BIOMAT 2014

International Symposium on
Mathematical and Computational Biology

BIOMAT 2014

International Symposium on
Mathematical and Computational Biology

Poznan, Poland 3 – 7 November 2014

edited by

Rubem P Mondaini
Federal University of Rio de Janeiro, Brazil

World Scientific

NEW JERSEY • LONDON • SINGAPORE • BEIJING • SHANGHAI • HONG KONG • TAIPEI • CHENNAI

Published by

World Scientific Publishing Co. Pte. Ltd.
5 Toh Tuck Link, Singapore 596224
USA office: 27 Warren Street, Suite 401-402, Hackensack, NJ 07601
UK office: 57 Shelton Street, Covent Garden, London WC2H 9HE

British Library Cataloguing-in-Publication Data
A catalogue record for this book is available from the British Library.

BIOMAT 2014
Proceedings of the International Symposium on Mathematical and Computational Biology

ISBN 978-981-4667-93-7

Printed in Singapore

Preface

The BIOMAT 2014 International Symposium has been held in the Conference Centre of the Stefan Banach International Mathematical Centre/ Institute of Mathematics of the Polish Academy of Sciences, at Bedlewo near Poznan, Poland, on November 03 -07, 2014. Senior representatives of the Polish Scientific Community have been in charge of the Local Organizing Committee. We thank the collaboration of Prof. Jerzy Tiuryn, Prof. Ryszard Rudnicki, Prof. Jacek Miekisz. Prof. Richard Kerner from France and his skills of being conversant with several languages from Eastern Europe has also given a fundamental cooperation in the task of administrative arrangements with our colleagues from Poland. He and Prof. Rafael Barrio from Mexico have also provided to all attendees some beautiful and unforgettable soirees of piano concerti in the Bedlewo Palace.

The participants came from twenty countries. From Europe (Belgium, France, Germany, Hungary, Italy, Poland, Romania, Russia, Ukraine,United Kingdom), from Americas (Argentina, Brazil, Canada, Colombia, Mexico, United States), from Africa (Algerie, South Africa), from Asia (India, Israel). In an informal atmosphere of continuous sessions, during five days, Plenary Speakers, authors of contributed papers and research students had the opportunity of discussing many research themes and to make commitments for future collaboration. The BIOMAT Consortium fine tradition of enhancing the scientific interdisciplinary activities of Biological and Mathematical Sciences of practitioners from developing countries has certainly succeed once more.

Financial support has been given by the Stefan Banach International Mathematical Centre and we thank our Polish colleagues for their kindness to provide some help consisting of some discount on the price of full pension accommodation in the Bedlewo Centre. Special thanks are due to the Director of the Bedlewo Centre, Slawomir Malecha and the expertise of his staff in Bedlewo. We also thank Prof. Eduardo Massad from University of São Paulo, Brazil, for the collaboration with the editorial costs of the BIOMAT 2014 book. We are indebted to the International Union of Biological Sciences (IUBS) and its Director, Dr. Nathalie Fomproix, from Paris, France, for the accommodation of some research students and young postdocs. The Society for Mathematical Biology (SMB) has also provided some support for participants through the request of Dr. Jane Heffernan from York University, Canada. On behalf of the BIOMAT Consortium, we thank very much all these institutions and their representatives. Their help has been essential in our offer of sixty fellowships to cover the expenses of full pension accommodation in Bedlewo.

The Editor of the BIOMAT International Series of Books and President of the BIOMAT Consortium-International Institute for Interdisciplinary Sciences, a non-profit international association of scientists and research students which is responsible by the organization of the BIOMAT Symposium series, would like to acknowledge the collaboration of his wife Carmem Lucia in the editorial work of the BIOMAT 2014 book. He is aware and very glad for the special dedication she usually gives in the final organization of the originals of the BIOMAT Series for publication after receiving all reports from reviewers and the submission of corrected versions by successful authors. He is also indebted to his research student Simão Coutinho de Albuquerque Neto who has made the adaptation of some accepted works to the requirements of the World Scientific LaTeX template.

Rubem P. Mondaini
President of the BIOMAT Consortium
Chairman of the BIOMAT 2014 Scientific Advisory Committee

Bedlewo near Poznan, Poland, November 2014

Editorial Board of the BIOMAT Consortium

Kristin Swanson	University of Washington, USA
Kerson Huang	Massachussets Institute of Technology, MIT, USA
Lisa Sattenspiel	University of Missouri-Columbia, USA
Louis Gross	University of Tennessee, USA
Ludek Berec	Biology Centre, ASCR, Czech Republic
Mariano Ricard	Havana University, Cuba
Michael Meyer-Hermann	Frankfurt Inst. for Adv. Studies, Germany
Nicholas Britton	University of Bath, UK
Panos Pardalos	University of Florida, Gainesville, USA
Peter Stadler	University of Leipzig, Germany
Pedro Gajardo	Federico Santa Maria University, Valparaíso, Chile
Philip Maini	University of Oxford, UK
Pierre Baldi	University of California, Irvine, USA
Ramit Mehr	Bar-Ilan University, Ramat-Gan, Israel
Raymond Mejía	National Institutes of Health, USA
Rebecca Tyson	University of British Columbia, Okanagan, Canada
Reidun Twarock	University of York, UK
Richard Kerner	Université Pierre et Marie Curie, Paris, France
Robijn Bruinsma	University of California, Los Angeles, USA
Rui Dilão	Instituto Superior Técnico, Lisbon, Portugal
Ruy Ribeiro	Los Alamos National Laboratory, New Mexico, USA
Seyed Moghadas	York University, Canada
Siv Sivaloganathan	Centre for Mathematical Medicine, Fields Institute, Canada
Suzanne Lenhart	University of Tennessee, USA
Timoteo Carletti	Facultés Universitaires Notre Dame de la Paix, Belgium
Vitaly Volpert	Université de Lyon 1, France
William Taylor	National Institute for Medical Research, UK
Zhijun Wu	Iowa State University, USA

Contents

Computational Biology

Multi-scale Models in Biological Sciences

Mathematical Morphology of Biological Structures

DYNAMICS OF Z-RING FORMATION IN LIPOSOMES*

R.A. BARRIO, C. VAREA

Instituto de Física, U.N.A.M., Apartado Postal 20-364, 01000 Mexico, D.F., Mexico

T. ALARCÓN

Centre de Recerca Matemàtica, Campus de Bellaterra and Departament de Matemàtiques, Universitat Autònoma de Barcelona, 08193 Bellaterra (Barcelona), Spain

C. B. PICALLO

Departament ECM, Facultat de Física, Universitat de Barcelona, E-08028 Barcelona, Spain

A. HERNANDEZ-MACHADO

Facultat de Física, Universitat de Barcelona, and Centre de Recerca Matemàtica, Campus de Bellaterra, Barcelona, Spain

We propose a model for the dynamics of the formation of rings of FtsZ on tubular liposomes which produce constriction on the corresponding membrane. The main characteristic of our model is the energy of interaction between the FtsZ protein and the membrane. This coupling is achieved through the protein increasing the spontaneous curvature energy. Our analysis is based on a linear dispersion relation, valid in general in the linear, short-time regime, which allows us to estimate the minimum concentration of FtsZ protein in order to induce constriction. Using this dispersion relation we are also able to predict the effect of increasing membrane rigidity on the constriction process. Our model allows to study the non-linear, long-term dynamics of the system where we observe coarsening of Z-rings on tubular liposomes.

*This work is supported by Conacyt (Mexico) through project no. 179616. A.H.M. gratefully acknowledges partial financial support from MINECO (Spain) under project FIS2013-47949-C2-1-P and DURSI project 2014SGR878. t.a. gratefully acknowledges MICINN for funding under grants MTM2010-18318-E, MTM2011-29342 and DURSI for funding under grant 2014SGR1307.

1. Introduction

Recent experimental results related to the origin of constrictive forces during cell division of rodlike bacteria have revealed that such forces are induced by the emergence of the so-called Z-rings. The major component of this structure is a filament formed by the FtsZ protein [1,2].

A popular experimental setup to study this phenomenon *in vitro* consists of using using a mixture of lipids and FtsZ protein. The lipids organise themselves into rodlike liposomes which resemble the structure of the bacterium membrane. The FtsZ protein, in turn, accumulates on this rod-shaped structure forming rings scattered over the length of the liposome. An important observation that results from these experiments is that the initially unstructured distribution of Z-rings evolves to a configuration in which the separation between two consecutive Z-rings is roughly equal to twice the radius of the liposome [1]. These experiments demonstrate that FtsZ suffices for assembling Z-rings and generate constriction forces on the liposome.

There is a discussion in the literature as to what are the mechanisms of force generation. One hypothesis postulates that there are lateral bonds which induce lateral attraction between protein filaments. If, in addition, the overlapping ends are capable of sliding over each other and reducing the radius of the liposome, constriction would emerge as the increased number of lateral bonds makes this configuration more stable. A very different hypothesis consists of assuming that FtsZ protofilaments are curved and when tethered to the membrane or liposome, produce a bending force [2].

These experimental results have motivated a number of modelling and theoretical studies examining both scenarios. The mechanism based on lateral bonding has been modelled and analysed by Hörger et al. using Langevin computer simulations [3,4].

Regarding the second scenario where bending energy is the force-generating mechanism, Hörger et al. [3,5] have proposed a model based on the Helfrich energy formalism with the addition of a constant spontaneous curvature term. This equilibrium model leads to the result that the constriction force is 50-100 pN in agreement with both *in vivo* and *in vitro* experimental results. Cytrynbaum et al. [6], based on previous models of constriction generation via hydrolysis of FtsZ-GTP [7,8,9], have proposed a method for calculating the value of the bending modulus consistent with experimental data for the separation between Z-rings using a Langevin equation for the movement of the rings.

It has been observed that the presence of a homogeneous distribution of FtsZ protein induces a regularly spaced constriction of the membrane of liposomes by the formation of Z-rings [1,10,2]. Although several models have explored this feature [7,8,11,12], the force generation mechanism of the Z-ring and the location where it acts remains unclear. We present a model to give an explanation for the dynamics of the patterning of Z-rings based on a simple bending energy model that captures the dynamical interplay between the protein and the membrane. Specifically, we propose a dynamic model where the spontaneous curvature depends on the local concentration of FtsZ protein. The vast number of experimental studies on FtsZ ring formation provides access to reliable data [13,10,2].

In this paper we formulate an analytical theory based on linearisation of the model equations which allows us to determine which parameters play a critical role in the process of constriction. We show that these parameters are the maximum protein concentration and the membrane rigidity. We also study the long-time behaviour of the system where non-linear effects take over. The numerical results indicate the presence of coarsening of Z-rings in long tubular liposomes. These coarsening gives rise to a pattern where rings are separated by a distance twice the radius of the liposome, in agreement with experimental results.

2. The Model

Here we would like to model the constriction of the vesicle membrane in the presence of an external agent, the protein. We suppose that the main contribution to the membrane energy is bending, following the well-known approach of Canham-Helfrich. We implement a Canham-Helfrich minimization scheme by means of a phase field that takes the value $\phi = 1$ inside the vesicle and $\phi = 1$ out of the vesicle. Hence, the points where $\phi = 0$ give the membrane location. The bending free energy is given by [14].

$$F_b = A_b \int_V \Phi_b^2 dV \tag{1}$$

and

$$\Phi_b = -\phi + \phi^3 - \epsilon^2 \nabla^2 \phi + C_0 \epsilon (1 - \phi^2) \tag{2}$$

where C_0 is the spontaneous curvature term that describes the natural trend of the vesicle to acquire a shape with a certain non-zero curvature. In our case, in the absence of FtsZ, the vesicle has no curvature and hence this term must be zero. However, as the protein concentrates on the membrane, this acquires curvature, giving rise to the known constricted shapes shown in figures (2) and (3) of references [1] and [2], respectively.

Therefore, this spontaneous curvature term should be proportional to a field u representing the concentration of protein FtsZ in the domain. As u should be positive, a non linear dependence on the FtsZ concentration is assumed in the form $C_0 = C_0(u) = \beta u^2$, where β measures the strength of the interaction between the protein and the membrane.

The free energy of the protein alone has two wells, at the fixed points $u = u_{\max}$ (maximum concentration of protein) and $u = u_{\min}$ (minimum concentration of protein, usually taken to be zero). Since we know that FtsZ acts only on the surroundings of the vesicle membrane the coupling of the protein and the vesicle should have two terms,

$$F_{sf} = \int_V (A_s V_s + A_f V_f) dV \tag{3}$$

$$V_s = (\phi^2 - 1)^2 (u - u_{\min})^2 (u - u_{\max})^2 + \lambda |\nabla u|^2 \tag{4}$$

$$V_f = \phi^2 (u - u_{\text{far}})^2 \tag{5}$$

where the term V_s only acts in the vicinity of the membrane ($\phi \approx 0$), and $\lambda|\nabla u|^2$ reflects the surface tension of the protein field. The term V_f becomes important far from the membrane ($\phi = \pm 1$), and thus, u_{far} represents the average value of protein concentration in the bulk of the system and can also be taken to be nil. Therefore, the total free energy of our vesicle-protein system is given by

$$F = \int_V (A_b \Phi_b^2 + A_s V_s + A_f V_f + \sigma |\nabla \phi|^2) dV \tag{6}$$

We also impose a restriction on the vesicle area (see [14]) that mimics the limited number of lipids that form the membrane. This is done by means of the Lagrange multiplier σ that is calculated by imposing surface area conservation of the membrane [15]. Both fields u and ϕ should be conserved, hence their evolution is given by Cahn-Hilliard (or model B) equations [16,17],

$$\frac{\partial \phi}{\partial t} = D_\phi \nabla^2 \left(\frac{\delta F}{\delta \phi} \right) \tag{7}$$

$$\frac{\partial u}{\partial t} = D_u \nabla^2 \left(\frac{\delta F}{\delta u} \right) \tag{8}$$

From the definition $\frac{\delta}{\delta \phi} F(\phi, \nabla\phi, \nabla^2\phi) = \frac{\partial F}{\partial \phi} - \nabla \frac{\partial F}{\partial \nabla \phi} + \nabla^2 \frac{\partial F}{\partial \nabla^2 \phi}$ we obtain the complete expressions for our dynamical system,

$$
\begin{aligned}
\frac{\partial \phi}{\partial t} &= D_\phi \nabla^2 \left[A_b(2\Phi_b(-1+3\phi^2-2\epsilon\beta u^2\phi)-2\epsilon^2\nabla^2\Phi_b)+ \right.\\
&\quad + A_s 4\phi(\phi^2-1)(u-u_{\text{min}})^2(u-u_{\text{max}})^2+\\
&\quad \left. + A_f 2\phi(u-u_{\text{far}})^2 - 2\sigma\nabla^2\phi \right]\\
\frac{\partial u}{\partial t} &= D_u \nabla^2 \left[A_b 4\epsilon\beta u(1-\phi^2)\Phi_b+ \right.\\
&\quad + A_s 2((u-u_{\text{min}})(u-u_{\text{max}})(2u-(u_{\text{min}}+u_{\text{max}}))(\phi^2-1)^2-\lambda\nabla^2 u)+\\
&\quad \left. + A_f 2\phi^2(u-u_{\text{far}}) \right]
\end{aligned} \tag{9}
$$

3. Linear Stability Analysis

In order to analyse the stability of a membrane with homogeneous distribution of protein, one can study the effect of small perturbations on a flat interface. In particular, we consider that perturbations take the form of plane waves: $\phi = \phi_0 e^{ikz+\omega t}$ and $u = u_0 e^{ikz+\omega t}$ around the membrane ($\phi = 0$) and the fixed point of maximum protein concentration (u_{max}) that we expect to be reached at the sites of FtsZ rings. We further assume that the amplitudes $\phi_0 \ll 1$ and $u_0 \ll 1$. Substituting these expression into the linearized version of Eqs. 7, 8 we obtain two different roots, $\omega_1(k)$ and $\omega_2(k)$, of the dispersion relation $\omega(k)$. One of them is shown to exhibit a

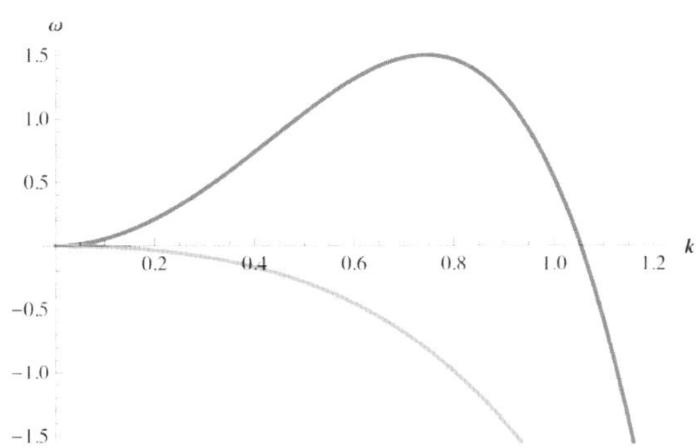

Figure 1. Dispersion relation showing an unstable branch which exhibits a global maximum for a well defined value of the wave number. Parameter values: $D_\phi = 4.5$, $D_u = 2.7$, $A_b = 2$, $A_s = 2$, $A_f = 0.2$, $\epsilon = 0.01$, $\beta = 100$, $u_{min} = 0$, $u_{max} = 1$, $\lambda = 1$, $u_{far} = 0$, and $\sigma = 1$.

region of values of k such that $\omega_1(k) > 0$ (darker line in Fig. 1), which indicates instability. The wavelength corresponding to the global maximum of $\omega_1(k)$ gives us the inverse of the characteristic length, l_c, scale of the separation between protein rings in our system. In Fig. 1 both the stable and unstable branches of the dispersion relation are shown for typical values of the parameters.

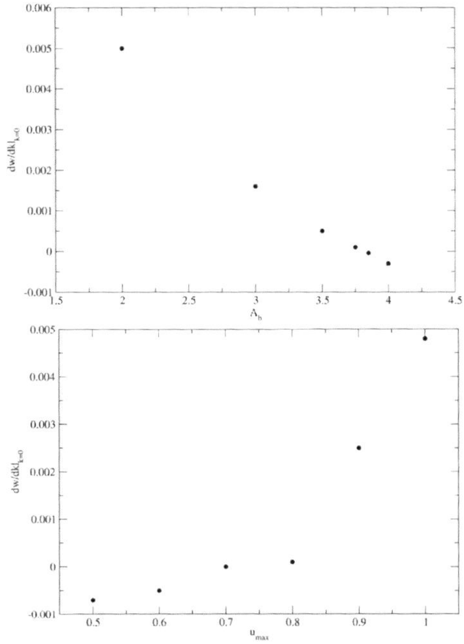

Figure 2. We analyse the onset of instability, which we identify with the change of sign (from negative to positive), of the function $\frac{dw}{dk}|_{k=0}$ as function of A_b (upper plot) and u_{max} (lower plot). The rest of parameter values are given in Fig. 1.

In Fig. 2 we analyse the dependence of the instability boundary. We define the instability boundary as the change of sign of $\frac{dw_1}{dk}|_{k=0}$ which determines the appearance of the unstable branch of the dispersion relation. We look at the behaviour of $\frac{dw_1}{dk}|_{k=0}$ as a function of two key parameters, u_{max}, the maximum concentration of FtsZ protein, and A_b, which measures the rigidity of the membrane, in order to ascertain how these parameters affect the onset of the constriction. We predict that there exists a minimum value of the protein concentration, measured in terms of the parameter u_{max}, below which there is no constriction. Regarding the dependence on the membrane rigidity, A_b, we observe that there is a maximum value above which constriction is prevented.

4. Numerical Study

We integrate numerically equations (7) and (8) with periodic boundary conditions using second-order finite differences for the spatial dependence and an Euler scheme for the time dependence. Since the standard second-order finite differences is a consistent finite difference method, the time step was chosen following the Courant-Friedrichs-Lewy stability criterion: $\Delta t \leq c\Delta x$, where c is a positive constant [18]. The initial condition corresponds to a cylindrical shape corresponding to the membrane of the vesicle. The protein is uniformly distributed over the integration domain. As time progresses,

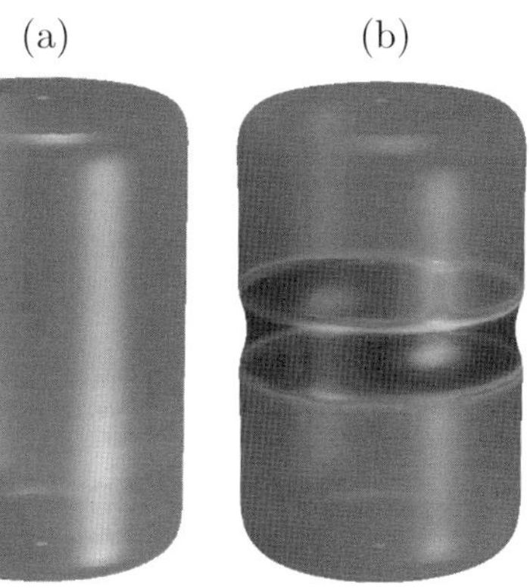

Figure 3. FtsZ ring formation in a cylindrical vesicle. Plot (a) corresponds to length 22, shorter than l_c and plot (b) to length $28 \approx l_c$.

the protein gets attached to the membrane forming rings. The short time behaviour is controlled by the linear dispersion relation, whereas the long-time behaviour depends on the length of the cylindrical membrane.

In agreement with our analytical results, short tubular vesicles, shorter than the length scale provided by the dispersion relation, l_c, show no ring formation and, therefore, no constriction (see Fig. 3(a)). For lengths of the order of l_c, which for the parameter values given in Fig. 1 is $l_c \simeq 28$ (measured in units of Δx), we observe that only one ring is formed located in the middle of the membrane (see Fig. 3(b)). Longer systems, such as tubular liposomes, whose length is several times l_c, show a richer behaviour. Initially, in agreement with the behaviour predicted by the linear analysis, a number of rings are formed, located at distances given by l_c. At longer times, a nonlinear regime ensues in which rings exhibit coarsening where rings move over the membrane and eventually fuse maintaining a distance between them of twice the radius (see Fig. 4).

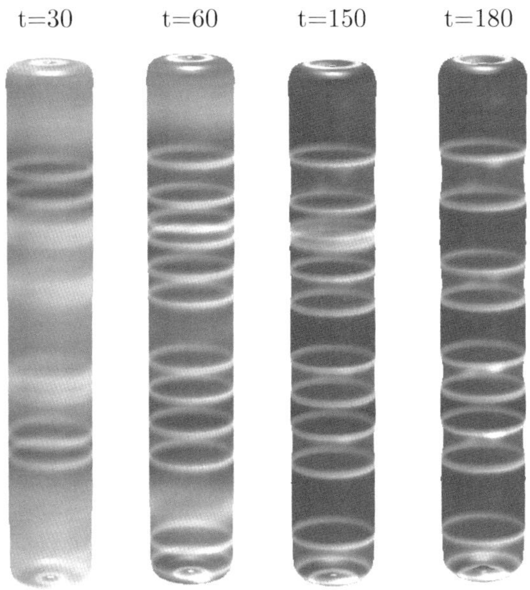

Figure 4. FtsZ ring formation as a function of time in a tubular domain of length $l \approx 4l_c$.

5. Conclusions and Discussion

We have proposed a model to study the dynamics of the constriction and ring formation on bacteria and tubular liposomes. This model is based on a free energy functional accounting for the bending energy of the membrane including spontaneous curvature. The spontaneous curvature is a function of the local concentration of FtsZ protein. Concentration of protein on the membrane is achieved by means of an extra contribution to the free energy, which penalises the presence of protein away from the membrane. The dynamical model has been studied analitically by means of a linear stability analysis in which we obtain that an instability which leads to a periodic pattern in the concentration of protein and the onset of constriction, eventually leading to the formation of Z-rings. The dispersion relation resulting from the linear analysis allows us to predict a lower threshold of FtsZ protein concentration to induce constriction. Similarly, we are able to predict an upper threshold for the membrane rigidity parameter to prevent constriction.

We have studied the non-linear, long time regime by means of numerical simulation. The non-linear regime is particularly relevant for longer systems (e.g. tubular liposomes several times l_c in length). In these systems, according to the results of the linear analysis, a periodic pattern of peaks in protein concentration is formed. As time progresses, this periodic structure goes through a process of coarsening which leads to the movement of rings over the membrane which eventually merge.

In the case of a closed cylindrical membrane, our model predicts a minimum length, l_c, below which the membrane would not be able to constrict. We have also determined that a lower threshold of protein concentration exists below which protein accumulation is not possible. This suggests that lowering the affinity between the protein and the membrane in an appropriate proportion, determined by our analysis, would result in a decrease of protein concentration on the membrane, hence, constriction is prevented.

We have further determined the existence of a upper threshold for the membrane rigidity above which constriction is not possible. This model prediction can be experimentally verified in *in vitro* experiments by increasing the rigidity of tubular liposome membranes by polimerisation [19].

References

1. M. Osawa, D. E. Anderson and and H. P. Ericson , *Science* **320**, 1729 (1980).
2. M. Osawa and H. P. Erickson,*Molecular Microbiology* **81**, 571 (2011).
3. I. Hörger et al., *Phys. Rev. E*, 011902 (2008)
4. I. Hörger et al., *Biophys. J.* , L81 (2008).
5. I. Hörger, F. Campelo, A. Hernńdez-Machado, and P. Tarazona, *Phys. Rev. E* **81**, 031922 (2010).
6. E. N. Cytrynbaum, Y. D. Li, J. F. Allard, and H. Mehrabian, *Phys. Rev. E*, 011902 (2012).
7. B. Ghosh and A. Sain, *Phys. Rev. Lett.* **101**, 178101 (2008).
8. J. F. Allard and E. N. Cytrynbaum, *Proceedings of the National Academy of Sciences* **106**, 145 (2009).
9. H. P. Erickson, *Proc. Natl Acad. Sci.* **107**, 9238 (2009).
10. H. P. Erickson, D. E. Anderson, and M. Osawa, *Microbiology and Molecular Biology Reviews* **74**, 504 (2010).
11. G. Lan et al., *Proceedings of the National Academy of Sciences* **106**, 121 (2009).
12. E. Fischer-Friedrich and N. Gov, *Physical Biology* **8**, 026007 (2011).

13. S. Wang, H. Arellano-Santoyo, P. A. Combs, and J. W. Shaevitz, *Proceedings of the National Academy of Sciences* (2010).
14. F. Campelo and A. Hernández-Machado, *Phys. Rev. Lett.* **99**, 088101 (2007).
15. C. Varea, R. Barrio, and A. Hernández-Machado, *Phys. Rev. E* **84**, 061922 (2011).
16. J. W. Cahn and J. E. Hilliard, *The Journal of Chemical Physics* **28**, 258 (1958).
17. A. Bray, *Advances in Physics* **43**, 357 (1994).
18. F. Campelo and A. Hernández-Machado, *Eur. Phys. J. Special Topics* **143**, 101 (2007).
19. H. Miyata and H. Hotani, *Proc. Natl. Acad. Sci.* **89**, 11547 (1992).

OPTIMAL CONTROL OF A COARSE-GRAINED MODEL FOR PROTEIN DYNAMICS

R.P. MONDAINI, S.C. DE ALBUQUERQUE NETO

Federal University of Rio de Janeiro, Centre of Technology, COPPE
Rio de Janeiro, Brazil, 21941-972, P.O. Box 68511

An elementary optimal control analysis based on Pontryagin's Maximum Principle is applied to the dynamics of protein molecular structure. The Protein Dynamics Evolution is chosen to be modeled by a coarse-grained model in the two cases of smallness of atomic masses and smallness of atomic mass differences of the backbone atoms.

1. Introduction

A Theory of Protein Formation and evolution would elucidate the necessary initial conditions for describing the molecular dynamics from primitive "protomolecules" to the present proteins and it would make previsions for the protein folding process. This theory would be able to describe all intermediary states which are used by Nature to circumvent "Levinthal's paradox"[1] In the absence of this theory, our previsions seem to be restricted to the derivation of the polymerization process leading to the formation of peptides of equal aminoacids. We can then write:

$$A_1 = H_2NC_\alpha HRCOOH \tag{1}$$

$$A_j = H(HNC_\alpha HRCO)_jOH \tag{2}$$

$$nA_1 = A_n + (n-1)H_2O \tag{3}$$

$$nA_j = A_{jn} + (n-1)H_2O \tag{4}$$

$$A_j + A_k = A_{j+k} + H_2O \tag{5}$$

$$nA_j + mA_k = A_{jn} + A_{km} + (n+m-2)H_2O = A_{jn+km} + (n+m-1)H_2O \tag{6}$$

In the real case of different aminoacids $A_1, A_2, A_3, \ldots\ldots\ldots, A_n$, with an ordering previously specified by Nature, "life" is not so easy, but we could have an insight of its organization by considering the necessary "funnel" intermediate states to be originated in the transcription of a problem of potential energy minimization corresponding to the interaction of each atom site with its nearest neighbors, into a Fermat Problem of minimizing the sum of powers of distances among these atom sites[2,3,4]. This could be inferred from a constrained Optimal Control Problem to be solved by methods of the Pontryagin's Maximum Principle[5,6]. We started from a coarse-grained model with a mass-spring force and we succeed at finding a well-posed non-bounded control problem. We are then able to decide the value of the power exponents and these will provide information about the effective dependence of the potential energy on the distances among atom sites. This is a global optimization problem approach and it is essentially different from the usual one of molecular dynamics, which starts from a phenomenological form of the field of force.

The polypeptide structure of fig. 1 will be taken as a paradigm to apply these Optimal control techniques to be introduced on section 3.

2. Methodology

We now assume a coarse-grained model for the dynamics of the atom sites C_α, N, C:

$$m_{C_{\alpha_j}} \overrightarrow{\ddot{\Delta r}_{C_{\alpha_j}}} - m_{N_j} \overrightarrow{\ddot{\Delta r}_{N_j}} = -\gamma(\overrightarrow{\dot{\Delta r}_{C_{\alpha_j}}} - \overrightarrow{\dot{\Delta r}_{N_j}}) + \overrightarrow{F_{C_{\alpha_j}}} - \overrightarrow{F_{N_j}} \tag{7}$$

$$m_{C_j} \overrightarrow{\ddot{\Delta r}_{C_j}} - m_{C_{\alpha_j}} \overrightarrow{\ddot{\Delta r}_{C_{\alpha_j}}} = -\gamma(\overrightarrow{\dot{\Delta r}_{C_j}} - \overrightarrow{\dot{\Delta r}_{C_{\alpha_j}}}) + \overrightarrow{F_{C_j}} - \overrightarrow{F_{C_{\alpha_j}}} \tag{8}$$

$$m_{N_{j+1}} \overrightarrow{\ddot{\Delta r}_{N_{j+1}}} - m_{C_j} \overrightarrow{\ddot{\Delta r}_{C_j}} = -\gamma(\overrightarrow{\dot{\Delta r}_{N_{j+1}}} - \overrightarrow{\dot{\Delta r}_{C_j}}) + \overrightarrow{F_{N_{j+1}}} - \overrightarrow{F_{C_j}} \tag{9}$$

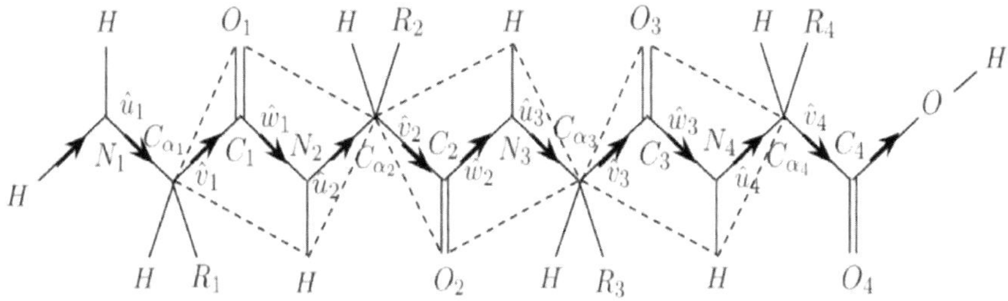

Figure 1.: Scheme for protein backbone modelling where $\hat{u_j}, \hat{v_j}, \hat{w_j}, j = 1, 2, ..., n$, are the unit vectors of $\overrightarrow{r_{C_{\alpha_j}}} - \overrightarrow{r_{N_j}}, \overrightarrow{r_{C_j}} - \overrightarrow{r_{C_{\alpha_j}}}, \overrightarrow{r_{N_{j+1}}} - \overrightarrow{r_{C_j}}$.

where $m_{C_{\alpha_j}}, m_{N_j}, m_{C_j}$ are the masses of the alpha-carbon atom sites, the nitrogen atom sites and the carbon atom sites of the carbonyl $C = O$, respectively; γ is the friction constant and (.) stands for $\frac{d}{dt}$.

The field of force is given by

$$\overrightarrow{F_{C_{\alpha_j}}} - \overrightarrow{F_{N_j}} = -\kappa(\overrightarrow{\Delta r_{C_{\alpha_j}}} - \overrightarrow{\Delta r_{N_j}}) + \overrightarrow{U_{C_{\alpha_j}, N_j}} \tag{10}$$

$$\overrightarrow{F_{C_j}} - \overrightarrow{F_{C_{\alpha_j}}} = -\kappa(\overrightarrow{\Delta r_{C_j}} - \overrightarrow{\Delta r_{C_{\alpha_j}}}) + \overrightarrow{U_{C_j, C_{\alpha_j}}} \tag{11}$$

$$\overrightarrow{F_{N_{j+1}}} - \overrightarrow{F_{C_j}} = -\kappa(\overrightarrow{\Delta r_{N_{j+1}}} - \overrightarrow{\Delta r_{C_j}}) + \overrightarrow{U_{N_{j+1}, C_j}} \tag{12}$$

where the first terms on the right hand sides stand for a mass-spring Hooke force and the second terms stand for the control vector functions. We can write,

$$\overrightarrow{\Delta r_{C_{\alpha_j}}} - \overrightarrow{\Delta r_{N_j}} = l_{C_{\alpha_j}, N_j} \hat{u}_j \tag{13}$$

$$\overrightarrow{\Delta r_{C_j}} - \overrightarrow{\Delta r_{C_{\alpha_j}}} = l_{C_j, C\alpha_j} \hat{v}_j \tag{14}$$

$$\overrightarrow{\Delta r_{N_{j+1}}} - \overrightarrow{\Delta r_{C_j}} = l_{N_{j+1}, C_j} \hat{w}_j \tag{15}$$

where we define the unit vectors $\hat{u}_j$, $\hat{v}_j$, $\hat{w}_j$ and $l_{C_{\alpha_j},N_j}$, $l_{C_j,C\alpha_j}$, l_{N_{j+1},C_j} are the variations of lengths of the bonds among the atoms C_{α_j}, N_{j+1}, C_j under the force field.

As the control functions are concerned, they should be forces which are able to drive the system to another equilibrium position. We also assume that these controls will drive the molecular vibrations only along the bonds among atoms. This will lead to:

$$\overrightarrow{U_{C_{\alpha_j},N_j}} = u_{C_{\alpha_j},N_j}\hat{u}_j \tag{16}$$

$$\overrightarrow{U_{C_j,C_{\alpha_j}}} = u_{C_j,C_{\alpha_j}}\hat{v}_j \tag{17}$$

$$\overrightarrow{U_{N_{j+1},C_j}} = u_{N_{j+1},C_j}\hat{w}_j \tag{18}$$

We should note that

$$m_{C_j} = m_{C_{\alpha_j}} = m_j \tag{19}$$

From eqs. (10)–(12), (13)–(15) and (16)–(18), eqs. (7)–(9), turn into

$$m_{C_{\alpha_j}}\overrightarrow{\Delta\ddot{r}_{C_{\alpha_j}}} - m_{N_j}\overrightarrow{\Delta\ddot{r}_{N_j}} = (-\gamma\dot{l}_{C_{\alpha_j},N_j} - \kappa l_{C_{\alpha_j},N_j} + u_{C_{\alpha_j},N_j})\hat{u}_j \tag{20}$$

$$m_j\ddot{l}_{C_j,C_{\alpha_j}} = -\gamma\dot{l}_{C_j,C_{\alpha_j}} - \kappa l_{C_j,C_{\alpha_j}} + u_{C_j,C_{\alpha_j}} \tag{21}$$

$$m_{N_{j+1}}\overrightarrow{\Delta\ddot{r}_{N_{j+1}}} - m_{C_j}\overrightarrow{\Delta\ddot{r}_{C_j}} = (-\gamma\dot{l}_{N_{j+1},C_j} - \kappa l_{N_{j+1},C_j} + u_{N_{j+1},C_j})\hat{w}_j \tag{22}$$

We also assume that the initial conditions on the temporal derivatives

$$\dot{\hat{u}}_j(t = t_0) = 0, \dot{\hat{v}}_j(t = t_0) = 0, \dot{\hat{w}}_j(t = t_0) = 0 \tag{23}$$

will be preserved during the controlled motion about the equilibrium positions, or

$$\begin{aligned}\dot{\hat{u}}_j(t_0 \le t \le t_0 + \Delta T) = 0\\ \dot{\hat{v}}_j(t_0 \le t \le t_0 + \Delta T) = 0\\ \dot{\hat{w}}_j(t_0 \le t \le t_0 + \Delta T) = 0\end{aligned} \tag{24}$$

where ΔT is the duration of the controls.

There are two cases:

1) Smallness of atomic masses[7] $m_j \approx 0, m_{N_j} \approx 0$:

$$\gamma \dot{l}_{C_{\alpha_j},N_j} = -\kappa l_{C_{\alpha_j},N_j} + u_{C_{\alpha_j},N_j} \tag{25}$$

$$\gamma \dot{l}_{C_j,C_{\alpha_j}} = -\kappa l_{C_j,C_{\alpha_j}} + u_{C_j,C_{\alpha_j}} \tag{26}$$

$$\gamma \dot{l}_{N_{j+1},C_j} = -\kappa l_{N_{j+1},C_j} + u_{N_{j+1},C_j} \tag{27}$$

2) Smallness of atomic mass differences $m_j - m_{N_j} = \Delta m_j \approx 0$:

$$m_j \ddot{l}_{C_{\alpha_j},N_j} = -\gamma \dot{l}_{C_{\alpha_j},N_j} - \kappa l_{C_{\alpha_j},N_j} + u_{C_{\alpha_j},N_j} \tag{28}$$

$$m_j \ddot{l}_{C_j,C_{\alpha_j}} = -\gamma \dot{l}_{C_j,C_{\alpha_j}} - \kappa l_{C_j,C_{\alpha_j}} + u_{C_j,C_{\alpha_j}} \tag{29}$$

$$m_j \ddot{l}_{N_{j+1},C_j} = -\gamma \dot{l}_{N_{j+1},C_j} - \kappa l_{N_{j+1},C_j} + u_{N_{j+1},C_j} \tag{30}$$

This last set of equations can also be written as:

$$\dot{l}_{C_{\alpha_j},N_j} = L_{C_{\alpha_j},N_j} \tag{31}$$

$$m_j \dot{L}_{C_{\alpha_j},N_j} = -\gamma L_{C_{\alpha_j},N_j} - \kappa l_{C_{\alpha_j},N_j} + u_{C_{\alpha_j},N_j} \tag{32}$$

$$\dot{l}_{C_j,C_{\alpha_j}} = L_{C_j,C\alpha_j} \tag{33}$$

$$m_j \dot{L}_{C_j,C_{\alpha_j}} = -\gamma L_{C_j,C_{\alpha_j}} - \kappa l_{C_j,C_{\alpha_j}} + u_{C_j,C_{\alpha_j}} \tag{34}$$

$$\dot{l}_{N_{j+1},C_j} = L_{N_{j+1},C_j} \tag{35}$$

$$m_j \dot{L}_{N_{j+1},C_j} = -\gamma L_{N_{j+1},C_j} - \kappa l_{N_{j+1},C_j} + u_{N_{j+1},C_j} \tag{36}$$

We assume that each triplet of variables $l_{C_{\alpha_j},N_j}$, $l_{C_j,C_{\alpha_j}}$, l_{N_{j+1},C_j} does not depend on j. The evolution in time of the lengths is then uniform along the protein backbone and they start from the values:

$$l_{C_{\alpha_j},N_j}(t_0) = l_{0_{C_{\alpha_j},N_j}}, l_{C_j,C_{\alpha_j}}(t_0) = l_{0_{C_j,C_{\alpha_j}}}, l_{N_{j+1},C_j}(t_0) = l_{0_{N_{j+1},C_j}} \tag{37}$$

Protein data will lead to

$$l_{0_{C_{\alpha_j},N_j}} = 0.153\mathring{A},\ l_{0_{C_j,C_{\alpha_j}}} = 0.132\mathring{A},\ l_{0_{N_{j+1},C_j}} = 0.147\mathring{A} \tag{38}$$

We also have

$$\begin{aligned} L_{C_{\alpha_j},N_j}(t_0) &= L_{0_{C_{\alpha_j},N_j}}, \\ L_{C_j,C_{\alpha_j}}(t_0) &= L_{0_{C_j,C_{\alpha_j}}}, \\ L_{N_{j+1},C_j}(t_0) &= L_{0_{N_{j+1},C_j}} \end{aligned} \tag{39}$$

The modelling will proceed by supposing these relations to be valid $\forall j$.

3. Pontryagin's Maximum Principle and a class of well-posed Control Problems

According to the Pontryagin's Maximum Principle[5,6], we look for piecewise continuous controls and their associated state variables in order to maximize an objective function such as

$$J[u_{jk}] = \phi(l_{jk}(t_F) + \int_{t_0}^{t_F} f(l_{jk}(t), u_{jk}(t))\mathrm{d}t \tag{40}$$

where

$$k = C_{\alpha_j}, N_j, \quad C_j, C_{\alpha_j}, \quad C_{N_{j+1}}, C_j, \quad \forall j \tag{41}$$

Let f be a homogeneous function of degree p in the state variables $l_{jk}(t)$ and of degree q in the control variables $u_{jk}(t)$, or

$$f(l_{jk}(t), u_{jk}(t)) = \sum_k \sum_{j=1}^{n} [(l_{jk}(t))^p + A_{jk}(u_{jk}(t))^q] \tag{42}$$

The payoff function[6] ϕ is also a homogenous function of degree r in the state variables,

$$\phi(l_{jk}(t)) = \sum_k \sum_{j=1}^{n} (l_{jk}(t))^r \tag{43}$$

and p, q, r, are positive integers.

The assumed uniformity of the triplets of bond lengths along the protein structure will lead to

$$l_{jk} = l_k; \quad A_{jk} = A_k; \quad u_{jk} = u_k, \quad \forall j \tag{44}$$

and the payoff function will turn into

$$\phi(l_k(t)) = n \sum_k (l_k(t))^r \tag{45}$$

From section 2, and the uniformity of the triplets of bond lengths of eq. (42) we write eqs. (25)–(27) and (31)–(36) as

$$\dot{l}_k = -\frac{\kappa}{\gamma} l_k + \frac{1}{\gamma} u_k \tag{46}$$

and

$$\dot{l}_k = L_k \tag{47}$$

$$\dot{L}_k = -\frac{\gamma}{m} L_k - \frac{\kappa}{m} l_k + \frac{1}{m} u_k \tag{48}$$

The Hamiltonian functions can then be written as:

$$H_0 = \sum_k \lambda_{l\,k}(t)\dot{l}_k(t) + n\sum_k((l_k(t))^p + A_k(u_k(t))^q) \tag{49}$$

and

$$H = \sum_k(\lambda_{l\,k}(t)\dot{l}_k(t) + \lambda_{L\,k}(t)\dot{L}_k(t)) + n\sum_k((l_k(t))^p + A_k(u_k(t))^q) \tag{50}$$

The Optimality conditions and the conditions for a minimum[6] corresponding to eqs. (44), (47) will be given by

$$0 = \frac{\partial H_0}{\partial u_k(t)} = \frac{\lambda_{l\,k}(t)}{\gamma} + nqA_k(u_k(t))^{q-1} \tag{51}$$

$$\frac{\partial^2 H_0}{\partial u_k^2(t)} = nq(q-1)A_k(u_k(t))^{q-2} > 0 \tag{52}$$

respectively.

The Adjoint equations for the Lagrange multiplier functions and the Transversality conditions[6] for this case can be written as

$$\dot{\lambda}_{l_k}(t) = -\frac{\partial H_0}{\partial l_k(t)} = \frac{\kappa}{\gamma}\lambda_{l_k}(t) - np(l_k(t))^{p-1} \tag{53}$$

$$\lambda_{l_k}(t_F) = \frac{\partial \phi}{\partial l_k(t)}(l_k(t_F) = nr(l_k(t_F))^{r-1} \tag{54}$$

From eqs. (49), (51), (52), we get the equations for the Control variables and their boundary conditions:

$$(q-1)\dot{u}_k(t) = \frac{\kappa}{\gamma}u_k(t) + \frac{p}{q}\frac{1}{\gamma A_k}(l_k(t))^{p-1}(u_k(t))^{2-q} \tag{55}$$

$$u_k(t_F) = (\frac{-r}{q\gamma A_k})^{\frac{1}{q-1}}(l_k(t_F))^{\frac{r-1}{q-1}} \tag{56}$$

respectively.

The case given by eqs. (45), (46) and (48) will lead to the same conditions for a minimum of eq. (50) and the corresponding Optimality conditions will be given by

$$0 = \frac{\partial H}{\partial u_k(t)} = \frac{\lambda_{L_k}(t)}{m} + nqA_k(u_k(t))^{q-1} \tag{57}$$

The Adjoint equations for this case can be written as

$$\dot{\lambda}_{l_k}(t) = -\frac{\partial H}{\partial l_k(t)} = \frac{\kappa}{m}\lambda_{L_k}(t) - np(l_k(t))^{p-1} \tag{58}$$

$$\dot{\lambda}_{L_k}(t) = -\frac{\partial H}{\partial L_k(t)} = -\lambda_{l_k}(t) + \frac{\gamma}{m}\lambda_{L_k}(t) \tag{59}$$

The Transversality conditions are the same as those of the first case for $\lambda_{l_k}(t)$, eq. (52), and for $\lambda_{L_k}(t)$, we have

$$\lambda_{L_k}(t_F) = \frac{\partial \phi}{\partial L_k(t)}(L_k(t_F)) = 0 \tag{60}$$

From eqs. (55), (56), (57), we can write the equations for the Control variables $u_k(t)$ and Adjoint variables $\lambda_{l_k}(t)$:

$$(q-1)\dot{u}_k(t) = \frac{\gamma}{m}u_k(t) + \frac{1}{nq}\frac{1}{mA_k}\lambda_{l_k}(t)(u_k(t))^{2-q} \tag{61}$$

$$\dot{\lambda}_{l_k}(t) = -n[p(l_k(t))^{p-1} + q\kappa A_k(u_k(t))^{q-1}] \tag{62}$$

The only remaining Transversality condition is given by

$$u_k(t_F) = 0 \tag{63}$$

After getting rid of the suffix k, we can write the dynamic equations as:

For small atomic masses $m \approx 0$:

$$\dot{l} = -\frac{\kappa}{\gamma}l + \frac{1}{\gamma}u \tag{64}$$

$$(q-1)\dot{u} = \frac{\kappa}{\gamma}u + \frac{p}{q}\frac{1}{\gamma A}l^{p-1}u^{2-q} \tag{65}$$

For small atomic mass differences $\Delta m \approx 0$:

$$\dot{l} = L \tag{66}$$

$$\dot{L} = -\frac{\gamma}{m}L - \frac{\kappa}{m}l + \frac{1}{m}u \tag{67}$$

$$\dot{\lambda}_l = -n(pl^{p-1} + q\kappa Au^{q-1}) \tag{68}$$

$$(q-1)\dot{u} = \frac{\gamma}{m}u + \frac{1}{nq}\frac{1}{mA}\lambda_l u^{2-q} \tag{69}$$

4. Dimensionless Equations

We now scale variables into eqs. (62), (63) according to

$$t = \alpha_t t^*, \quad l = \alpha_l l^*, \quad u = \alpha_u u^* \tag{70}$$

There are no additional dimensionless parameters. The coefficients $\alpha_t, \alpha_l, \alpha_u$ will be given by:

$$\alpha_t = \frac{\gamma}{\kappa} \tag{71}$$

$$\alpha_l^{p-q} = A\kappa^q \tag{72}$$

$$\alpha_u^{p-q} = A\kappa^p \tag{73}$$

The equations (62) and (63), after getting rid of the asterisks, will turn into:

$$\dot{l} = -l + u \tag{74}$$

$$(q-1)\dot{u} = u + \frac{p}{q} l^{p-1} u^{2-q} \tag{75}$$

The same procedure is adopted for eqs. (64)–(67). This will lead to

$$t = \beta_t t^*, \quad l = \beta_l l^*, \quad u = \beta_u u^*, \quad L = \beta_L^*, \quad \lambda_l = \beta_{\lambda_l} \lambda_l^* \tag{76}$$

There are at most three dimensionless additional parameters. We choose only one and this will be given by

$$a = \frac{\gamma}{\sqrt{\kappa m}} \tag{77}$$

and the coefficients above could be given by

$$\beta_t = \sqrt{\frac{m}{k}} \tag{78}$$

$$\beta_l^{p-q} = A\kappa^q \tag{79}$$

$$\beta_u^{p-q} = A\kappa^p \tag{80}$$

$$\beta_L^{p-q} = A\kappa^{\frac{3p-q}{2}} m^{\frac{p-q}{2}} \tag{81}$$

$$\beta_{\lambda_l}^{p-q} = A^{p-1}\kappa^{\frac{2pq-(p+q)}{2}} m^{\frac{p-q}{2}} \tag{82}$$

From eqs. (74) and (64)–(67), after dropping the asterisks, we can write:

$$\dot{l} = L \tag{83}$$

$$\dot{L} = -aL - l + u \tag{84}$$

$$\dot{\lambda}_l = -npl^{p-1} - nqu^{q-1} \tag{85}$$

$$(q-1)\dot{u} = au + \frac{1}{nq}\lambda_l u^{2-q} \tag{86}$$

5. Some Elementary States and Their Controls

Analytical solutions can be easily found for $p = 1, q = 2$ and $p = 2, q = 2$. Other couples of values for $p > 1$ and $q > 2$ could also lead to interesting dynamics and control processes but they should be searched by a numerical solution approach. From figs. 2, 3, we can make the comparison of states and controls for the cases $m \approx 0$ and $\Delta m \approx 0$. The minimum $m \approx 0$ problems, corresponding to fig. 2 below are solved by $A > 0$ which means $u(t_F) < 0$ for $r \geq 1$. From eqs. (50) and (54), we have $0 < t < min(t_{1F}, t_{2F})$ where $l_1(t_{1F}) = 0$, $l_2(t_{2F}) = 0$, since we should have $l(t) \geq 0$. For the minimum $\Delta m \approx 0$ problem, corresponding to fig. 3 below, this has also to be solved by $A > 0$ from eqs. (50), (61) and we have $u_1(t_{1F}) = 0$, $u_2(t_{2F}) = 0$. A detailed analysis on the restriction of final time values, should be done from the application of the free terminal time techniques of Optimal Control.

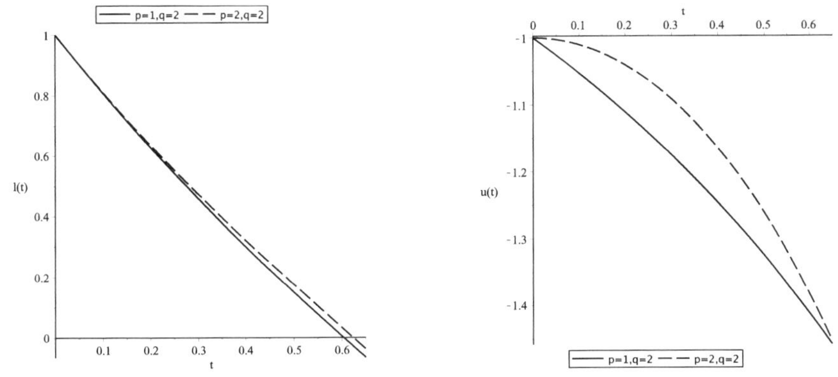

(a) (b)

Figure 2.: $m \approx 0$. (a) Comparison of states, $l_1(t), l_2(t)$. (b) Comparison of controls, $u_1(t), u_2(t)$. $l_1(0) = 1, l_2(0) = 1, u_1(0) = -1, u_2(0) = -1, t_{1F} = 0.603\ s,\ t_{2F} = 0.623\ s$.

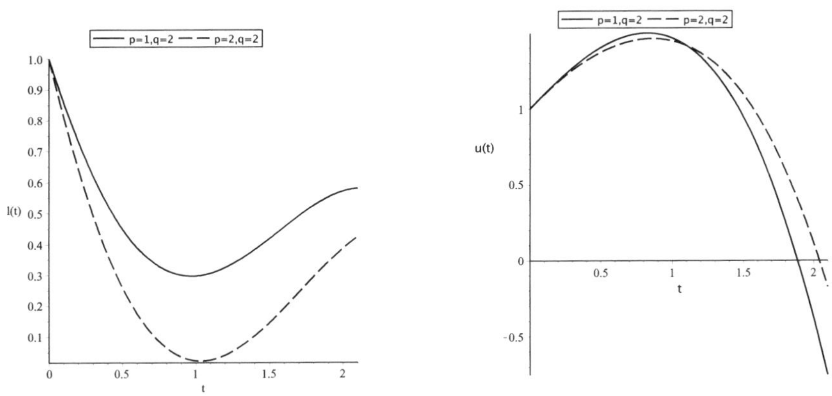

(a) (b)

Figure 3.: $\Delta m \approx 0$. (a) Comparison of states, $l_1(t), l_2(t)$. (b) Comparison of controls, $u_1(t), u_2(t)$. $n = 100, a = 1, l_1(0) = 1, l_2(0) = 1, L_1(0) = -1.5, L_2(0) = -2, \lambda_{l_1}(0) = 0, \lambda_{l_2}(0) = 0, u_1(0) = 1, u_2(0) = 1, t_{1F} = 1.886\ s,\ t_{2F} = 2.040\ s$.

6. Concluding Remarks

Proteins will perform the "Symphony of Life"[8,9] as musical instruments. A detailed study of protein structure and function should be undertaken in order to clarify all the aspects of this phenomenon[10,11]. Each protein vibrates in a series of low fundamental frequencies which favour the protein interaction via docking, the formation of molecular aggregates and the upsurge of life. Unfortunately, viruses could also be formed this way but we think that viruses can be destroyed by special combination of these frequencies. It would be worthwhile to search for the frequency distribution on the molecular virus structure as well as to identify the virus frequencies which do not overlap too much with those of their host cells. The understanding of this process and the formulation of a well posed and bounded optimal control problem are expected to be the first step towards the proposal of a non-invasive surgical treatment. In a forthcoming work the potential energy of an anharmonic Morse potential will replace the quadratic approximation leading to a mass-spring force used here. A consideration of bounded optimal control techniques will lead to a new subsequent analysis to be published elsewhere.

7. Acknowledgements

The authors are indebted to an anonymous referee for all suggested improvements of the present work as well as for all enlightening comments on it.

References

1. C. Levinthal, "Are there Pathways to Protein Folding?". J. Chimie Physique, 65(1968)44-45.
2. R.P. Mondaini. A Connection between Atom sites and Amide planes in Protein Structures - In: BIOMAT 2009 International Symposium on Mathematical and Computational Biology. World Scientific Co. Pte. Ltd., 2010.
3. R.P. Mondaini. The Euclidean Steiner Ratio and the measure of Chirality of biomacromolecules. Genetics and Molecular Biology 27 (2004) 658 - 664.
4. R.P.Mondaini. The Steiner Tree Problem and its application to the modeling of Biomolecular Structures - In: Mathematical Modelling of Biosystems, Applied Optimization series, 102, Springer Verlag, 2008.
5. L.S. Pontryagin. The Mathematical Theory of Optimal Processes. Gordon and Breach Science Publishers, 1987.
6. S. Lenhart, J.F. Workman. Optimal Control Applied Models. Chapman & Hall/CRC. Mathematical & Computational Biology,2007.
7. Y. Arkun, M.Gur. Combining Optimal Control Theory and Molecular Dynamics for Protein Folding. PLOS One 7 (2012) 1 - 8.
8. G. Acbas, K.A. Niessen, E.H. Snell, A.G. Markelz. Optical measurements of

Long-range Protein vibrations. Nature Communications 5 (2014) article no. 3076.

9. D.A. Turton, H. M. Senn,T.Harwood, A.J. Lapthorn, E.M. Ellis, K. Wyne. Terahertz underdamped Vibrational Motion governs protein-ligand binding in solution. Nature Communications 5 (2014) article no. 3999.
10. B.Tidor, M. Karplus. The contribution of Vibrational Entropy to Molecular Association. J. Mol. Biol. 238 (1994) 405 - 414.
11. M. Karplus, J. Kuriyan. Molecular Dynamics and Protein Function. Proc. Natl. Acad. Sci. USA 102 (2005) 6679 - 6685.

TIME ASYMMETRY OF CROSS-CORRELATION FUNCTIONS AS A SIGNATURE OF NON EQUILIBRIUM STEADY STATES

A. LEMARCHAND[1,2] AND C. BIANCA[1,2]

[1] *Sorbonne Universités, UPMC Univ Paris 06, UMR 7600, Laboratoire de Physique Théorique de la Matière Condensée, 4, place Jussieu, case courrier 121, 75252 Paris cedex 05, France*
E-mail: anle@lptmc.jussieu.fr

[2] *CNRS, UMR 7600 LPTMC, Paris, France*

The complex spatio-temporal structures that appear in biological systems require far-from-equilibrium conditions which lead to the circulation of reaction fluxes. We investigate in which respect time asymmetry of cross-correlation functions of concentration fluctuations may be exploited to determine non vanishing reaction fluxes at the cellular level. Biochemical mechanisms associated with nontrivial dynamical behaviors such as bistability and oscillations of different natures are studied. Using simulations of the master equation as a reference, we show that, far from a bifurcation, Langevin approach provides a reliable tool to compute analytical expressions for the cross-correlation functions of concentration fluctuations in a non equilibrium steady state. However, the relation between cross-correlation functions and reaction flux sensitively depends on the nonlinearities of chemical dynamics. In the absence of an in-depth knowledge of reaction mechanism, the design of a universal protocol for determining a reaction flux from cross-correlation functions remains a delicate task.

1. Introduction

The interplay between nonlinearity and departure from equilibrium is at the origin of complex spatio-temporal organization in biochemical systems [1,2]. The diversity of the structures developing at the cellular level is related to the nonlinearities of dynamics, and organization occurs at the cost of energy or matter consumption. Sustained exchanges with the exterior, which drive the system of interest in far-from-equilibrium conditions, are necessary to the emergence of organized structures. Adenosine triphosphate (ATP) offers a well-known example of biochemical species whose concentration is regulated and that plays a crucial role in maintaining a biological system in a non equilibrium state. The evaluation of the departure from equilibrium in biology is therefore an essential task. In the absence of reaction flux, all the chemical reactions obey detailed balance and time reversibility is observed. Recent progress in non equilibrium statistical physics have been achieved and time asymmetry of the correlation functions of concentration fluctuations has been shown to be a signature of far-from-equilibrium systems [3,4,5,6,7,8,9,10,11,12].

We consider chemical mechanisms involving two variable concentrations and possessing a stable non equilibrium steady state. When the deterministic dynamics of such systems is linear, the difference of cross-correlation functions of concentration fluctuations has been shown to be universally related to the reaction flux circulating into the system in the non equilibrium steady state [13,14]. We propose to assess whether the determination of the cross-correlation functions can be used to evaluate the reaction flux in the biologically relevant cases of nonlinear deterministic dynamics.

The paper is organized as follows. In the next Section, the method used to derive an analytical expression of the cross-correlation functions from Langevin-type equations is given and the result obtained for linear dynamics is reminded [13,15]. In Section 3, a nonlinear model of enzymatic catalysis is studied and the approximate expression of the cross-correlation functions deduced from the Langevin approach is compared to simulations of the chemical master equation, considered as the reference [15]. Then, chemical systems exhibiting oscillations of different natures are considered. We give the results obtained for the Brusselator model in the vicinity of a Hopf bifurcation [16] in Section 4 and for a system having a saddle-node infinite period bifurcation in Section 5. The possibility to design a method for reaction flux evaluation from the cross-correlation functions in the absence of knowledge of the considered dynamics is discussed in Section 6 which concludes this work.

2. Cross-Correlation Functions Near a Steady State

We consider a chemical mechanism involving two species X and Y of variable concentrations X and Y. At the macroscopic scale, the rate laws are described by ordinary differential equations that are often nonlinear. The considered systems are supposed to possess at least a stable steady state (X_s, Y_s). The chemical system may be maintained far from equilibrium due to sustained exchanges of matter or energy with the exterior. The Langevin approach to chemical kinetics offers the simplest way to incorporate internal fluctuations into the description of small systems [17]. The Langevin equations are simply obtained by adding a Langevin force with zero mean to the deterministic equations. The variances of the Langevin forces have well-known expressions depending on the rate constants and the concentrations X and Y [18,19]. The linearized Langevin equations around the steady state (X_s, Y_s) may be written as:

$$\frac{dx}{dt} = ax + by + \xi_x(t) \tag{1}$$

$$\frac{dy}{dt} = cx + dy + \xi_y(t) \tag{2}$$

where $x = X - X_s$, $y = Y - Y_s$ and the four real parameters a, b, c, and d depend on the rate constants of the considered mechanism. The variances of the Langevin forces $\xi_x(t)$ and $\xi_y(t)$ are evaluated at (X_s, Y_s) and denoted by:

$$\langle \xi_x(t)\xi_x(t')\rangle = F_{xx}\delta(t'-t), \tag{3}$$

$$\langle \xi_x(t)\xi_y(t')\rangle = F_{xy}\delta(t'-t), \tag{4}$$

$$\langle \xi_y(t)\xi_y(t')\rangle = F_{yy}\delta(t'-t). \tag{5}$$

For discussing the stability of the steady state, we compute the eigenvalues $\lambda_\pm$ of the matrix $\mathbf{M}$ associated with Eqs. (1,2):

$$\mathbf{M} = \begin{pmatrix} a & b \\ c & d \end{pmatrix} \tag{6}$$

which are solutions of the characteristic polynomial:

$$\lambda^2 - \lambda(a+d) + ad - bc = 0. \tag{7}$$

The solutions $\lambda_\pm$ can be real or complex numbers.

The new variables $q(t)$ and $r(t)$ in the basis of the eigenvectors are related to $x(t)$ and $y(t)$ according to:

$$\begin{pmatrix} x(t) \\ y(t) \end{pmatrix} = \begin{pmatrix} p_{xx} = \frac{b}{\sqrt{b^2+(\lambda_+-a)^2}} & p_{xy} = b\sqrt{b^2+(\lambda_- - a)^2} \\ p_{yx} = \frac{\lambda_+ - a}{\sqrt{b^2+(\lambda_+-a)^2}} & p_{yy} = \frac{\lambda_- - a}{\sqrt{b^2+(\lambda_- - a)^2}} \end{pmatrix} \begin{pmatrix} q(t) \\ r(t) \end{pmatrix} \tag{8}$$

where p_{xx}, p_{yx} are the coordinates of the eigenvector associated with the eigenvalue λ_+ and p_{xy}, p_{yy} are the coordinates of the eigenvector associated with λ_-.

The correlation functions are easily determined for the new variables and the change of basis matrix defined in Eq. (8) is used to compute the time correlation functions of the concentration fluctuations between times t and t'. In the long time limit, the correlation functions depend only on the time delay $\tau = t' - t$ and the difference $I(\tau)$ of cross-correlation functions is given by [15]:

$$\begin{aligned} I\ (\tau) &= \langle x(0)y(\tau)\rangle - \langle x(\tau)y(0)\rangle \,\text{with}\, x = X - X_s,\ y = Y - Y_s \\ &= -\frac{(p_{yx}p_{yy}F_{xx} - (p_{xx}p_{yy} + p_{xy}p_{yx})F_{xy} + p_{xx}p_{xy}F_{yy})(e^{\lambda_+\tau} - e^{\lambda_-\tau})}{(p_{xx}p_{yy} - p_{xy}p_{yx})(\lambda_+ + \lambda_-)} \end{aligned} \tag{9}$$

When the initial dynamics is linear, the difference of cross-correlation functions is proportional to the reaction flux J, which measures the departure from equilibrium [13,15]:

$$I(\tau) = J\,\frac{e^{\lambda_+\tau} - e^{\lambda_-\tau}}{\lambda_+ - \lambda_-} \tag{10}$$

At equilibrium, the reaction flux J vanishes and consequently, one has $I(\tau) = 0$. When the system is maintained in a non equilibrium steady state, the time symmetry of the cross-correlation functions is lost. Our aim is to check the validity of Eq. (10) in the case of nonlinear chemical mechanisms. Three typical behaviors are investigated: bistability [15], oscillations arising after a Hopf bifurcation [16], and oscillations resulting from a saddle node infinite period bifurcation.

3. Bistable Enzymatic System

In this Section, we consider a modified version of the well-known Michaelis-Menten mechanism for enzymatic catalysis, in which the autocatalytic production of species X by the reverse reaction 1 has been introduced, leading to a possible bistable behavior of the system [15]:

$$3\text{X} + \text{S} \underset{k_{-1}}{\overset{k^*_{+1}}{\rightleftharpoons}} \text{Z} + 2\text{X} \tag{11}$$

$$\text{Z} \underset{k_{-2}}{\overset{k_{+2}}{\rightleftharpoons}} \text{Y} \tag{12}$$

$$\text{Y} \underset{k^*_{-3}}{\overset{k_{+3}}{\rightleftharpoons}} \text{X} + \text{P} \tag{13}$$

The concentrations of species S and P are supposed to be constant due to appropriate exchanges with the surrounding. The notation $k_{+1} = k^*_{+1}S$ and $k_{-3} = k^*_{-3}P$ for apparent rate constants is employed. The laws of chemical kinetics lead to the following deterministic equations for the two variables X and Y:

$$\frac{dX}{dt} = -k_{+1}X^3 + k_{-1}ZX^2 - k_{-3}X + k_{+3}Y \tag{14}$$

$$\frac{dY}{dt} = k_{-3}X - k_{+3}Y + k_{+2}Z - k_{-2}Y \tag{15}$$

where the variable Z may be eliminated using the conservation relation $X + Y + Z = C$, where C is constant. Depending on the parameter values, the system may admit one or three stationary states. In what follows, we set the values of $k_{\pm 1}$, $k_{\pm 2}$, k_{+3} and use k_{-3} as a convenient control parameter. The steady values X_s of the concentration of species X as k_{-3} varies are given in Fig. 1. Bistability is observed for k_{-3} in the range $[0.468, 1.196]$. The reaction flux J, which measures the departure from detailed balance, obeys:

$$J = k_{+1}(X_s)^3 - k_{-1}Z_s(X_s)^2 = k_{+2}Z_s - k_{-2}Y_s = k_{+3}Y_s - k_{-3}X_s \tag{16}$$

where (X_s, Y_s, Z_s), with $Z_s = C - X_s - Y_s$, is a stable stationary state value.

For this system, the four parameters a, b, c and d introduced in Eqs. (1,2) are given by:

$$a = -3k_{+1}(X_s)^2 + k_{-1}(2(C - Y_s)X_s - 3(X_s)^2) - k_{-3}, \tag{17}$$

$$b = k_{+3} - k_{-1}(X_s)^2, \tag{18}$$

$$c = k_{-3} - k_{+2}, \tag{19}$$

$$d = -(k_{+2} + k_{-2} + k_{+3}) \tag{20}$$

and the variances of the Langevin forces by:

$$F_{xx} = k_{+1}(X_s)^3 + k_{-1}Z_s(X_s)^2 + k_{-3}X_s + k_{+3}Y_s \tag{21}$$

$$F_{xy} = -(k_{-3}X_s + k_{+3}Y_s) \tag{22}$$

$$F_{yy} = k_{+2}Z_s + k_{-2}Y_s + k_{-3}X_s + k_{+3}Y_s \tag{23}$$

In the case of the enzymatic system, the explicit calculation of the difference of cross-correlation functions using Eq. (9) leads to [15]:

$$I(\tau) = J\left(1 - \frac{4k_{-3}}{a + d}\right)\frac{e^{\lambda_+\tau} - e^{\lambda_-\tau}}{\lambda_+ - \lambda_-} \tag{24}$$

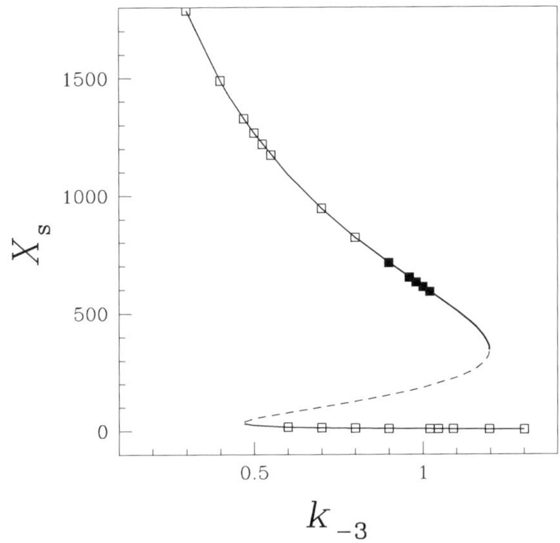

Figure 1. Bifurcation diagram associated with the enzymatic system. The lines give the stationary values X_s of species X concentration, solutions of the nonlinear deterministic equations (Eqs. (14,15)), versus rate constant k_{-3}. The other parameters are fixed at $k_{+1} = 10^{-6}$, $k_{-1} = 4 \times 10^{-6}$, $k_{+2} = 0.5$, $k_{-2} = 0.1$, $k_{+3} = 10^{-3}$. The solid lines are associated with the two stable branches of steady states. The thick line of the upper branch corresponds to focuses associated with complex eigenvalues and the thin lines to nodes associated with real eigenvalues. The dashed line is the unstable branch. The squares give the mean stable steady state values deduced from the solution of the master equation (Eq. (25)) for $N = 10000$ particles and an ensemble average over 10^7 realizations. The solid squares are associated with oscillating correlation functions and the open squares, with non-oscillating correlation functions.

where the reaction flux J is given by Eq. (16), the eigenvalues $\lambda_\pm$ are solutions of Eq. (7) and the two parameters a and d are defined in Eqs. (17,20).

The validity of the linearized Langevin approach is checked by comparing the results given by Eq. (24) and a direct computation of the difference of cross-correlation functions from kinetic Monte Carlo simulations of the master equation for the probability of the numbers N_X, N_Y and N_Z of molecules of species X, Y and Z, respectively[17,18,20]. The master equation associated with the mechanism described in Eqs. (11-13) reads:

$$
\begin{aligned}
&\frac{dP(N_X, N_Y, N_Z)}{dt} = \\
&= k_{+1}[(N_X+1)N_X(N_X-1)P(N_X+1, N_Y, N_Z-1) \\
&-N_X(N_X-1)(N_X-2)P(N_X, N_Y, N_Z)] \\
&+k_{-1}[(N_Z+1)(N_X-1)(N_X-2)P(N_X-1, N_Y, N_Z+1) \\
&-N_Z N_X(N_X-1)P(N_X, N_Y, N_Z)] \\
&+k_{+2}[(N_Z+1)P(N_X, N_Y-1, N_Z+1) - N_Z P(N_X, N_Y, N_Z)] \\
&+k_{-2}[(N_Y+1)P(N_X, N_Y+1, N_Z-1) - N_Y P(N_X, N_Y, N_Z)] \\
&+k_{+3}[(N_Y+1)P(N_X-1, N_Y+1, N_Z) - N_Y P(N_X, N_Y, N_Z)] \\
&+k_{-3}[(N_X+1)P(N_X+1, N_Y-1, N_Z) - N_X P(N_X, N_Y, N_Z)] \quad (25)
\end{aligned}
$$

with $N = N_X + N_Y + N_Z$. We use the algorithm introduced by Gillespie [21,22] to simulate the master equation. Each chemical process of the nonlinear mechanism has an assigned rate $r_{\pm i}$, with $i = 1, 2, 3$, equal to its transition probability in the master equation. For example, when the system contains N_X particles of species X, the transition probability of the forward reaction 1 is given by $r_{+1} = k_{+1}N_X(N_X-1)(N_X-2)$. The total rate $r_{tot} = \sum_{i=1}^{3} r_{\pm i}$ is computed. Starting from a given state (N_X, N_Y) and $N_Z = N - N_X - N_Y$, we randomly choose a process and a waiting time θ sampled from the distribution $r_{tot}\exp(-r_{tot}\theta)$. Time is increased by θ, the new population values and the transition rates are updated according to the changes associated with the chosen reaction. Hence, a stochastic trajectory is generated in the phase space (N_X, N_Y). The quantity $I(\tau)$ defined in Eq. (9) is computed using an ensemble average over a large number of stochastic trajectories.

The results of the three stochastic approaches are compared in Figs. 2, 3 for two values of the control parameter k_{-3}. The difference of cross-correlation functions $I(\tau)$ is calculated in the framework of Langevin approach, using Eq. (10) in the hypothesis of linear deterministic dynamics and Eq. (24) for the nonlinear enzymatic system. The two Langevin approaches are compared to the simulations of the master equation, considered as a reference. As shown in Fig. 2 for a sufficiently small value of k_{-3}, the three approaches are in good agreement. According to Fig. 3 obtained for a larger value of k_{-3}, the Langevin approach applied to nonlinear dynamics remains in good agreement with the results of the master equation and they both depart from the expression of $I(\tau)$ established for linear dynam-

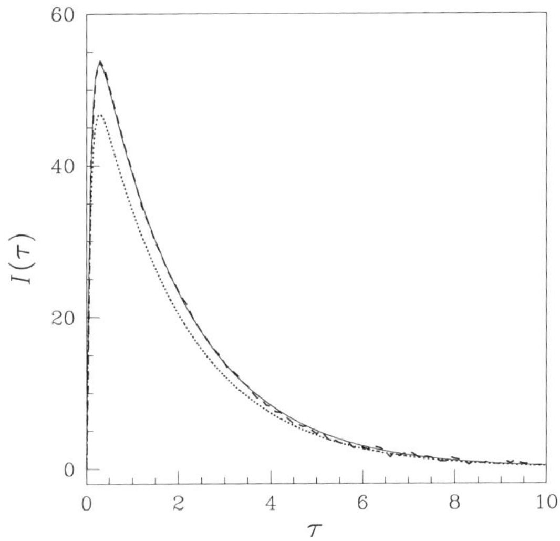

Figure 2. Enzymatic system. Difference of temporal cross-correlation functions $I(\tau) = \langle x(0)y(\tau)\rangle - \langle x(\tau)y(0)\rangle$ versus time lag τ for $k_{-3} = 0.4$ in the domain of monostability associated with a high X_s value. The black dotted line refers to the result of Eq. (10) deduced from Langevin approach to a system with assumed linear dynamics. The thin solid line refers to the result of Eq. (24) deduced from Langevin approach to nonlinear dynamics with autocatalysis. The dashed line refers to the results of the direct simulation of the master equation (Eq. (25)) associated with nonlinear dynamics. The statistics is performed over 10^7 stochastic trajectories. The other parameter values are given in the caption of Fig. 1.

ics. The difference between the results of the Langevin approach to linear dynamics and the analogous approach applied to nonlinear dynamics increases as k_{-3} increases, in agreement with the analytical expression of the correction term, $-\frac{4k_{-3}}{\kappa_1+\kappa_4}$, appearing in Eq. (24). The analytical result given in Eq. (24) can be considered as a good approximation of the correlation functions around the non equilibrium steady state deduced from kinetic Monte Carlo simulations of the master equation. Some differences between the two approaches are discussed in reference [15] in the close vicinity of the bifurcation between monostability and bistability.

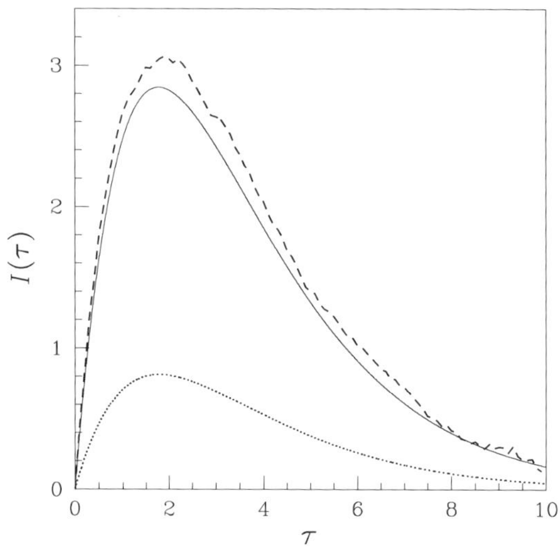

Figure 3. Enzymatic system. Difference of temporal cross-correlation functions $I(\tau) = \langle x(0)y(\tau)\rangle - \langle x(\tau)y(0)\rangle$ versus time lag τ for $k_{-3} = 0.7$ in the bistable domain around the node associated with a lower X_s value. See the caption of Fig. 2 for the meaning of the symbols and the parameter values.

4. Brusselator Model Close to Hopf Bifurcation

We consider the following chemical mechanism, inspired by the Brusselator model [23,24] which may in addition obey detailed balance under appropriate conditions [16]:

$$\mathrm{A} \underset{k_{-1}}{\overset{k_{+1}^{*}}{\rightleftharpoons}} \mathrm{X} \tag{26}$$

$$2\mathrm{X} + \mathrm{Y} \underset{k_{-2}}{\overset{k_{+2}}{\rightleftharpoons}} 3\mathrm{X} \tag{27}$$

$$\mathrm{B} + \mathrm{X} \underset{k_{-3}^{*}}{\overset{k_{+3}^{*}}{\rightleftharpoons}} \mathrm{C} + \mathrm{Y} \tag{28}$$

where the concentration of species A, B, and C is constant. The deterministic dynamics is governed by the following system of ordinary differential equations:

$$\frac{dX}{dt} = k_{+1} - k_{-1}X + k_{+2}X^2Y - k_{-2}X^3 - k_{+3}X + k_{-3}Y \quad (29)$$

$$\frac{dY}{dt} = -k_{+2}X^2Y + k_{-2}X^3 + k_{+3}X - k_{-3}Y \quad (30)$$

where we have set $k_{+1} = k^*_{+1}A$, $k_{+3} = k^*_{+3}B$, and $k_{-3} = k^*_{-3}C$ for apparent rate constants.

The steady state (X_s, Y_s) of the system reads:

$$X_s = \frac{k_1}{k_{-1}} \quad (31)$$

$$Y_s = X_s\frac{k_{+3} + k_{-2}X_s^2}{k_{-3} + k_{+2}X_s^2} \quad (32)$$

and the reaction flux, which vanishes at equilibrium, can be written as:

$$J = k_{+2}X_s^2Y_s - k_{-2}X_s^3 = k_{+3}X_s - k_{-3}Y_s \quad (33)$$

The four parameters a, b, c and d of Eqs. (1,2), which define the linear stability operator, obey:

$$a = -k_{-1} + 2k_{+2}X_sY_s - 3k_{-2}X_s^2 - k_{+3}, \quad (34)$$

$$b = k_{-3} + k_{+2}X_s^2, \quad (35)$$

$$c = -k_{-1} - a, \quad (36)$$

$$d = -b \quad (37)$$

The solution of the characteristic polynomial defined in Eq. (7) are complex and written in the following form: $\lambda_\pm = \alpha \pm i\beta$. A Hopf bifurcation occurs for:

$$a = b \quad (38)$$

For $a - b > 0$, the focus (X_s, Y_s) is unstable and the attractor of the system is a limit cycle associated with oscillations of period $2\pi/\beta$.

In the case of the Brusselator, the variances of the Langevin forces are given by:

$$F_{xx} = k_{+1} + k_{-1}X_s + k_{+2}X_s^2Y_s + k_{-2}X_s^3 + k_{+3}X_s + k_{-3}Y_s, \quad (39)$$

$$F_{xy} = -(k_{+2}X_s^2Y_s + k_{-2}X_s^3 + k_{+3}X_s + k_{-3}Y_s), \quad (40)$$

$$F_{yy} = k_{+2}X_s^2Y_s + k_{-2}X_s^3 + k_{+3}X_s + k_{-3}Y_s. \quad (41)$$

Finally, the expression of the difference of cross-correlation functions may be written as [16]:

$$I(\tau) = \frac{4k_{-1}}{a-b}\, J \exp(\alpha\tau)\, \frac{\sin(\beta\tau)}{\beta} \quad (42)$$

The master equation for the probability of the number of particles N_X and N_Y of species X and Y is written for the chemical scheme given in Eqs. (26-28) [16] and numerically solved using the same lines as for the enzymatic system [21,22].

The two Langevin approaches to the cross-correlation functions and the numerical solutions of the master equation are compared in Figs. 4 and 5 for two values of the control parameter k_{+3}. Sufficiently far from the Hopf bifurcation as shown in Fig. 4, the analytical result given in Eq. (42) following Lanvegin approach to nonlinear dynamics perfectly agrees with the simulations of the master equation, whereas the estimation provided by Eq. (10) for an assumed linear dynamics sensitively differs. Figure 5

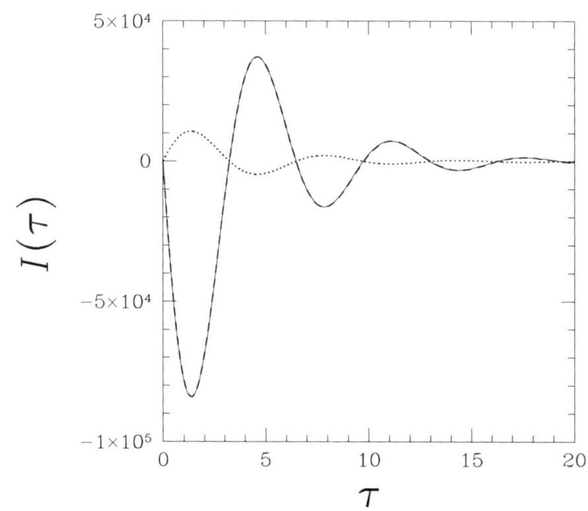

Figure 4. Brusselator model. Difference of temporal cross-correlation functions $I(\tau) = \langle x(0)y(\tau)\rangle - \langle x(\tau)y(0)\rangle$ versus time lag τ for $k_{+3} = 1.5$ in the domain of stability of the focus (X_s, Y_s). The other parameters are fixed at $k_{+1} = 10^4$, $k_{-1} = 1$, $k_{+2} = 10^{-8}$, $k_{-2} = 10^{-11}$, and $k_{-3} = 10^{-3}$. The solid line refers to the result of Eq. (42) deduced from Langevin approach. The dashed line refers to the results of the direct simulation of the master equation. The statistics is performed over 10^6 stochastic trajectories. The dotted line refers to the result of Eq. (10) valid when assuming linear deterministic dynamics for the same values of the reaction flux J and the eigenvalues $\lambda_\pm$ as for nonlinear dynamics.

shows the results of the three stochastic approaches for parameter values close to the Hopf bifurcation associated with $a = b$ according to Eq. (38). In these critical conditions, for which Eq. (42) predicts the divergence of $I(\tau)$, the Langevin approach does not correctly provide the amplitude of oscillation and the three approaches disagree. We will come back on the possible discrepancies between Langevin approach and master equation in the vicinity of a bifurcation in Section 6. In the next Section, we study

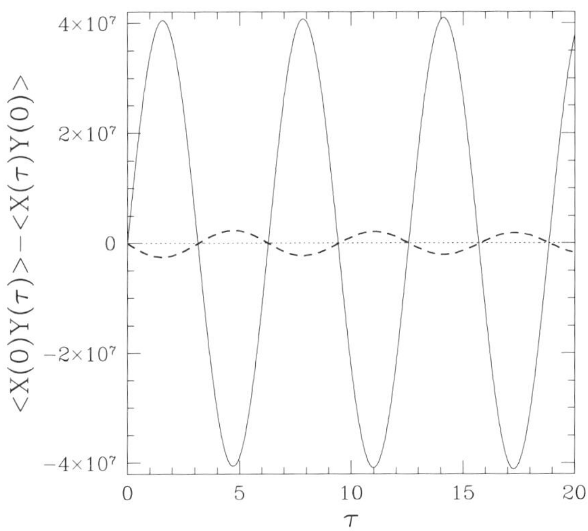

Figure 5. Brusselator model. Same caption as in Fig. 4 for $k_{+3} = 2$, close to the Hopf bifurcation occurring for $k_{+3}^c = 2.008$. The statistics is performed over 10^3 trajectories.

the behavior of the difference of cross-correlation functions $I(\tau)$ as another type of bifurcation leading to oscillations arises.

5. Chemical System Close to a Saddle-Node Infinite Period Bifurcation

We use a slightly modified version of the two-variable chemical mechanism proposed by Erban *et al* [25] in order to investigate the vicinity of a saddle-node infinite period (SNIPER) bifurcation:

$$\mathrm{A} \overset{k^*_{+1}}{\longrightarrow} \mathrm{Y} \overset{k_{+2}}{\longrightarrow} \mathrm{X} \overset{k_{+3}}{\longrightarrow} \mathrm{B} \tag{43}$$

$$\mathrm{B} + 2\mathrm{X} \underset{k_{-4}}{\overset{k^*_{+4}}{\rightleftharpoons}} 3\mathrm{X} \tag{44}$$

$$\mathrm{A} + \mathrm{X} + \mathrm{Y} \overset{k^*_{+5}}{\longrightarrow} \mathrm{X} + 2\mathrm{Y} \tag{45}$$

$$2\mathrm{X} + \mathrm{Y} \overset{k_{+6}}{\longrightarrow} 2\mathrm{X} + \mathrm{B} \tag{46}$$

where we have introduced two species of constant concentrations A and B. Apparent rate constants can be written as $k_{+1} = k^*_{+1}A$, $k_{+4} = k^*_{+4}B$, and $k_{+5} = k^*_{+5}A$. The macroscopic equations governing the dynamics of species X and Y are given by:

$$\frac{dX}{dt} = k_{+2}Y - k_{+3}X + k_{+4}X^2 - k_{-4}X^3 \tag{47}$$

$$\frac{dY}{dt} = k_{+1} - k_{+2}Y + k_{+5}XY - k_{+6}X^2Y \tag{48}$$

The stationary states are defined as the intersections of the nullclines $\frac{dX}{dt} = 0$ and $\frac{dY}{dt} = 0$. Figures 6, 7 give two typical behaviors obtained as the control parameter k_{+1} increases and crosses the critical value $k^c_{+1} \simeq 488$ associated with the saddle-node infinite period bifurcation. As shown in Fig. 7 for $k_{+1} = 440$ above the SNIPER bifurcation, the system has three steady states, an unstable node, a saddle and a stable node (X_s, Y_s). In Fig. 8 for $k_{+1} = 520$, beyond the saddle-node infinite period bifurcation, the system has a single unstable node (UN) surrounded by a stable limit cycle. Contrary to the (supercritical) Hopf bifurcation, which gives birth to a limit cycle with vanishing radius, the saddle-node infinite period bifurcation leads to an already well-formed limit cycle from a homoclinic trajectory. Such a bifurcation can be found in models of cell-cycle regulation, for example to account for a transition between two phases of mitosis [26]. Above the SNIPER bifurcation, we define the reaction flux J at the stable steady state (X_s, Y_s) by:

$$J = k_{+1} + k_{+5}X_sY_s = k_{+3}X_s - k_{+4}(X_s)^2 + k_{-4}(X_s)^3 + k_{+6}(X_s)^2Y_s \tag{49}$$

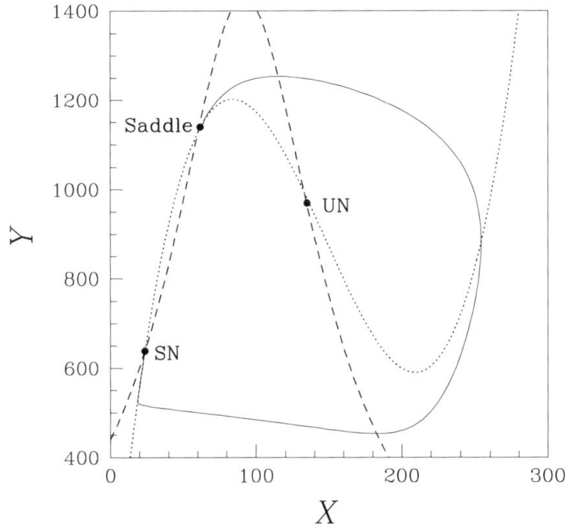

Figure 6. System with SNIPER bifurcation. Nullclines (dashed line and dotted line) obtained for $k_{+1} = 440$, above the saddle-node infinite period bifurcation. The other parameters are fixed at $k_{+2} = 1$, $k_{+3} = 33$, $k_{+4} = 0.275$, $k_{-4} = 0.000625$, $k_{+5} = 0.015$, $k_{+6} = 0.00008125$. The system possesses three steady states, a stable node (SN), a saddle and an unstable node (UN). The thin solid line gives one trajectory connecting the saddle and the stable node.

The four parameters a, b, c and d defining the linear operator in Eqs. (1,2) are given by:

$$a = -k_{+3} + 2k_{+4}X_s - 3k_{-4}(X_s)^2, \tag{50}$$

$$b = k_{+2}, \tag{51}$$

$$c = k_{+5}Y_s - 2k_{+6}X_sY_s, \tag{52}$$

$$d = -k_{+2} + k_{+5}X_s - k_{+6}(X_s)^2 \tag{53}$$

The variances of the Langevin forces are:

$$F_{xx} = k_{+2}Y_s + k_{+3}X_s + k_{+4}(X_s)^2 + k_{-4}(X_s)^3, \tag{54}$$

$$F_{xy} = -k_{+2}Y_s, \tag{55}$$

$$F_{yy} = k_{+1} + k_{+2}Y_s + k_{+5}X_sY_s + k_{+6}(X_s)^2Y_s. \tag{56}$$

Then Eqs. (7,8) are used to compute the eigenvalues of the linear operator and the change of basis matrix. Finally, Eq. (9) gives the difference

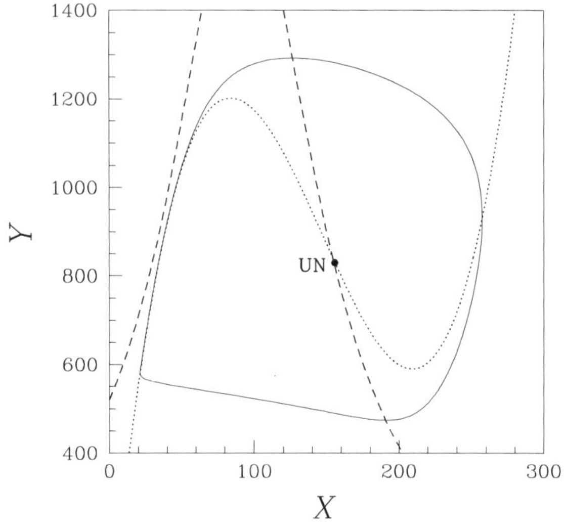

Figure 7. System with SNIPER bifurcation. Same caption as in Fig. 6 for $k_{+1} = 520$, beyond the saddle-node infinite period bifurcation. The system possesses a single unstable node (UN) and a stable limit cycle (solid line).

of cross-correlation function $I(\tau)$. For the sake of conciseness, we do not explicitly write the master equation associated with the SNIPER model, which can be easily written and numerically solved using Gillespie algorithm [21,22].

The results are given in Figs. 8, 9 for $k_{+1} = 440$ far from the SNIPER bifurcation and for $k_{+1} = 487.6$, close to the critical value k^c_{+1}. Exactly as in the case of the Hopf bifurcation, the analytical prediction of the Langevin approach to nonlinear dynamics (Eq. (9)) is validated by the simulations of the master equation as shown in Fig. 8, sufficiently far from the SNIPER bifurcation. In the case of the SNIPER bifurcation, Eq. (10), obtained for linear dynamics, is not able to predict the sign of the flux, i.e. if the chemostats are solicitated to provide species A and absorb species B or the contrary. Close to the SNIPER bifurcation (see Fig. 9), the master equation leads to a deep minimum for the difference of cross-correlation functions whereas the Langevin approaches predict a maximum of small amplitude. Hence, for critical parameter values, the sign of the reaction flux is not correctly predicted by the Langevin approaches even when consider-

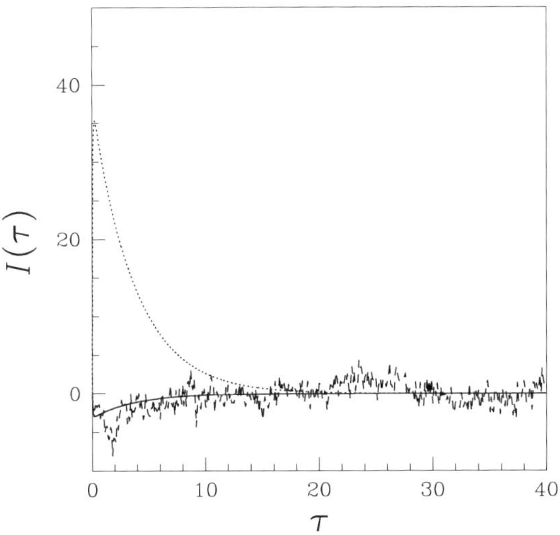

Figure 8. System with SNIPER bifurcation. Difference of temporal cross-correlation functions $I(\tau) = \langle x(0)y(\tau)\rangle - \langle x(\tau)y(0)\rangle$ versus time lag τ for $k_{+1} = 440$ above the saddle-node infinite period bifurcation. The dotted line refers to the result of Eq. (10) deduced from Langevin approach to the system associated with linear deterministic dynamics. The solid line refers to the result of Eq. (9) deduced from Langevin approach to the nonlinear mechanism given in Eqs. (43-46). The dashed line refers to the results of the direct simulation of the master equation associated with Eqs. (43-46). The other parameter values are given in the caption of Fig. 6. The statistics is performed over 10^5 stochastic trajectories.

ing nonlinear dynamics. It is worth to note that the Langevin approaches are local, since the stochastic differential equations are linearized around the steady state even in the case of nonlinear dynamics. The limit cycle formed through the SNIPER bifurcation directly corresponds to a large excursion of the phase space. Clearly, the behavior of the fluctuations is less and less correctly captured by a local stochastic analysis as the bifurcation gets closer.

6. Discussion and Conclusion

Time asymmetry of fluctuations observed in far-from-equilibrium systems [3] incited us to design a method to evaluate a reaction flux in a chemical or biological system with at least two variable concentrations. The dif-

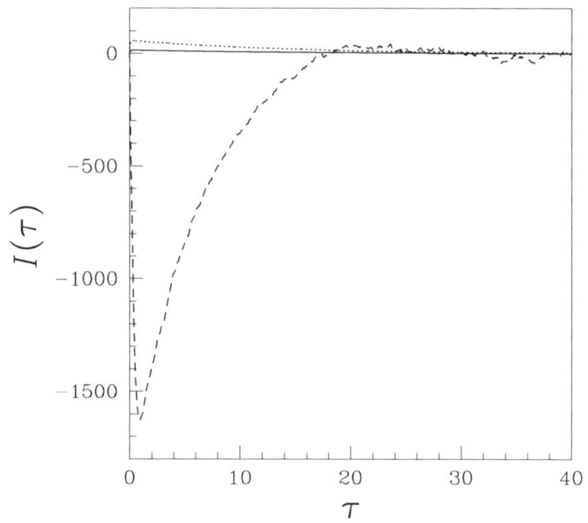

Figure 9. System with SNIPER bifurcation. Same caption as in Fig. 8 for $k_{+1} = 487.6$, close to the saddle-node infinite period bifurcation.

ference $I(\tau)$ of time cross-correlations of concentration fluctuations can be experimentally determined in small biological systems using, for example, tagged fluorescent species and fluctuation correlation spectroscopy (FCS) [14,27]. When the deterministic rate equations for concentrations are linear, the reaction flux can be deduced from the slope of $I(\tau)$ for vanishing time delay τ, without any further knowledge of the system of interest [13].

We have determined approximate, analytical expressions of the difference of time cross-correlations of concentration fluctuations using Langevin-type equations associated with biologically relevant nonlinear chemical mechanisms. Systems exhibiting bistability and periodic oscillations of different natures have been studied. The results have been compared with simulations of the master equation, considered as a reference. For non critical parameter values, the predictions of the Langevin approach to nonlinear dynamics perfectly agree with the results of the master equation.

However, discrepancies between the two approaches are observed in the vicinity of a bifurcation. Indeed, the Langevin equations are deduced from the master equation by a truncation of the Kramers-Moyal expansion to the two first terms, the drift and the diffusion term [17]. The Kramers-Moyal expansion results from a Taylor expansion of the master equation, when the stochastic processes are assumed to involve only small jumps of the random variables [17]. A bifurcation is associated with vanishing real parts of the eigenvalues of the deterministic operator and, consequently, small real parts of the drift term. Hence, terms of order higher than two in the Kramers-Moyal expansion may be not negligible close to a bifurcation and Langevin approach may fail. This phenomenon is observed in the case of the Hopf bifurcation. In the case of the saddle-node infinite period bifurcation, another effect magnifies the discrepancies between the Langevin approach and the master equation. Actually, we use linearized Langevin equations around the steady-state whereas the SNIPER bifurcation would require a global analysis due to the emergence of a directly well-developed attractor.

In the generic case of non critical parameter values, the estimation of reaction flux from the determination of correlation function remains delicate if the chemical system of interest is entirely unknown. Indeed, the Langevin approach to nonlinear dynamics shows that the amplitude of $I(\tau)$ is proportional to the reaction flux but also to a specific combination of parameters which depends on the nonlinearities of dynamics. Assuming linear dynamics and applying the simple, general expression of cross-correlation functions obtained in this case often lead to very bad estimations of the reaction flux. Even the sign of the flux may be wrong.

To harvest the determination of correlation function for reaction flux estimation in a given chemical system, it is therefore essential, first, to identify the reaction mechanism [28], then, to evaluate the associated rate constants [29], and, finally, to compute the specific expression of the difference of time cross-correlation functions for this mechanism. If these demanding requirements may be fulfilled and provided that the system is far from a bifurcation, the stochastic differential equations of Langevin type governing the fluctuating dynamics of concentrations provides a reliable, analytical formula relating the reaction flux and the correlations of fluctuations in the vicinity of a stable steady state. From a theoretical point of view, the interplay between fluctuations and nonlinearities of deterministic dynamics is subtle and leads to specific expressions for the time cross-correlations of concentration fluctuations, that depend on the details of dynamics.

Acknowledgments

We thank l'Agence Nationale de Recherche (ANR T-KiNet project) for support.

References

1. A. Goldbeter, Biochemical oscillations and cellular rhythms (Cambridge, Cambridge) 1996.
2. J. D. Murray, Mathematical biology: I An introduction (Springer, New York) 2002.
3. U. M. B. Marconi, A. Puglisi, L. Rondoni and A. Vulpiani, *Phys. Rep.* **461**, 111 (2008).
4. D.J. Evans, E.G.D. Cohen and G.P. Morriss, *Phys. Rev. Lett.* **71**, 2401 (1993).
5. G. Gallavotti and E. G. D. Cohen, *J. Stat. Phys.* **80**, 931 (1995) .
6. D. J. Evans and D. J. Searles, *Adv. Phys.* **52**, 1529 (2002).
7. P. Gaspard *J. Chem. Phys.* **120**, 8898 (2004).
8. W. Min, L. Jiang, J Yu, S.C. Kou, H. Qian and X. S. Xie, *Nano Lett.* **5**, 2373 (2005).
9. H. Ge, M. Qian and H. Qian, *Phys. Rep.* **510**, 87 (2012).
10. C. Paneni, D. J. Searles and L. Rondoni, *J. Chem. Phys.* **124**, 114109 (2006).
11. J. R. Gomez-Solano, A. Petrosyan, S. Ciliberto, R. Chetrite and K. Gawedzki, *Phys. Rev. Lett.* **103**, 040601 (2009).
12. C. Paneni, D. J. Searles and L. Rondoni, *J. Chem. Phys.* **128**, 164515 (2008).
13. W. J. Heuett and H. Qian, *J. Chem. Phys.* **124**, 044110 (2006).
14. D. R. Sisan, D. Yarar, C. M. Waterman and J. S. Urbach, *Biophys. J.* **98**, 2432 (2010).
15. C. Bianca and A. Lemarchand, *J. Chem. Phys.* **140**, 224105 (2014).
16. C. Bianca and A. Lemarchand, submitted.
17. N. G. van Kampen, Stochastic processes in physics and chemistry (North Holland, Amsterdam) 1983.
18. C. W. Gardiner, Handbook of Stochastic Methods (Springer, Berlin) 1985.
19. D. T. Gillespie, *J. Chem. Phys.* **113**, 297 (2000).
20. G. Nicolis and I. Prigogine, Self-Organization in Nonequilibrium Systems (Wiley, New York) 1977.
21. D. T. Gillespie, *J. Phys. Chem.* **81**, 2340 (1977).
22. D. T. Gillespie, A. Hellander and L. R. Petzold, *J. Chem. Phys.* **138**, (2013) 170901.
23. I. Prigogine and R. Lefever, *J. Chem. Phys.* **48**, 1695 (1968).
24. P. Gaspard, *J. Chem. Phys.* **117**, 8905 (2002).
25. R. Erban, S. J. Chapman, I. G. Kevrekidis and T. Vejchodsky, *SIAM J. Appl. Math.* **70**, 984 (2009).
26. J. Tyson, A. Csikasz-Nagy and B. Nowak, *BioEssays* **24**, 1095 (2002).
27. H. Qian and E. L. Elson, *Proc. Natl. Acad. Sci. USA* **101**, 2828 (2004).
28. A. Lemarchand, H. Berthoumieux, L. Jullien and C. Gosse, *J. Phys. Chem.* **A116**, 8455 (2012).

29. F. Closa, C. Gosse, L. Jullien and A. Lemarchand, *J. Chem. Phys.* **138**, 244109 (2013).

MODELING SUSTAINABLE DEVELOPMENT FOR DECISION TAKING

D. ANGULO
Consiglio Nazionale delle Ricerche, Istituto dei Sistemi Complessi
via Madonna del Piano 10, 50019 Sesto Fiorentino, Italy
Universita degli Studi di Firenze, Dipartamento di Sistemi e Informatica
Viale Morgagni, 65, 50134 Firenze - Italy
E-mails:david.angulo@fi.isc.cnr.it, dangulog@unal.edu.co

G. OLIVAR AND F. ANGULO
Department of Electrical and Electronics Engineering & Computer Sciences
Universidad Nacional de Colombia, sede Manizales
Cra 27 No. 64-60, Manizales, Colombia
Emails: dangulog@unal.edu.co, golivart@unal.edu.co, fangulog@unal.edu.co

This paper presents a system of ordinary differential equations in order to explain each one of the sustainable development dimensions (resource, economy and social growth) and their dependance with population growth. Simulations show that system is highly dependant on social and technological parameters, which induce different types of bifurcations (Branching Points and Hopf). Analysis conclude that for a sustainable scheme, several sustainability actions must be introduced in the system in order to keep equilibrium between each one of the dimensions.

Keywords: Sustainable Development, Dynamical systems, Decision Taking.

1. Introduction

Sustainable development (SD) has been defined as the need to provide necessity of actual generation, without compromising ability of future generations to provide their owns [1]. Under this assumption, multidisciplinary studies have been carried out in order to achieve this goal. In this vein, many mathematic studies have been developed in the framework of modeling each one of the dimensions that regard to the Sustainable Development, all of this applied to decision making science, prevention and social control.

Studies of natural resources dynamics were carried out as a variation of Lotka-Volterra model by Brander and Taylor for Eastern Island dynamics [2]. Several authors introduced modifications so the model could be more realistic [3,4]. On the other hand, economical variable was studied as an important part of the SD schemes by several authors [5,6] most of them applying Cobb-Douglas to the evolution of economy. More complex systems were developed by Prskawetz et al. in [7,8] where human capital and technology are new state variables that account for the social dimension of the SD scheme. Most of these studies were validated with countries and regions with primitive economies (resource extraction, manufacturing, etc.) with very good approaches.

The aim of this work is to couple the 3 dimensions that SD contemplates, namely Social Development, Economy Growth, Resource Management along with their dependance with Population Dynamics, in a unique system of differential equations. Paper is organized as follows. Section 2 develops the dynamical model in a modular way, presenting resource, population, economy and social development dynamics. Simulations of the smooth system, analysis of the phase diagrams and temporal response of state variables are shown in Section 3 together with continuation of equilibrium points and one and two parameters bifurcation diagrams. Finally, conclusions and some prospects of future work are shown in Section 4.

2. Model

2.1. *Population-Resources Dynamics*

The dynamical model follows the guidelines proposed by Brander-Taylor model of population-resources dynamics in Eastern Island. According to this, we can describe the variation over time of the resource stock (S) and the population (L) as shown in equation (1).

$$\begin{cases} \dot{L} = (b - d + \phi\alpha\beta S)L \\ \dot{S} = \left(\rho\left(1 - S/k_1\right)\left(S/k_2 - 1\right) - \alpha\beta L\right)S \end{cases} \tag{1}$$

Growth rate of S depends on a resource regeneration function minus the extraction flow function which is proportional to the available resource in the environment and population. Moreover, population growth rate is related to a natural death rate assuming that there is no resource at all, plus a fertility function associated to the extraction as well. Parameters k_1, k_2 and ρ define the regeneration function; β refers to the percentage of population dedicated to extraction; α is a technological parameter and ϕ tells us the demand of resource in order to keep a positive population growth [3].

2.2. *Economy Dynamics*

The Cobb-Douglas production function is a widely used functional that expresses the relationship between an output and its inputs in economy. Formerly proposed by Knut Wicksell and tested with statistical evidence of USA economy by Charles Cobb and Paul Douglas in 1928 [9].

The proposed functional has the form

$$\Gamma = \gamma K^{q_1} B^{q_2} \tag{2}$$

Where K denotes the capital input, B denotes the labor input (population involved in production), γ is a constant called factor productivity and Γ denotes the total productivity. Exponents q_1 and q_2 refers to elasticities in economy. By using Cobb-Douglas production function we can define a simple economy model of the type:

$$\dot{K} = -\delta K + \gamma K^{q_1} (\alpha\beta LS)^{q_2} - \lambda L \tag{3}$$

We have made several assumptions in this simplified model of economy. First, only the secondary sector (resource extraction and manufacturing) accounts for the economy growth. The labor involved in production is equivalent to extraction flow, and capital growth is reduced by a depreciation factor δ and a consumption function linearly dependent to L.

2.3. *Social Development Dynamics*

Quantification of social development is not a simple task, since there are not mesurable units that can describe it. However many authors have associated social development with human capital, knowledge and education. Social development (H) shall increase with economy investment but shall decrease when population reaches critical values (overpopulation or underpopulation). Under these assumptions change ratio of H takes the form:

$$\dot{H} = \frac{HL}{1 + e^{-HK(1-H)}} \left(L/L_m - 1\right)\left(1 - L/L_M\right) \tag{4}$$

This expression guarantees an approximation to the range of H ($H \in (0,1]$) in such way that when H reaches certain values (1 in our case), its growth is very small. This model of social growth is very similar to a well known development indicator, namely Human Development Index. Since H is closely related to education and knowledge, we can make α (technological parameter) a linear function of H. Finally we obtain the non-linear system

$$\begin{cases} \dot{L} = (b - d + \phi\alpha(H)\beta S)\, L \\ \dot{S} = (\rho\,(1 - S/k_1)\,(S/k_2 - 1) - \alpha(H)\beta L)\, S \\ \dot{K} = -\delta K + \gamma K^{q_1}(\alpha(H)\beta LS)^{q_2} - \lambda L \\ \dot{H} = \dfrac{HL}{1 + e^{-HK(1-H)}}\,(L/L_m - 1)\,(1 - L/L_M) \end{cases} \tag{5}$$

where $\alpha(H) = \alpha_0 - \Delta\alpha(1 - 2H)$. α_0 is taken as the original parameter used by Brander and Taylor in order to guarantee a survivance of population and resource stock, and $\Delta\alpha$ is defined as the maximum variation of technology, assuming that technological development is bounded.

3. Simulations

Simulations were carried out during 1×10^6 time units to guarantee the steady state of the system. In order to analyze the steady state on phase diagrams, only last 5×10^5 steps were taken for the graphics. However, entire temporal response is shown so the transitory state of variables could be studied. Since technological function ($\alpha(H)$) is an important issue in sustainability, parameter $\Delta\alpha$ was chosen as a significative parameter for bifurcation diagrams. Remainder parameters are considered invariant and their values are shown in Table I.

$S(0)$	12000	b	1
$L(0)$	200	d	1.1
$K(0)$	300	q_1	0.25
$H(0)$	0.1	q_2	0.75
ρ	0.025	α_0	1×10^{-4}
$K1$	700	γ	1.8×10^{-4}
$K2$	12000	δ	1×10^{-3}
L_m	2700	L_M	3700
ϕ	0.55	β	0.29

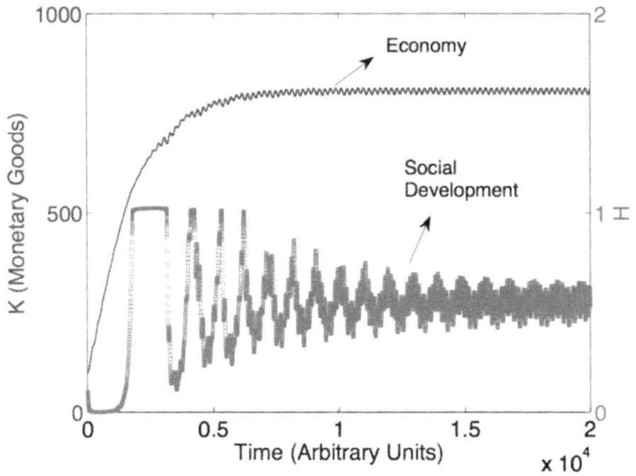

(a) Economy and Social Development Vs Time

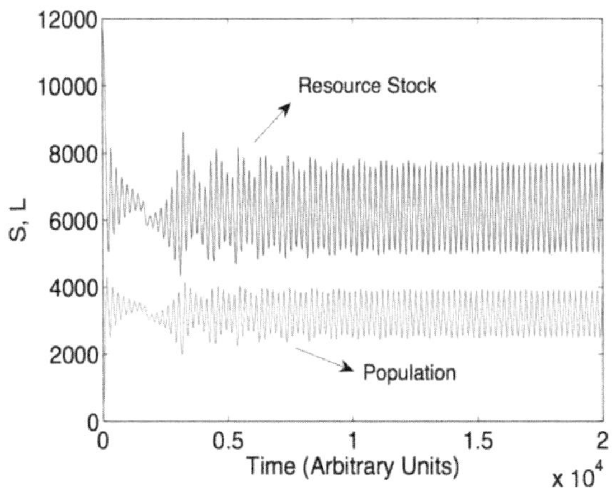

(b) Population and Resource Stock Vs Time

Figure 1. Temporal Response of State Variables. $\Delta\alpha = 5 \times 10^{-6}$.

Figures 1(a) and 1(b) show transient dynamics of all of the state variables. First 15000 units of time present a very rich non-linear dynamic due to the fast growth in economy and social development. Nevertheless, the growth in population makes that consumption function increases, so economy establishes around an equilibrium value. As economy is kept around the equilibrium, social development only depends on population dynamics, so we observe several oscillations of H due to L oscillations. These changes always occur around the defined values of overpopulation and underpopulation L_M and L_m respectively. After 20000 steps system reaches the steady state presenting periodic solutions.

Figures 2(a) to 3(d) show 2 dimensional projections of the system dynamical behavior of equation (5) for two different values of $\Delta\alpha$. We observe periodic orbits in all projections due to the variation of knowledge-education function α closely related to H. An increase in the value of $\Delta\alpha$, increases dramatically the transient behavior and produces period doubling as shown in Figures 3(a) to 3(d).

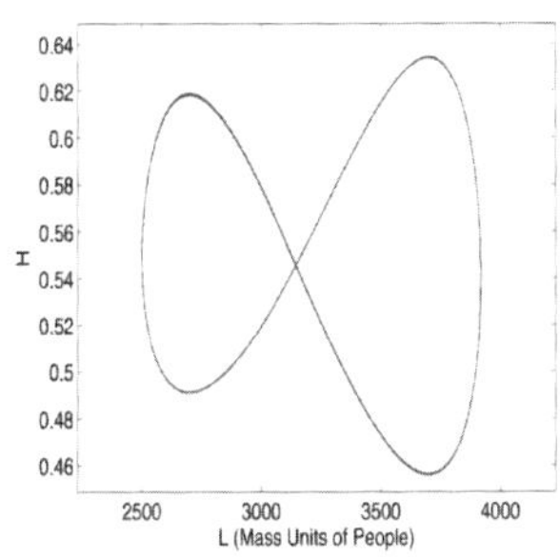

(a) Social Development Vs Population

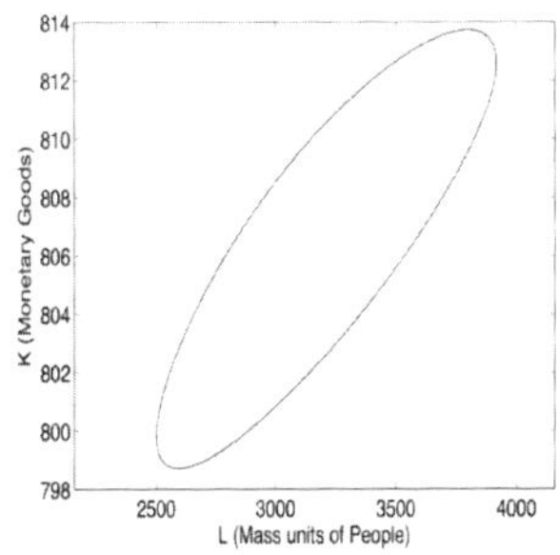

(b) Economy Vs Population

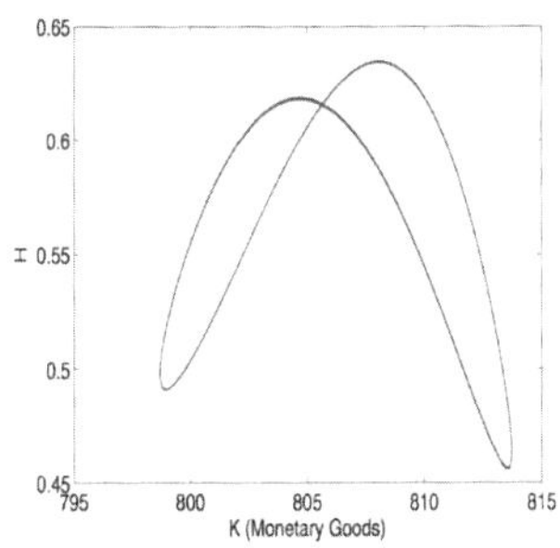

(c) Social Development Vs Economy

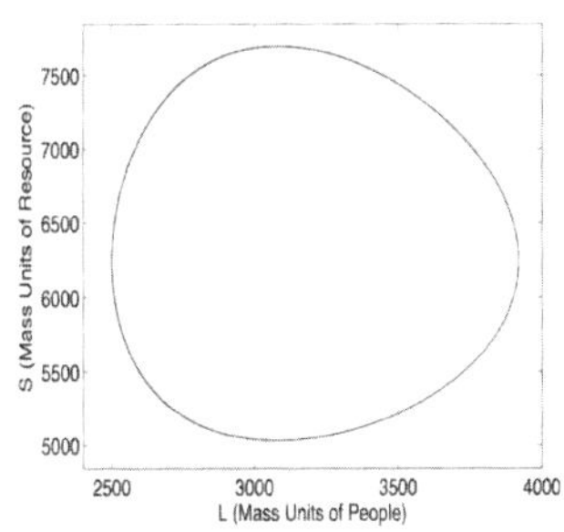

(d) Resource Stock Vs Population

Figure 2. 2-D Phase Diagrams of State Variables. $\Delta\alpha = 5 \times 10^{-6}$. Time steps to reach the steady state: 20000

Three manifolds of equilibrium were found in our system, where variable H may take any value (line of equilibrium), and several equilibrium points dependant on $\Delta\alpha$. A stable equilibrium was chosen in order to compute

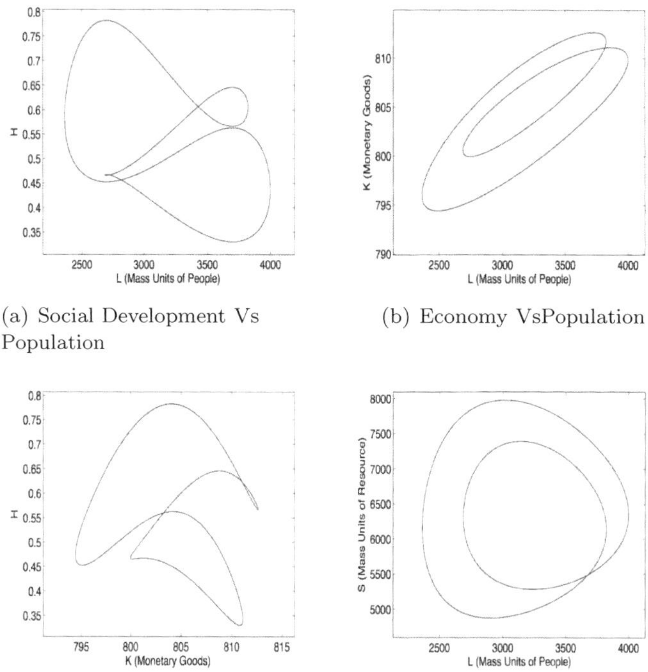

(a) Social Development Vs Population

(b) Economy VsPopulation

(c) Social Development Vs Economy

(d) Resource Stock Vs Population

Figure 3. 2-D Phase Diagrams of State Variables. $\Delta\alpha = 1 \times 10^{-5}$. Time steps to reach the steady state: 500000

the continuation of equilibria by using continuation software MATCONT. Bifurcation diagrams in Figures 4(a) and 4(b) show different bifurcations for variations of technological parameter $\Delta\alpha$.

The path of equilibrium points of variable L presents Branch Points (BP) bifurcations, the values of L where BP's are present are L_M and L_m. In the same vein, equilibrium branch on $H = 0$ presents the same Branch Points at L_M and L_m, in both positive and negative values of H although only non-negative bifurcation is represented in Figure 4(b). Observe that non-trivial equilibrium is achieved by following any of the unstable branches on BP. However, these values of non-trivial equilibrium are not possible to obtain because of the high values of $\Delta\alpha$.

A Hopf bifurcation can be observed at $\Delta\alpha = 1.26 \times 10^{-6}$, codimension 2 bifurcation diagram was computed taking parameters L_M and L_m Vs $\Delta\alpha$

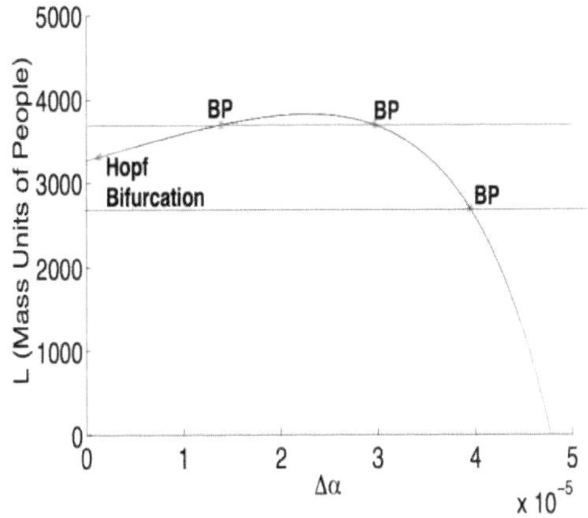

(a) Population Vs $\Delta\alpha$

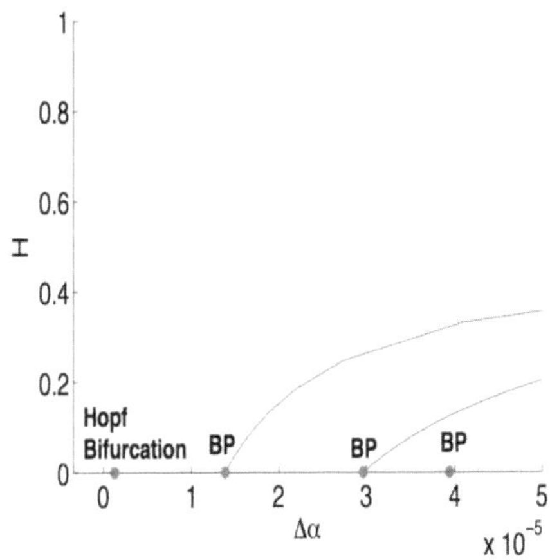

(b) Social Development Vs $\Delta\alpha$

Figure 4. Bifurcation Diagram of Population and Social Development. $\Delta\alpha = 5 \times 10^{-6}$

as shown in Figure 5. Hopf branch is only present for the specific value of $\Delta\alpha = 1.2 \times 10^{-6}$ so its presented as a vertical line. As overpopulation and underpopulation parameters are changed Hopf-Hopf and Zero-Hopf codimension-2 bifurcations may arise in a very small neighborhood which clearly demonstrates the importance of the carrying capacity of a region.

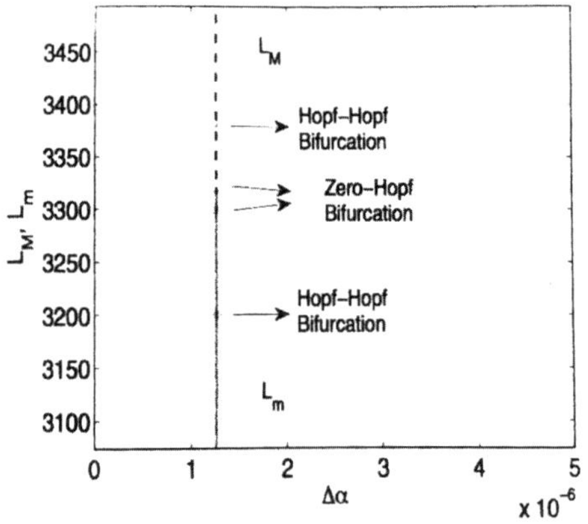

Figure 5. Codimension 2 Bifurcations. Continuation of Hopf bifurcation was carried out by taking L_M and $\Delta\alpha$ as parameters. Similar analysis was made taking L_m and $\Delta\alpha$

4. Conclusions

Simulations of our dynamical system showed rich non-linear behavior. Technological parameter seems to play a determinant parameter in complex behavior of the model, allowing appearance of different codimension 1 bifurcations such as Branch Points and Hopf. In the same way, codimension 2 bifurcations (Zero-Hopf and Hopf-Hopf) are consequence of the variation of population parameters L_M and L_m. Simulations with selected parameters do not show non-trivial unless technological perturbation $\Delta\alpha$ takes very high (unreal) values around. This is an interesting fact because it can be concluded that taking real values of parameters, a dynamical sustainable model is always an oscillating system. Future work is proposed in order to study the non-smooth behavior that can affect the economy, introducing control mechanisms in consumption or depreciation when the income per capita (K/L) reaches values below some crisis parameters. This assumptions can lead to a rich analysis of non-smooth type behavior such as sliding, as well as non-smooth bifurcations [10]. Detailed discussion about socioeconomic interpretations of the model, together with significance and repercussions of codimension 2 bifurcations, is being developed by the authors as future work as well.

References

1. WCED, *Oxford University Press, New York* (1987).

2. J. A. Brander and M. Scott Taylor, *The American Economics Review* **88**, (1998).
3. S. D'Alessandro, *Ecological Economics* **62**, (2007).
4. R. Reuveny and C. S. Decker, *Ecological Economics* **35**, (2000).
5. J.C.V. Pezzey and M.A. Toman, *Resources for the Future* **DP**, (2002).
6. M. De Lara and L. Doyen, *Sustainable Management of Natural Resources Mathematical Models and Methods* **Springer-Verlag**, (2008).
7. A. Prskawetz, G. Steinmann and G. Feichtinger, *Mathematical Population Studies* **7**, (2000).
8. A. Prskawetz, G. Feichtinger, M. Luptacik, A. Milik, F. Wirl, F. Hof and W. Lutz *European Journal of Population* **14**, (1999).
9. C.W. Cobb and P.H. Douglas *The American Economic Review* **18**, (1928).
10. Yu.A. Kuznetsov, *Elements of Applied Bifurcation Theory* **Springer-Verlag**, (2004).

WILD HERBIVORES IN FORESTS: FOUR CASE STUDIES*

GIORGIO SABETTA, EMMA PERRACCHIONE AND EZIO VENTURINO†
Dipartimento di Matematica "Giuseppe Peano",
Università di Torino,
via Carlo Alberto 10, 10123 Torino, Italy
email: ezio.venturino@unito.it

To model the interaction of herbivores with natural resources, we consider a three population system with a top predator population, i.e. the herbivores, and two prey populations, grass and trees, following the ideas first presented for the case of urban parks management. We apply the model for the study the specific situation of four other natural mountain parks in Northern Italy, three located in the Eastern Alps, two of which in the Dolomites and one in the Julian Alps, and one in the Maritime Alps, Northwest Italy. The simulations, based on actual data gathered from contacts with rangers and parks administrators, field samplings and published material, provide useful information on the behavior of the vegetation-wild herbivores interactions and the possible medium-long term evolution of these ecosystems. At the same time they show that these ecosystems are in a very delicate situation, for which the animal populations could become extinguished in case of adverse environmental conditions. A sensitivity analysis and the determination of the so-called *sensitivity surfaces* support our findings and indicate some possible preventive measures to the park admistrators.

1. Introduction

The management of wild parks poses difficult questions to the administrators. The mutual relationships that arise between the various animal and vegetation populations living in them constitute a very complex network of interactions. To intervene in such situations by trying to modify one or some of the parameter values related to these interactions may lead to consequences that are hard to assess and might even be the opposite of the desired outcomes.

In this paper we use a previously built dynamical system model for the

*This work has been partially supported by "Metodi numerici in teoria delle popolazioni" of the Dipartimento di Matematica "Giuseppe Peano".
†Corresponding author

interaction of one top predator, here identified as the herbivores population, and two prey, namely the grass and the trees, to study the evolution in the medium term of a few recently created natural parks in Northern Italy, [17].

In spite of the fact that the first models of this kind were built in order to assess the damages that the vegetation, in particular the trees, suffer from overexploitation by the herbivores, [18], the outcome of this investigation points out that, in reality and quite surprisingly, it is the herbivore population that at present is the most endangered one, especially in view of the assumption made in the modelling process. Indeed, we have deliberately excluded possible further negative effects on the herbivores due to their natural predators. This assumption should probably be removed in further studies, as wolves apparently begin to populate at least part of the Western Alps, [3], while as mentioned later on, even the bears have reappeared in the Eastern Alps. But introducing the predators will most likely only worsen the present situation. To assess the herbivores population risk of extinction, we carry out a sensitivity analysis on the model parameters. Moreover, we reconstruct numerically the sensitivity surfaces which enable us to suggest some measures in order to possibly prevent extinction.

The paper is organized as follows. In the next Section, for easiness of the reader, we briefly outline the model already introduced in [17,18]. Section 3 contains the analysis of the four parks taken in consideration, namely: the Dolomiti Bellunesi Park, the Dolomiti Friulane Park, the Alpi Marittime Park and finally the Prealpi Giulie Park. Section 4 contains the sensitivity analysis and in the final Section we discuss these whole ecosystems survival possibilities at the coexistence level, i.e. with all the population thriving.

2. Background on the Model

For the benefit of the reader we recall here the model introduced in [17] that will be used in the subsequent sections for the analysis of the various specific situations and its main features.

Let H, G and T represent respectively the herbivores, grass and trees populations of the environment in consideration. Apart from the obvious ecological fact that the two types of vegetation are different, in that grass grows fast but has a low carrying capacity for surface unit, while trees grow slowly but they contain a large biomass, the reason for considering trees in this context is due to the phenomenon of debarking, that occurs especially when resources are scarce. In such situation, herbivores searching for alternative food tend to bite off stripes of barks from the trees. They thus interrupt the canals that from the roots go up to the leaves to take there the nutrients absorbed from the ground. In this way they damage the tree, sensibly diminishing the clorophyll production up to totally preventing it from occurring. This in due time leads to the tree death.

The model, [17,18], is a classical predator with two prey system, in which the resources are consumed following a concave response function, usually called the Beddington-De Angelis function, [2,9]. It has the feature of expressing somewhat the concept of feeding switching, [19,20,13,14,15], for which herbivores, as stated above, turn to the second resource when the main one is scarce. The Beddington-De Angelis function prevents the herbivores from consuming more than the available amount of grass even if their population becomes very large. Similarly if there is a huge amount of grass the per capita quantity eaten by the herbivores cannot exceed their per capita maximal consumption, α^{-1}. Similarly, β is the inverse of the herbivores maximal consumption of trees. Letting r_1 and r_2 denote the grass and trees growth rates and K_1 and K_2 their respective carrying capacities, μ the metabolic rate of herbivores, c and g the half saturation constants, $e \leq 1$ and $f \leq 1$ the conversion factors of food into new herbivores biomass and a and b the daily feeding rates due to grass and tress, respectively, the model thus reads as follows

$$\dot{H} = -\mu H + ae\frac{H\ G}{c + H + \alpha\ G} + bf\frac{H\ T}{g + H + \beta\ T + \alpha\ G}$$

$$\dot{G} = r_1 G\left(1 - \frac{G}{K_1}\right) - a\frac{H\ G}{c + H + \alpha G} \tag{1}$$

$$\dot{T} = r_2 T\left(1 - \frac{T}{K_2}\right) - b\frac{H\ T}{g + H + \beta\ T + \alpha\ G}.$$

All parameters are nonnegative. K_1, K_2, c and g are measured in biomass, $e \leq 1$, $f \leq 1$, α and β are pure numbers, μ, r_1, r_2, a and b are rates.

This model has a few equilibria. Coexistence can only be assessed via numerical simulations, in view of the high nonlinearities appearing in (1). The origin is always unstable, preventing ecosystem collapse. Of the remaining possible equilibria, we mention here only the herbivore-free point $E_3 = (0, K_1, K_2)$ and the forest-free equilibrium $E_4 = (H_4, G_4, 0)$ because they play a role also in this investigation. In fact, the stability of equilibrium with loss of woods, representing a severe damage for the environment, motivated the earlier investigations [17,18]. Here we will need their feasibility and stability conditions given by, [17,18]: stability of E_3

$$ae\frac{K_1}{c+\alpha\ K_1}+bf\frac{K_2}{g+\alpha\ K_1+\beta\ K_2}<\mu\ ; \tag{2}$$

feasibility of E_4

$$G_4>\frac{c\mu}{ae-\mu\alpha}\ ; \tag{3}$$

stability of E_4:

$$r_2\,(g+H_4+\alpha\ G_4)<H_4\ . \tag{4}$$

3. Parameter estimation and simulations

In this Section we discuss the parameters that are common to all parks. The remaining parameters, namely the carrying capacities K_1 and K_2 and the grass and trees half saturation constants c and g, are park-dependent and will be discussed in next subsections.

To estimate the parameters we refer to [12,17,18]. Specifically, in [12] tables providing estimation of the annual net primary production of several environments are shown. The latter allow to estimate the growth rates of grass and trees, that in all the considered natural parks are about $r_1 = 0.01$ and $r_2 = 0.0006$. In such estimate we take into account that the growth period of grass and trees can be estimated to be about 120 days every year.

To set the parameter $\mu = 0.03$ we refer to [17]. It means that an herbivore with no food available dies in about 30 days. This is consistent with similar mammals not accustomed to a lethargic period.

The parameter α, i.e. the inverse of the per capita maximal consumption of grass, can be approximated as a percentage of the herbivore itself; this percentage is about $4\%-5.5\%$, [18]. In the following we fix $\alpha = 0.05^{-1}$. Even if the parameter β is the analogous of α for trees, its estimation is different. In fact when an herbivore switches its attention to a tree it may cause the death of the whole tree even if it takes a small piece of barks. In [18], in case of sheep, β is estimated to be 1. However the tree death in case of our herbivores does not occur with likelihood higher than the one due to sheep because a wild herbivore peels off vertical stripes of bark (and not circular stripes of barks at the bottom of the tree). This difference implies that the communication between roots and leaves is not totally interrupted. As a consequence, remarking that β is the inverse of the per capita maximal consumption of trees by herbivores, we set $\beta = 8$.

We remark that e and f are the herbivores assimilation coefficients of grass and trees, respectively. Since grass represents the preferred resource and moreover herbivores can survive by eating grass alone, it means that an adequate amount of grass can satisfy the metabolic needs of herbivores, i.e. $e >> f$. Similarly, taking into account the above consideration $a >> b$ must hold. In the following we set $e = 0.605$, $f = 0.001$, $a = 0.98$ and $b = 0.002$ as reference values.

In our simulations all the above parameters have a fixed value, since they are about the same in each park. Anyway, to better assess the possible uncertainties in the parameters estimation, a sensitivity analysis is carried out in Section 4.

A common remark for the subsequent analysis is the fact that the whole ecosystems dynamics is analysed in the absence of wolf and other herbivores predators and of possible human activities and interferences.

Further, the herbivores-free equilibrium E_3 in all parks is unstable. The left-hand side of (2) has indeed the values 0.0302 for the Dolomiti Bellunesi Park and for the Dolomiti Friulane Park, and 0.0301 for the Alpi Marittime and the Prealpi Giulie, while the right hand side is always $\mu = 0.03$. The equilibrium E_4 stating the extinction of the forests is feasible in all the parks. In fact the right-hand side of (3) and the minimum of G_4 in two centuries have respectively the values 6.78×10^5 and 1.83×10^6 for the Dolomiti Bellunesi, 1.05×10^6 and 2.52×10^6 for the Dolomiti Friulane, 4.63×10^5 and 1.19×10^6 for the Alpi Marittime, 1.56×10^5 and 3.43×10^5 for the Prealpi Giulie parks. However E_4 is never stable, condition (4) is not verified. For each case we find indeed the following values of the left and right hand sides of (4): Dolomiti Bellunesi: 6.9×10^5 and 3.2×10^5; Dolomiti Friulane: 5.2×10^5 and 2.7×10^5; Alpi Marittime: 2.9×10^5 and 2.1×10^5; Prealpi Giulie: 1.13×10^5 and 8.7×10^4. Nevertheless, we will see later in assessing the sensitivity surface that in spite of these results, the possibility of herbivores extinction is real. No other equilibrium exists at a stable level with one or more vanishing populations.

3.1. *Dolomiti Bellunesi National Park*

The Dolomiti Bellunesi National Park was established in 1990. It is located in the Veneto Region (NE Italy) in the territory of 15 municipalities, with a surface of about 32000 ha. of which 8000 ha. of grass and about 19000 of forests; in the remaining part there are rivers, lake, pastures and rocks, which are excluded from our study. About 16000 ha. comprise 8 naturals reserves belonging to the biogenetic reserve of the Council of Europe. The territory includes medium and high mountain areas with altitudes between 400 and 2565 meters above sea level, the highest peak being the Schiara mountain.

There are some protected areas near the park that contribute to construct a large biogeographic network. The park includes a great environmental variety, allowing many animal species to find suitable living and reproducing conditions. Nowadays the park harbors about 115 birds species, 20 amphibian and reptile species, about 100 butterfly species and 50 beetle species. Here some insect species are present that are found nowhere else in the world.

In this park there is a relevant herbivore population. Several species are present, estimated by the park rangers, among which about 3000 chamois (*Rupicapra rupicapra*), whose average weight is about 50 Kg., about 2000 roes (*Capreolus capreolus*), their average weight being around 25 Kg., 250 mouflons (*Ovis aries*), with an average weight of 35 Kg. and 300 deers (*Cervus elaphus*), each one on average weighting about 200 Kg. These last species occupy the most elevated areas. They exhibit seasonal migrations, climbing upwards in the summer and descending in the valley during the winter.

Let us recall here that the biomass unit is given in tons for all the model populations, the hectares given in the Tables for measuring the amount of grass and trees are suitably converted into biomass tons. On the basis of the above data and following [12] for the conversions, we find the carrying capacities and the half saturation constants to be $K_1 = 3469640.64$, $K_2 = 15695993.39$, $c = 101862.16$ and $g = 1001229580.18$. We set the initial conditions using the rangers data and the park vegetation distribution:

$$H(0) = 268.750 \qquad G(0) = 2313093.76 \qquad T(0) = 1046399.56. \qquad (5)$$

We ran our Matlab code to assess the system's behavior in the medium term timespan. Our simulations show that under undisturbed conditions, the system could reach a stable coexistence equilibrium in about 150 years.

There is a continuous increase of the herbivores population until they reach the equilibrium where they become 647.44 times the initial condition. Due to a initial slow increase of herbivores, the grass population has a fast initial increase. After that, when herbivores grows a little faster, the grass population starts to decrease until the equilibrium where it becomes about the 79% of the initial condition. The trees population has as imilar behavior. The latter has an initial growth, a little slower than the grass

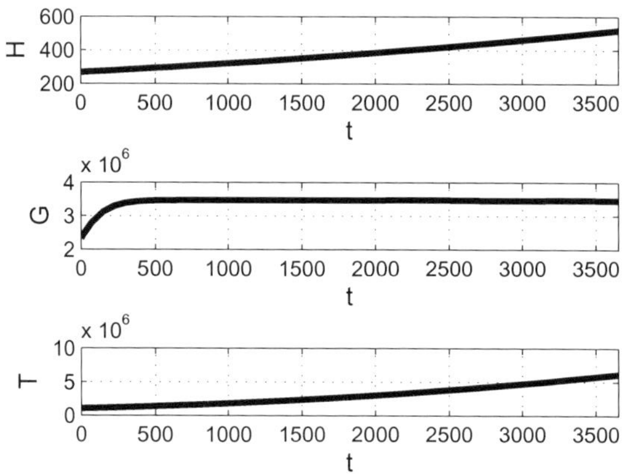

Figure 1. Dolomiti Bellunesi's system evolution over ten years with initial conditions set as in Equation (5).

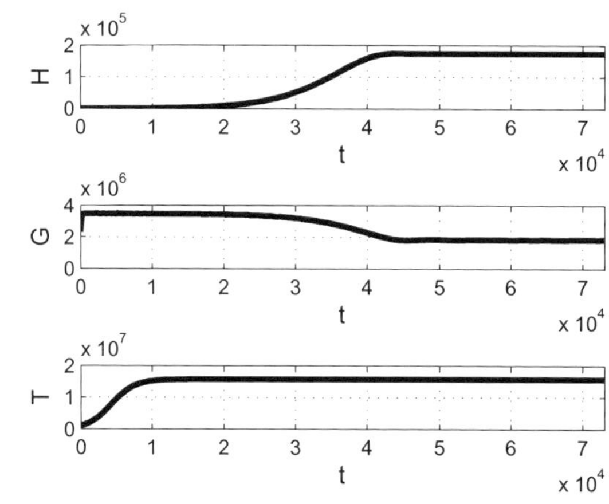

Figure 2. Dolomiti Bellunesi's system evolution over two centuries with initial conditions set as in Equation (5).

population. After that, it stabilizes at the equilibrium. See Figure 1 for the short range and Figure 2 for the medium-long term.

3.2. *Dolomiti Friulane Natural Park*

The park, established in 1996, is inserted in the mountains above the high plains of the Friuli-Venezia Giulia region (NE Italy).

The territory has high geological, environmental and natural interest, with a high degree of wilderness, particularly evident due to the absence of many communication roads. Three major tectonic lines are present in this area. The widespread presence, up to several thousands of years ago, of ancient glaciers characterized its geomorphological aspect. In much more recent times, we recall the 1963 catastrophe due to the great landslide of Monte Toc into the artificial lake of Vajont. It constitutes a unique example of colossal landslide.

The park extends for 37275.49 hectares, its composition is shown in Table 1, the acronym **CLC code** stands for the **CORINE-Land Cover code**.

Table 1. Dolomiti Friulane Natural Park environmental composition.

CLC code	**Denomination**	**Area (ha)**	**%**
122	Railways, roads and technical infrastructures	1.07	0.003%
131	Mining areas	2.56	0.01%
3113	Mixed forest of broad-leaf (*Fraxinus, Orno-ostrietum*)	576.01	1.5%
3115	Beech forest (*Fagus sylvatica L.*)	9247.13	24.8%
3122	Pine forest (*Pinus nigra, Pinus nigra laricio, Pinus sylvestris, Pinus Heldreichii*)	2751.61	7.4%
3123	Fir forest (*Picea abies, Abies alba*)	2020.34	5.4%
3124	Larch (*Larix*) and pine (*Pinus cembra*) forest	474.02	1.3%
3131	Conifers and broad-leaf forest	2957.55	7.9%
3211	Continuous meadows	2931.26	7.9%
3212	Discontinuous meadows	891.00	2.4%
322	Moorland and bushed land	8370.44	22.5%
332	Rocks	3367.80	9.0%
333	Thin vegetation areas	3060.52	8.2%
411	Internal swamp	1.77	0.005%
511	Rivers, drains and waterways	474.99	1.3%
512	Ponds	147.42	0.4%
	Total	37275.49	100%

The park can be regarded as wild, as there are no pastures due to the poor anthropization, a further sign of this fact being the presence of the golden eagle. There are about 2500 chamois (*Rupicapra rupicapra*), 300 deer (*Cervus elaphus*), 800 roes (*Capreolus capreolus*) and a large herd of rock goat (*Capra ibex*), with an average weight of about 85 Kg. The latter population consists nowadays of at least 196 individuals versus the 239 of 2008, but in spite of this it presently appears to be in expansion, due to an increase of the colonization and an increase of reproductive capabilities. In the last three years an increase of about 10% in the population of chamois (*Rupicapra rupicapra*) has been observed, probably as a rebound with respect to the decrease of the previous years: 2586 individuals in 2006, 2476 in 2007 and 2373 in 2008.

For the simulations we use the following parameters based on the data above: $K_1 = 4385112.18$, $K_2 = 10464154.82$, $c = 155246.86$ and $g = 804938520.99$, with initial conditions:

$$H_0 = 221.660 \qquad G_0 = 2923408.12 \qquad T_0 = 6976103.21. \tag{6}$$

Again, from our simulation we observe that the herbivores population grows until the equilibrium is reached in about 150 years, see Figure 3. There herbivores attain a value which is about 965 times the initial condition. Grass and trees populations have a faster growth in the first period, due to a slow growth of the herbivores population. After this period, grass and trees populations reach the equilibrium. Specifically, after an initial surge, grass population decreases until the equilibrium becoming 85% of the initial condition.

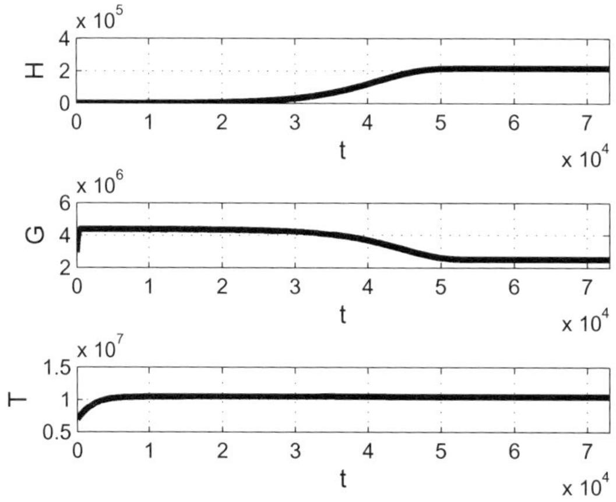

Figure 3. Dolomiti Friulane's system evolution over two centuries with initial conditions set as in Equation (6).

3.3. *Alpi Marittime Natural Park*

This park was established only in 2009. It is located in the south-west part of Piedmont region, NW Italy. The territory extends for about 19751.23 ha. subdivided as shown in Table 2.

According to a 2012 survey, in this park live about 3828 chamois (*Rupicapra rupicapra*), and 591 rock goats (*Capra ibex*), this datum going back to a census of the year 2003. Also a few individuals of roes (*Capreolus capreolus*), deer (*Cervus elaphus*) and mouflons (*Ovis aries*), have been spotted, but never accurately surveyed. In absence of better data, to account for their contributions too, in our simulations we arbitrarily set their overall population at 100 individuals with an average weight of about 85 kilograms.

Table 2. Alpi Marittime Natural Park environmental composition.

Denomination	**Area (ha)**
Fir forest (*Abies alba, Picea abies*)	170.48
Mixed forests (*Acer, Fraxinus, Tilia*)	277.55
Bushes	626.60
Pioneer and invasion woodlands (*Populus, Betula, Corylus avellana*)	429.05
Chestnut forest *Castanea Sativa Miller*	42.42
Fagus sylvatica L.	5099.38
Larch (*Larix*) and pine (*Pinus cembra*) forest	1088.48
Shrubbery	1522.21
Meadows	2011.35
Pines forest (*Pinus mugo*)	3082.2
Rocked meadows	6042.34
Pastures	1309.4
Oak forests (*Quercus petraea, Quercus pubescens*)	988.30
Reafforestation	864.60
Total	19751.230

In this situation, the carrying capacities and the half saturation constants turn out to have the following values: $K_1 = 2184427.67$, $K_2 = 5720905.98$, $c = 69587.52$ and $g = 395981556.29$. Moreover we set the next initial conditions as follows:

$$H_0 = 209.635 \qquad G_0 = 1456285.16 \qquad T_0 = 3813937.32. \tag{7}$$

The pattern already discovered of a initial fast growth of grass and trees populations matched by a corresponding slow increase of the herbivores population in the first period occurs here as well. After this period again we have a faster growth of the herbivores and consequently their stabilization at the equilibrium. At the same time we have a decrease of grass until the

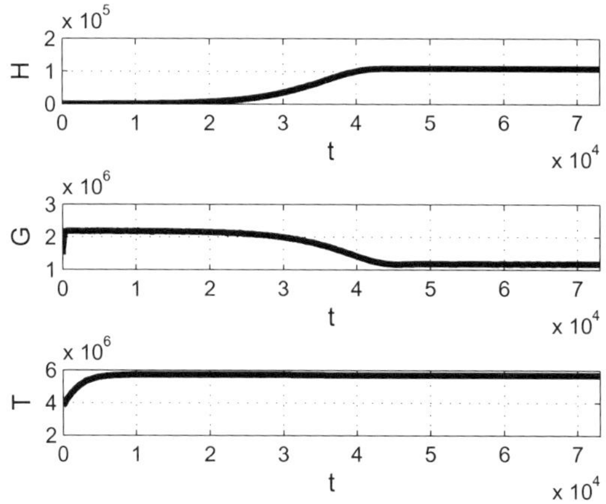

Figure 4. Alpi Marittime's system evolution over two centuries with initial conditions set as in Equation (7).

equilibrium, where the herbivores are about 519 times the initial condition and grass is about the 81% the initial condition.

3.4. *Prealpi Giulie Natural Park*

This park was established in 1996 in the Friuli-Venezia Giulia region (NE Italy). It is an area of about 10000 ha., specific because in it the Mediterranean, the Illyrian and the Alpine biogeographic regions come in contact. Their characteristics combine to form an ecosystem with extraordinary biodiversity. The composition of the park is given in Table 3.

Table 3. Prealpi Giulie Natural Park environmental composition.

Denomination	**Area (ha)**
Meadows	1691.15
Evolving forests	722.53
Thin vegetation areas	723.94
Agricultural fields	23.63
Conifer forests	218.65
Broad-leaf forests	3186.00
Mixed forests	1885.60
Moorland and bushed land	157.43
Rocks and cliff	795.98
Sands	30.48
Total	9435.40

In the park wildlife species of southern, Mediterranean, and Eastern European origin thrive. In recent years even the presence of the brown bear and of the lynx has been established. About 100 species of birds have been observed, among them there are several predators (eagle owl, tawny owl, boreal owl, golden eagle). As for the herbivores population, at present it is composed of about 30 deer (*Cervus elaphus*), 257 ibex (*Capra ibex*) and of chamois (*Rupicapra rupicapra*). The chamois population had the following evolution: 227 individuals in 2008, 249 in 2009, 153 in 2010, with a rebound to 383 in 2011 and 492 in 2012. From this and using Table 3, in the simulations we employ the following values: $K_1 = 576651.52$, $K_2 = 3572655.97$, $c = 23537.55$ and $g = 316851220.79$ and initial conditions:

$$H_0 = 52.445 \qquad G_0 = 384434.35 \qquad T_0 = 6012.78. \tag{8}$$

The system qualitatively behaves similarly to the former ones.

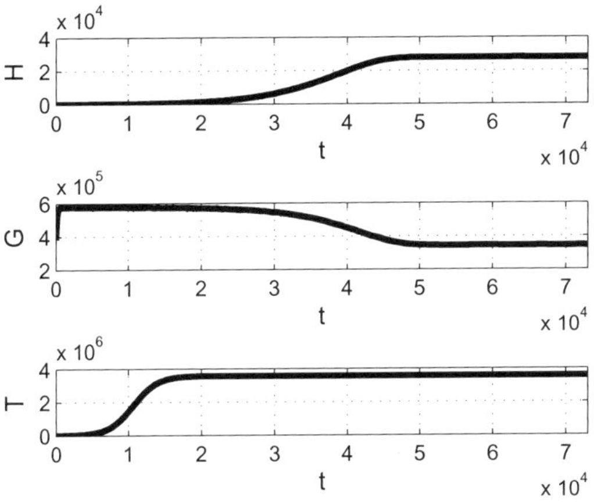

Figure 5. The Prealpi Giulie system evolution over two centuries with initial conditions set as in Equation (8).

4. Sensitivity Analysis

In this Section we perform the sensitivity analysis on (1), as an illustrative example focusing only on the Dolomiti Bellunesi's system, for sake of brevity. We have indeed found that all the parks show a similar behavior, therefore the conclusions on the model qualitative behavior are very close also for the parameter sets of all the remaining parks.

4.1. *Approximation of sensitivity surfaces*

Sensitivity analysis allows us to rank the parameters with respect to their influence on the ecosystem. Precisely we rank them on the basis of the magnitude of the maximum norm, the details being described in Section 5. The results indicate that the parameters most affecting the system's final configuration are μ, e and α. In particular the herbivores' population level appears to be very sensitive and under the threat of a high risk of extinction. In Figure 6 left, the surface shows the value attained by the herbivores as function of the parameters μ and e after 4 years. The surface has been reconstructed with state-of-the-art techniques, [5,6,7]. The dot represents the situation in the present ecosystem conditions. The right frame shows that it is very close to the line beyond which the herbivores population would vanish. Thus, under possible ecosystem parameter changes, induced for instance by climatic variations or random environmental disturbances, the present situation has some potential risks for an evolution toward an extinction of the herbivores population in the short time span.

In the three-dimensional parameter space (μ, e, α) Figure 7 represents instead the separatrix of the basins of attraction of the coexistenc equilibrium E^* and of the herbivores-free equilibrium E_3. Again the actual value of the herbivores population, represented by the dot, is very close to this surface, although at present beloging to the not endangered region. But clearly changes in the environmental conditions that might lead even to relatively small parameter perturbations may well push it into the region where the herbivores would be doomed.

4.2. *Parameters sensitivity analysis*

We now estimate the influence on the populations involved in the model of the parameters, to find those that most affect the herbivores evolution. At first we determine the sensitivity equations, differentiating the original equations with respect to the parameters of interest, [8,10]. Thus, letting $\boldsymbol{\lambda}$ the vector of the p parameters, f_i, $i = 1, 2, 3$ the right-hand sides of equations (1) and $P_i = H, G, T$, $i = 1, 2, 3$, the sensitivity system for $j =$

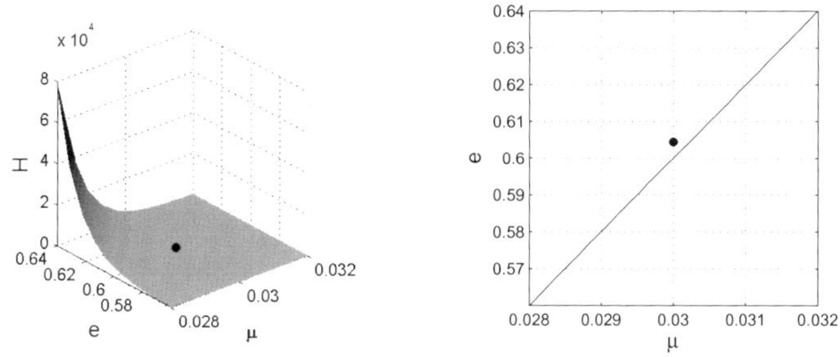

Figure 6. Dolomiti Bellunesi's system. Left frame: the surface represents the equilibrium value of the herbivores as function of the parameters μ and e at the end of 4 years. Note that the surface flattens onto the horizontal coordinate plane; Right frame: the curve separates the region in which the surface has a nonzero level, top left, from the region in which the herbivores population vanishes, bottom right. The dot represents the herbivores population level in the present conditions of the ecosystem. Its closeness to the range in which it would vanish is evident.

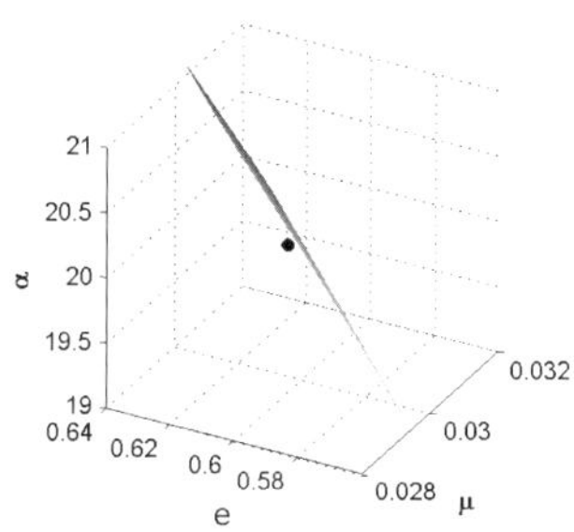

Figure 7. Dolomiti Bellunesi's system. In the parameter space (μ, e, α) we show the surface separating the basins of attraction of the coexistence equilibrium E^*, bottom left, from the one of the herbivore-free equilibrium E_3, top right. The dot represents the actual situation. Again, small parameter variations may push it into the other basin of attraction, thereby entailing the herbivores extinction.

$1, \ldots, p$ becomes

$$\frac{d}{dt}\left(\frac{\partial P_i(\boldsymbol{\lambda}, t)}{\partial \lambda_j}\right) = \sum_{k=1}^{3} \frac{\partial f_i(\boldsymbol{P}, \boldsymbol{\lambda}, t)}{\partial P_k} \frac{\partial P_k(\boldsymbol{\lambda}, t)}{\partial \lambda_j} + \frac{\partial f_i(\boldsymbol{P}, \boldsymbol{\lambda}, t)}{\partial \lambda_j}, \tag{9}$$

with initial conditions:

$$\frac{\partial P_i(\boldsymbol{\lambda}, 0)}{\partial \lambda_j} = 0, \quad i = 1, 2, 3, \quad j = 1, \ldots, p\ .$$

The semi-relative solutions, obtained by multiplying the sensitivity solutions by the chosen parameters, are shown in Figure 8. They indicate the size of the change in the population state as function of the parameters changes. For instance, let us suppose that for the population H we change μ into μ_1; the semirelative solution is the differential of H with respect to μ_1 in the direction μ. This represents a weight allowing the ranking of the parameters with respect to their influence over the model dynamics, [1]. In general the chosen parameter is perturbed by doubling or halving it. When possible we follow such an approach: but in Figure 8 the parameter e is halved, while we set $\mu = 0.028$ and α at its minimum, i.e. $\alpha = 18$ (4%).

These changes have a considerable impact on the herbivores, Figure 6. Even if the influence of all parameters gradually becomes negligible, the large peaks indicate that the system is really sensitive: a smaller μ, as a smaller α, has a positive effect on the herbivores population, while a decrease of e leads to a negative influence on the latter population, consistently with Figure 6, see especially the right frame, and Figure 7. Since the scales on the vertical axis of Figure 8 (top left) are very large for the herbivores population, we zoom-in the time interval in which the change of the parameters shows the most significant influence for this population, Figure 8 (bottom right).

Moreover, referring again to Figure 8 (bottom right), we approximately get the expected percentage change from a parameter perturbation in the time interval in which its influence is largest. Specifically, we can compute the logarithmic sensitivity solutions for $P_i = H, G, T$, for $i = 1, 2, 3$, observing that, [4], for $j = 1, \dots, p$:

$$\frac{\partial \log(P_i(\boldsymbol{\lambda}, t))}{\partial \log(\lambda_j)} = \frac{\lambda_j}{P_i(\boldsymbol{\lambda}, t)} \frac{\partial P_i(\boldsymbol{\lambda}, t)}{\partial \lambda_j}. \tag{10}$$

Computing (10) at the end of the observation period shown in Figure 8 (bottom right), we find that the percentage change with respect to μ and α is about 14% and 16%, respectively, while the one with respect to e is greater than 100% and it thus leads to the extinction.

We now turn to the parameters pair β and f, since they are related to trees as α and e are related to grass. By doubling them, we get the semi-relative sensitivity solutions and the logarithmic sensitivity solutions shown in Figure 9 (top and bottom, respectively). Consistently with the previous behavior, doubling β, as doubling α, has a small negative impact on the herbivores (less than 1%), while doubling f leads to a positive effect (about 95%).

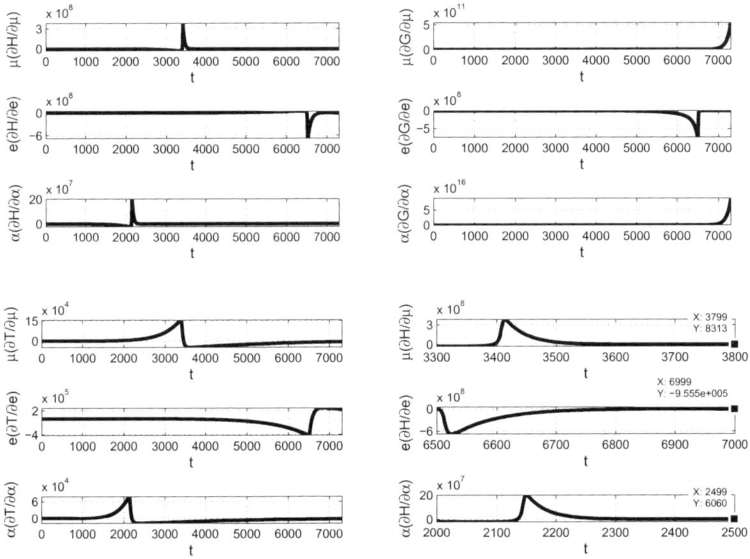

Figure 8. Dolomiti Bellunesi's system. Semi-relative solutions for the population H (top left), G (top right) and T (bottom left) with respect to the parameters μ, e and α. In the bottom right frame we zoom-in the time interval where the parameters change has the most significant impact on herbivores.

Consistently with the fact that grass is the preferred resource for the herbivores, the above analysis indicates that the parameter e affects the dynamics more than f. We thus expect a similar behavior from the sensitivity analysis with respect to the parameters a and b. This is confirmed in Figure 10. By considering the vertical scales of the two top frames in Figure 10, even if in both cases doubling the parameters leads to a positive effect, evidently the influence of b in the dynamics is negligible. Consistently with such consideration, from the logarithmic sensitivity solutions we can state that the positive increment, due to b, for the herbivores population is less than 1%, while the one due to a is orders of magniture larger.

To conclude the analysis wet double the parameters r_1 and r_2. Since they represent the grass and trees growth rates, we expect a positive effect on the population of herbivores. The latter have a larger benefit from grass than from trees and thus, consistently with Figure 11 (bottom), the percentage change with respect to r_1 and r_2 at the end of the observation period is about 400% and 20%, respectively.

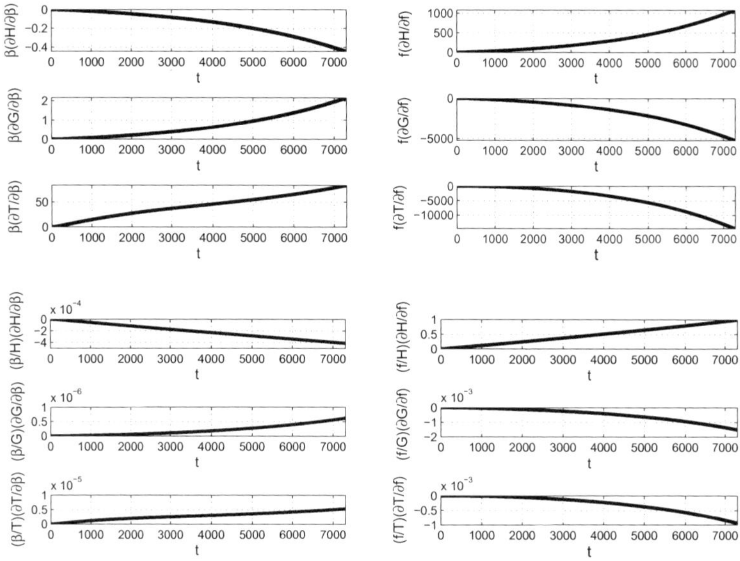

Figure 9. Dolomiti Bellunesi's system. Semi-relative solutions with respect to the parameters β and f (top) and the logarithmic sensitivity solutions (bottom).

4.3. *Sensitivity on the initial conditions*

The initial conditions for the populations G and T were set as the two thirds of the carrying capacities K_1 and K_2, respectively, as in [17,18], while the initial condition of the population H is known. We now validate this choice. For $P_i = H, G, T$, $i = 1, 2, 3$, the sensitivity equations with respect to the initial conditions for $j = 1, 2, 3$ are, [10]:

$$\frac{d}{dt}\left(\frac{\partial P_i(\boldsymbol{P}^0, t)}{\partial P_j^0}\right) = \sum_{k=1}^{3} \frac{\partial f_i(\boldsymbol{P}, \boldsymbol{P}^0, t)}{\partial P_k} \frac{\partial P_k(\boldsymbol{P}^0, t)}{\partial P_j^0}, \tag{11}$$

with initial conditions:

$$\frac{\partial P_i(\boldsymbol{P}^0, 0)}{\partial P_j^0} = 1, \quad i, j = 1, 2, 3.$$

Figure 12 (top left and right, respectively) contains the results of doubling and halving the initial condition $H(0)$. Consistently with our analysis, since the system does not show multi-stability, $H(0)$ does not affect the final configuration of the model, but it influences the time needed to reach the equilibrium. Comparing the two frames, by doubling $H(0)$ the system

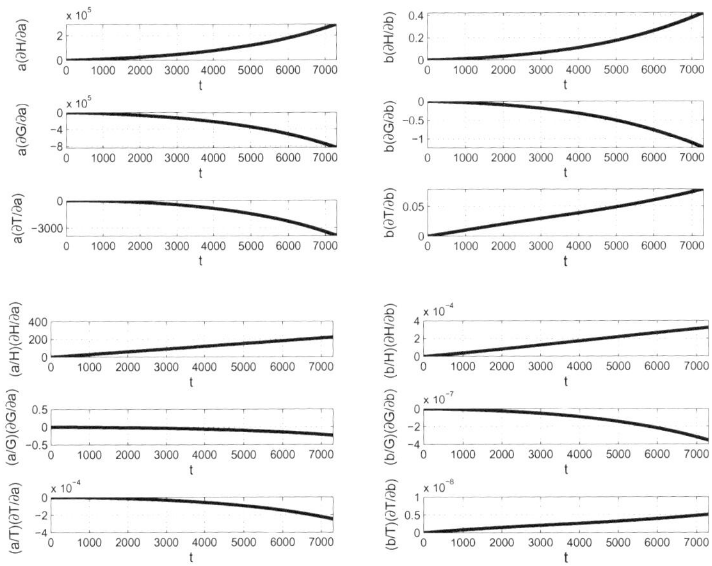

Figure 10. Dolomiti Bellunesi's system. Semi-relative solutions with respect to the parameters a and b (top) and the logarithmic sensitivity solutions (bottom).

stabilizes at time $t \approx 5 \times 10^4$, while by halving it, it settles later, at time $t \approx 6 \times 10^4$. Furthermore, from Figure 12 (bottom frames), the initial conditions $G(0)$ (bottom left) and $T(0)$ (bottom right) do not influence the system dynamics, neither its asymptotic behavior. In fact the sensitivity solutions occur essentially at the same time whenever $G(0)$ and $T(0)$ are doubled or halved.

5. A Caveat on the Ecosystems Preservation

Qualitatively, all the simulations performed in Section 3, by estimating the parameters with techniques which are widely used in literature [12,17], show very similar results for all the four ecosystems in consideration. But note that the results obtained for the Dolomiti Friulane Natural Park and for the Dolomiti Bellunesi National Park are very similar also from the quantitative point of view. This remark could be attributed to the similar extension and to the similar initial erbivores biomass. It is likely that by considering wolves and other herbivores predators, as well as possible human intervention, the results might differ somewhat.

Assessing and comparing more closely all the simulations results shown

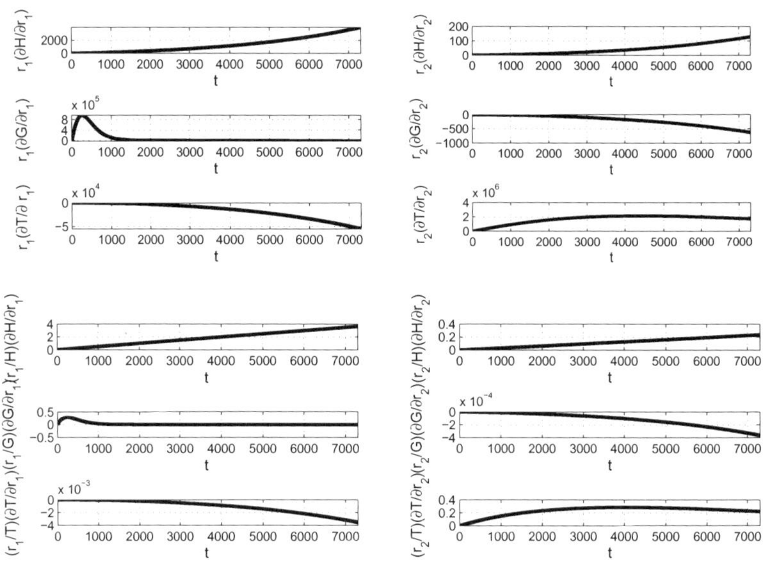

Figure 11. Dolomiti Bellunesi's system. Semi-relative solutions with respect to the parameters r_1 and r_2 (top) and the logarithmic sensitivity solutions (bottom).

in Section 3 it is to be further remarked that in all these ecosystems there seems no immediate high risk of extinction for the herbivores in the actual situation. This result is substantiated by the equilibrium analysis. While for coexistence we have to rely only on simulations, for the herbivores-free equilibrium $E_3 = (0, K_1, K_2)$ we have the stability conditions explicitly, (2). As stated in Section 3 above, the equilibrium is not stable. Moreover we can also exclude any other bistability cases. This result is consistent with the sensitivity analysis we carried out on initial conditions. In fact, a perturbation on such initial conditions does not affect the qualitative model behavior. Excluding bistability situations implies also that the equilibrium with no forests E_4 is always unstable. This is consistent with the fact that, differently from the context of sheep [17], the trees damage due to wild herbivores is less significant than the one due to sheep.

Therefore, from this analysis there seems to be no risk of extinction for herbivores in the present conditions. On the other hand, in Section 4 we pointed out that the system is really sensitive with respect to small perturbations of several parameters and such perturbations can instead lead to the extinction. By reconstructing the sensitivity curves and surfaces we can

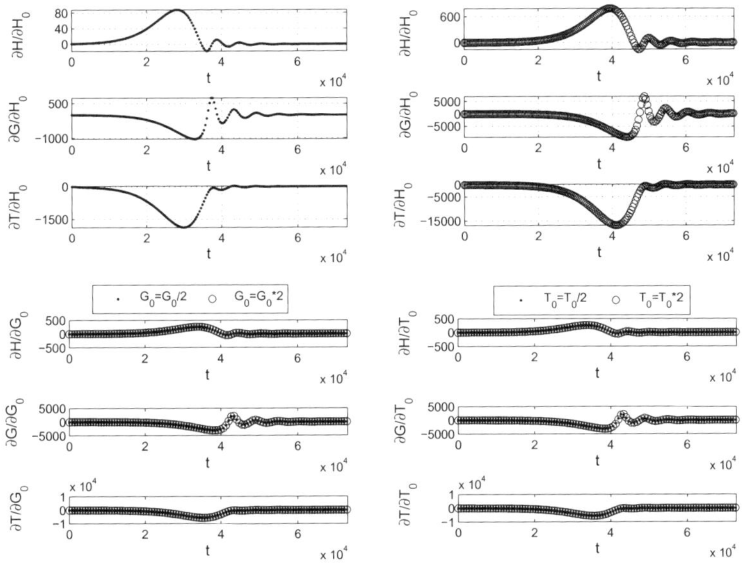

Figure 12. Dolomiti Bellunesi's system. Sensitivity solutions referred to the initial condition $H(0)$ (top: left by halving it; right: by doubling it) and referred to $G(0)$ and $T(0)$ (bottom left and right, respectively).

deduce that the system (1) is really sensitive with respect to the parameters μ, e and α. Specifically, for small perturbations of such parameters a transcritical bifurcation between the coexistence equilibrium point and the herbivores-free equilibrium occurs. Furthermore, since from the estimation of parameters the current ecosystem state is really close to the separatrix a small perturbation can drive the ecosystem into the region where herbivores extinction occurs.

In order to possibly prevent the ecosystem to fall into the unwanted region the strategy is therefore to move away from the separatrix surface.

Now using sensitivity analysis we rank the parameters with respect to their influence on the ecosystem populations. To do so, we compare the infinity norms of the semi-relative herbivores solutions, by defining for $j = 1, \ldots, p$

$$\left\| \frac{\partial H}{\partial \lambda_j} \right\|_\infty = \max_{t \geq 0} \left| \frac{\partial H(\boldsymbol{\lambda}, t)}{\partial \lambda_j} \right|. \tag{12}$$

Roughly speaking the parameters are sorted by comparing the vertical scales of the sensitivity solutions plots, the largest ones indicating the predominant influences. Thus from Figures 8-11, we can surely state that the parameters which most affect the system are μ, e and α, as previously claimed.

Having identified the parameters most influencing the herbivores population, we can now try to modify them so as to drive the ecosystem into a safer position. Specifically, from a mathematical point of view, a decrease of μ and α, combined with an increase of e, leads to a benefit for the population H. We can try to directly act on μ and α. Recalling that μ is the mortality rate, building safety niches during winter for instance surely leads to a decrease of the herbivores mortality rate. Moreover, by planting grass more nutritious than the existing one and/or by using fertilizers we can try to increase the nutrient assimilation coming from grass, leading to an increase of α^{-1}. In this way, mathematically speaking, in order to prevent extinction in the $\mu - e - \alpha$ parameter space we move away the ecosystem state from the separatrix.

Acknowledgments

The authors thank the parks rangers: Elisa Ramassa, (Gran Bosco di Salbertrand), Laura Martinelli (Alpi Marittime) E. Vettorazzo (Dolomiti Bellunesi) Stefano Santi (Prealpi Giulie) Eugenio Granziera (Dolomiti Friulane) for providing data and useful information.

References

1. H. T. Banks, D. M. Bortz, *J. Math. Biol.* **50**, 6 (2004), 607.
2. J. Beddington, *J.Anim. Ecol.* **51**, 331 (1975).
3. M. Borgia, *Luna Nuova*, Avigliana, Italy, 2003.
4. D. M. Bortz, P. W. Nelson, *Bulletin of Mathematical Biology* **66**, 1009 (2004).
5. R. Cavoretto, S. Chaudhuri, A. De Rossi, E. Menduni, F. Moretti, M. C. Rodi, E. Venturino, *Numerical Analysis and Applied Mathematics ICNAAM 2011*, T. Simos, G. Psihoyios, Ch. Tsitouras, Z. Anastassi (Editors), AIP Conf. Proc. **1389**, 1220 (2011); doi: 10.1063/1.3637836.
6. R. Cavoretto, A. De Rossi, E. Perracchione, E. Venturino, *International Journal of Computer Mathematics*, to appear (2015).
7. R. Cavoretto, A. De Rossi, E. Perracchione, *arXiv:1409.5576v1*, (2014).
8. V. Comincioli, *Metodi numerici e statistici per le scienze applicate (Numerical and statistical methods for applied sciences)*, Università degli studi di Pavia, 2004.
9. D. De Angelis, R. Goldstein, R. O'Neill, *Ecol.* **56**, 881 (1975).
10. S. L. Ernstberge, *Sensitivity Methods for Dynamical Sistems*, ProQuest, UMI Dissertation Publishing, (2011).

11. G. E. Fasshauer, *Meshfree Approximations Methods with* MATLAB, World Scientific Publishers Co., Inc., River Edge, NJ (2007).
12. T. Fujimori, *Ecological and Silvicultural Strategies for Sustainable Forest Managment*, Elsevier - Amterdam, Nederland (2001).
13. Q.J.A. Khan, E.V. Krishnan, M.A. Al-Lawatia, *Z. Angew. Math. Mech.* **82**, 125 (2002).
14. Q. J. A. Khan, E. Balakrishnan, G. C. Wake, *Bull. Math. Biol.* **66**, 109 (2004).
15. S. Palomino Bean, A. C. S. Vilcarromero, J. F. R. Fernandes, O. Bonato, *TEMA Tend. Mat. Apl. Comput.* **7**, 317 (2006).
16. G. Pulina, R. Bencini, *Dairy Sheep Nutrition*, CABI Publishing, Cambridge MA, USA (2004).
17. L. Tamburino, E. Venturino, *Int. J. Comp. Math.* **89**, 1808 (2012).
18. L. Tamburino, V. La Morgia, E. Venturino, *Computational Ecology and Software* **2**, 26 (2012).
19. M. Tanksy, *J. Theor. Biol.* **70**, 263 (1978).
20. A. C. S. Vilcarromero, S. Palomino Bean, J. F. R. Fernandes, O. Bonato, *Proceedings Of Congreso Latino Americano de Biomatematica* (Alab-V Elaem) **X**, (2001).

EFFECT OF VIRAL DISEASE IN A DIFFUSIVE PLANKTON SYSTEM

NANDA DAS*
Maulana Azad College,
8, Rafiahmed Kidwai Road, Kolkata 700013, India
E-mail: nandadas72@gmail.com

SAMARES PAL
Department of Mathematics,
University of Kalyani, Nadia, India
E-mail: samaresp@yahoo.co.in

We consider a mathematical model in virus infected phytoplankton and zooplankton system with the help of reaction-diffusion equations that studies the bifurcation. The analytical explanation provide for understanding phytoplankton dynamics with virus infection in phytoplankton population. This model has a nonlinear disease transmission rate and Holling type-II response function. Our aim is to provide a qualitative analysis of Hopf-bifurcation mechanisms, taking rate of infection as bifurcation parameter, and to study effect of disease due to spatial diffusion. The analytical findings are supported by the numerical results. It is observed that the coexistence of all classes depend on the rate of diffusion.

1. Introduction

Plankton dynamics has become familiar in marine ecological research as phytoplankton plays a key role at the bottom of the marine food web. To study plankton dynamics different mathematical models have been proposed by the researchers depending on field studies. In plankton ecology, bloom formation is the common phenomena and the termination of plankton bloom may be accelerated by the viral infection [1]. Also some evidences are found in [2] that the viral infection acts as a potential regulator of oceanic plankton populations. The importance of viruses for marine plankton ecology has been acknowledged in several publications [3,4,5,6,7].

*The work of Dr. Nanda Das (Corresponding Author) is supported by the University Grants Commission (U G C), India (Grant no. PSW-37/12-13(ERO), dated 05.02.2013).

At first Segel and Jackson [8] applied Turing's model to a predator-prey system. After that diffusive instability plays an important role in the study of ecological ecosystem models [9,10]. In those studies, they considered that all the system parameters are independent of time and obtain the criterion for diffusive instability of the system. In the marine system (particularly in the ocean), the diffusivity can vary with time. Also horizontal currents (which depend on the depth) in the sea mixed with the vertical currents by the mixing process of the sea. As a result the horizontal dispersion occurs [11] and the horizontal diffusivity of phytoplankton (prey) and that of zooplankton (predator) are different and vary with time [12]. The complex dynamics of cross-diffusion SI epidemic model was studied by Wang [13] and obtained the conditions of the local and global stability of the nonnegative constant steady states. He also proved that the existence and nonexistence of the positive nonconstant steady states, guarantees the existence of the stationary patterns.

In this study, we consider a prey-predator model in marine plankton system with virus infection in prey (phytoplankton). The growth of prey population is assumed to follow the logistic law and disease transmitted over prey population according to nonlinear incidence rate. The predator (zooplankton) predate over susceptible prey with Holling Type-II rate function and over infected prey according to the mass action law. In this model we incorporate diffusion and analyze its stability near equilibrium. We modify the model of same author [14] by including the grazing rate of predator according to Holling Type-II rate function.

1.1. *Biological assumption and Mathematical model*

A mathematical model with three state variables is considered based on the following biological assumptions:

A1: At any instant t the prey(phytoplankton) and predator(zooplankton) population be N and Z respectively. The viral infection divides the prey population into susceptible prey S and infected prey I such that $N = S + I$.

A2: In the absence of vial infection, the growth of susceptible prey is according to the logistic law, $\frac{dS}{dt} = rS(1 - \frac{S}{k}),\quad 0 < S < k$ where $k(> 0)$ and $r(> 0)$ are carrying capacity and per capita growth rate respectively.

A3: The growth of infected prey depends only on the disease transmission rate. The rate of disease transmission is according to the nonlinear incidence rate $\dfrac{\lambda I}{1 + \lambda\delta I}$, where λ is average number of contacts sufficient for disease transmission and δ is the handling time for each prey [15].

A4: The predator population Z grazes on susceptible prey according to Holling-II functional response, where as it grazes over infected prey according to the mass action law and both population has positive effect on the growth rate of predator. The decay of predator population depends only on mortality rate constant d.

A5: Let D_1, D_2, D_3 are the self diffusion coefficients of susceptible prey, infected prey and predator population respectively.

The mathematical model based on the above biological assumptions is

$$\begin{aligned}
\frac{\partial S}{\partial t} &= S\left[r\left(1-\frac{S}{k}\right)-\frac{\lambda I}{1+\lambda\delta I}-\frac{\alpha Z}{a+S}\right]+D_1\nabla^2 S,\\
\frac{\partial I}{\partial t} &= I\left[\frac{\lambda S}{1+\lambda\delta I}-\mu-\beta Z\right]+D_2\nabla^2 I,\\
\frac{\partial Z}{\partial t} &= Z\left[\theta\left(\frac{\alpha S}{a+S}+\beta I\right)-d\right]+D_3\nabla^2 Z.
\end{aligned} \tag{1}$$

The parameters used in the model (1) are positive with their respective biological meaning as mentioned below:
r, the per capita growth rate; k, carrying capacity; λ, disease transmission rate; δ, handling time for each prey; α, grazing rate of susceptible prey; β, grazing rate of infected prey; a, half saturation coefficient; μ, natural death rate of infected prey; θ, conversion rate factor; d, natural death rate of predator.

We analyze the model system (1) with initial conditions

$$S(x,0)>0, I(x,0)>0, Z(x,0)>0$$

and (for $R>0$) zero flux boundary conditions

$$\left.\frac{\partial S}{\partial x}\right|_{x=0}^{x=R}=\left.\frac{\partial I}{\partial x}\right|_{x=0}^{x=R}=\left.\frac{\partial Z}{\partial x}\right|_{x=0}^{x=R}=0.$$

1.2. *Mathematical Study of the model without diffusion*

At first, we consider the model (1) without diffusion. Therefore the model (1) with assumption $D_1 = D_2 = D_3 = 0$ reduced to

$$\frac{dS}{dt} = S\left[r\left(1-\frac{S}{k}\right) - \frac{\lambda I}{1+\lambda\delta I} - \frac{\alpha Z}{a+S}\right] = f_1,$$
$$\frac{dI}{dt} = I\left[\frac{\lambda S}{1+\lambda\delta I} - \mu - \beta Z\right] = f_2, \tag{2}$$
$$\frac{dZ}{dt} = Z\left[\theta\left(\frac{\alpha S}{a+S} + \beta I\right) - d\right] = f_3.$$

1.2.1. *Existence and bounds of solution*

Rewrite the system (2) in form of vector equation $\frac{dX}{dt} = F(X)$, where $X = [\ S,\ I,\ Z\ ]^T$ and $F = [f_1, f_2, f_3]^T$. If $X(0) = X_0 = (\ S_0,\ I_0,\ Z_0\ ) \in \mathcal{R}^3_+ = \{(\ S,\ I,\ Z\ ) :\ S \geq 0, I \geq 0, P \geq 0\}$, then it is easy to check that $\{f_i(X)\}_{X_i=0} \geq 0$. Therefore by Nagumo's lemma [16] any solution of the system with $X_0 \in \mathcal{R}^3_+$, say $X(t) = X(t; X_0)$, is such that $X(t) \in \mathcal{R}^3_+$ for all $t > 0$. Hence all solutions of the system (2) exist in the region $\mathcal{R}^3_+$, remain non-negative for all $t > 0$.

Now we prove that all the solutions of the system (2) are uniformly bounded by the following theorem:

Theorem 1.2.1. *The solution of the system (2) is uniformly bounded in*

$$\Omega = \left\{(\ S\ ,I\ ,Z\ ) : 0 \leq S(t) + I(t) + \frac{1}{\theta}Z(t) \leq \frac{rk}{\eta} + \epsilon,\ \textit{for}\ \ \textit{all}\ \epsilon > 0\right\}.$$

Proof: Consider the function $U(t) = S(t) + I(t) + Z(t)$.
Obviously $0 \leq S(t) \leq k$.
Now $\frac{dU}{dt} = rS(1-\frac{S}{k}) - \mu I - \frac{d}{\theta}z \leq rk - \{rS + \mu I + d\frac{Z}{\theta}\}$.
Let $\eta = \min\{r,\ \mu,\ d\}$, then $\frac{dU}{dt} + \eta U \leq rk$.
Now by the theory of differential inequality (Birkhoff and Rota [17]) we have,

$$0 \leq U(t) \leq \frac{rk}{\eta}(1-e^{-\eta t}) + U(0)e^{-\eta t}.$$

As $t \to \infty$, then $0 \leq U(t) < \frac{rk}{\eta}$. Hence $U(t)$ is a bounded quantity.

Thus all the solutions of the system (2) are uniformly bounded in the region

$$\Omega = \left\{ (\ S,\ I,\ P\ ) : 0 \leq S(t) + I(t) + \frac{1}{\theta} Z(t) \leq \frac{rk}{\eta} + \epsilon,\ \text{for all}\ \ \epsilon > 0 \right\}.$$

1.2.2. *Equilibrium Analysis*

• The equilibria $E_0(\ 0,\ 0,\ 0\ )$ and $E_1(\ k,\ 0,\ 0\ )$ exist for all parametric values.

• The equilibrium $E_{12}(\ \bar{S},\ \bar{I},\ 0\ )$ exists if $0 < \bar{S} < k$, where

$$\bar{S} = \frac{1}{2r\lambda\delta} \left\{ k\lambda(r\delta - 1) + \sqrt{\lambda^2 k^2 (r\delta - 1)^2 + 4rk\mu\lambda\delta} \right\} \text{ and } \bar{I} = \frac{r}{\mu k} \bar{S}(k - \bar{S}).$$

• The planer equilibrium $E_{13}(\ \tilde{S},\ 0,\ \tilde{Z}\ )$ exists if $\theta\alpha k > d(a + k)$, where

$$\tilde{S} = \frac{ad}{\theta\alpha - d} \text{ and } \tilde{Z} = \frac{ar\theta}{k} . \frac{\theta\alpha k - d(a + k)}{(\theta\alpha - d)^2}.$$

• The positive equilibrium $E^*(\ S^*,\ I^*,\ Z^*\ )$ exists if

$$\theta\alpha k < d(a + k) \ \text{ and } \ \lambda S^* > \mu(1 + \lambda\delta I^*) \text{ hold},$$

where $I^* = \dfrac{ad + (d - \theta\alpha)S^*}{\theta\beta(a + S^*)}$, $Z^* = \dfrac{\lambda S^* - \mu(1 + \lambda\delta I^*)}{\beta(1 + \lambda\delta I^*)}$ and S^* is the positive root of the equation in S given by

$$\frac{r}{k}(S - k) + \frac{\lambda ad + \lambda(d - \theta\alpha)S}{a(\theta\beta + \lambda\delta d) + (\theta\beta + \lambda\delta d - \theta\lambda\delta\alpha)S}$$

$$+ \frac{\theta\lambda\alpha\beta S^2 + \alpha\{a\theta\lambda\beta - \mu(\theta\beta + \lambda\delta d - \theta\lambda\delta\alpha)\}S - a\mu\alpha(\theta\beta + \lambda\delta d)}{(a + S)\{a\beta(\theta\beta + \lambda\delta d) + \beta(\theta\beta + \lambda\delta d - \theta\lambda\delta\alpha)S\}} = 0.$$

1.2.3. *Stability Analysis*

For the stability analysis of the system (2), we have to compute the variational matrix $V(S, I, Z)$ at each equilibrium and check the stability.

Result 1 The equilibrium point $E_0(\ 0,\ 0,\ 0\ )$, is always unstable saddle.

Result 2 If $\lambda k < \mu$ and $\theta\alpha k < d(a + k)$ then $E_1(\ k,\ 0,\ 0\ )$ is locally asymptotically stable.

Result 3 If $\theta(\dfrac{\alpha\bar{S}}{a + \bar{S}} + \beta\bar{I}) < d$ then $E_{12}(\ \bar{S},\ \bar{I},\ 0\ )$ is locally asymptotically stable.

Result 4 If $E_{13}(\tilde{S}, 0, \tilde{Z})$ exists and $r(a + \tilde{S})^2 > \alpha k\tilde{Z}$ then E_{13} is locally asymptotically stable if $\lambda\tilde{S} < \lambda + \beta\tilde{Z}$.

1.2.4. *Local stability of E^* and Hopf bifurcation*

The variational matrix at $E^*(\ S^*,\ I^*,\ Z^*\ )$ of the system (2) is

$$V(E^*) = \begin{pmatrix} -m_{11} & -m_{12} & -m_{13} \\ m_{21} & -m_{22} & -m_{23} \\ m_{31} & m_{32} & -m_{33} \end{pmatrix},$$

where $m_{11} = \dfrac{rS^*}{k} - \dfrac{\alpha S^* Z^*}{(a+S^*)^2}, \quad m_{12} = \dfrac{\lambda S^*}{(1+\lambda\delta I^*)^2}, \quad m_{13} = \dfrac{\alpha S^*}{a+S^*},$

$$m_{21} = \frac{\lambda I^*}{1+\lambda\delta I^*}, \quad m_{22} = \frac{\lambda^2 \delta S^* I^*}{(1+\lambda\delta I^*)^2}, \quad m_{23} = \beta I^*,$$

$$m_{31} = \frac{a\theta\alpha z^*}{(a+S^*)^2}, \quad m_{32} = \theta\beta z^*, \quad m_{33} = 0.$$

Theorem 1.2.2. *Let the interior point $E^*(\ S^*,\ I^*,\ Z^*\ )$ of the system (2) exists, then E^* is locally asymptotically stable if the following conditions hold, (i) $r(a+S^*)^2 > \alpha k Z^*$, (ii) $(1+\lambda\delta I^*)^2 c\beta Z^* > \lambda\alpha S^*$.*

Proof:

The characteristic equation at the interior positive equilibrium E^* is $x^3 + a_1 x^2 + a_2 x + a_3 = 0$,

where $a_1 = m_{11} + m_{22}$,

$$a_2 = m_{11}m_{22} + m_{13}m_{31} + m_{12}m_{21} + m_{23}m_{32},$$

$$a_3 = m_{11}m_{23}m_{32} + m_{21}m_{13}m_{32} + m_{13}m_{31}m_{22} - m_{31}m_{12}m_{23}.$$

Now $a_1a_2 - a_3 = m_{11}^2 m_{22} + m_{11}m_{22}^2 + m_{11}m_{13}m_{31} - m_{13}m_{21}m_{32} + m_{11}m_{12}m_{21} + m_{12}m_{21}m_{22} + m_{12}m_{23}m_{31} + m_{22}m_{23}m_{32} > 0$.

Thus, if the condition stated in the theorem holds, then all the Routh-Hurwitz criteria

$$(i)\ a_1 > 0\ \ (ii)\ a_3 > 0\ \ (iii)\ a_1a_2 - a_3 > 0$$

are satisfied, and the system (2) is locally asymptotically stable around the positive equilibrium point. Hence the theorem.

Theorem 1.2.3. *If the rate of infection λ crosses a critical value λ^*, the system enters into Hopf bifurcation around E^* if (i) $a_1(\lambda^*) > 0$, (ii) $D_2(\lambda^*) = a_1(\lambda^*)a_2(\lambda^*) - a_3(\lambda^*) = 0$, (iii) at $\lambda = \lambda^*$, $\dfrac{d}{d\lambda}\{D_2(\lambda)\} \neq 0$.*

Proof: We assume that the steady state E^* is locally asymptotically stable. We take λ as the bifurcation parameter. If there exists a critical value λ^*

such that (i) $a_1(\lambda^*) > 0$, (ii) $a_1(\lambda^*)a_2(\lambda^*) - a_3(\lambda^*) = 0$, (iii) and at $\lambda = \lambda^*$, $\frac{d}{d\lambda}\{a_1(\lambda)a_2(\lambda) - a_3(\lambda)\} \neq 0$ then for the occurrence of Hopf-bifurcation at $\lambda = \lambda^*$, the characteristic equation must be of the form

$$\{\xi^2(\lambda^*) + a_2(\lambda^*)\}\{\xi(\lambda^*) + a_1(\lambda^*)\} = 0. \tag{3}$$

This equation has three roots

$$\xi_1(\lambda^*) = +i\sqrt{a_2(\lambda^*)},\ \ \xi_2(\lambda^*) = -i\sqrt{a_2(\lambda^*)} \text{ and } \xi_3(\lambda^*) = -a_1(\lambda^*).$$

For the occurrence of Hopf bifurcation at $\lambda = \lambda^*$, we need to verify the transversality condition

$$\left[\frac{d}{d\lambda}\{Re\xi(\lambda)\}\right]_{\lambda=\lambda^*} \neq 0.$$

For all λ, the roots are in general of the form

$$\xi_1(\lambda) = \eta(\lambda) + i\zeta(\lambda),\ \xi_2(\lambda) = \eta(\lambda) - i\zeta(\lambda),\ \xi_3(\lambda) = -a_1(\lambda).$$

Now to verify the transversality condition, we substitute $\xi(\lambda) = \eta(\lambda) \pm i\zeta(\lambda)$, into equation (1.2.1) and calculate the derivative.

$$P(\lambda)\eta'(\lambda) - Q(\lambda)\zeta'(\lambda) + M(\lambda) = 0, \tag{4}$$
$$P(\lambda)\eta'(\lambda) + Q(\lambda)\zeta'(\lambda) + N(\lambda) = 0, \tag{5}$$

where
$$\begin{aligned} P(\lambda) &= 3\eta^2(\lambda) + 2a_1(\lambda)\eta(\lambda) + a_2(\lambda) - 3\zeta^2(\lambda),\\ Q(\lambda) &= 6\eta(\lambda)\zeta(\lambda) + 2a_1(\lambda)\zeta(\lambda),\\ M(\lambda) &= \eta^2(\lambda)a_1'(\lambda) + a_2'(\lambda)\eta(\lambda) + a_3'(\lambda) - a_1^2(\lambda)\zeta^2(\lambda),\\ N(\lambda) &= 2\eta(\lambda)\zeta(\lambda)a_1'(\lambda) + \zeta(\lambda)a_2'(\lambda). \end{aligned}$$

Again we note that $\eta(\lambda^*) = 0$, $\zeta(\lambda^*) = \sqrt{a_2(\lambda^*)}$.
Therefore $P(\lambda^*) = -2a_2(\lambda^*)$, $Q(\lambda^*) = 2a_1(\lambda^*)\sqrt{a_2(\lambda^*)}$,
$M(\lambda^*) = a_3'(\lambda^*) - a_1'(\lambda^*)a_2(\lambda^*)$, $N(\lambda^*) = a_2'(\lambda^*)\sqrt{a_2(\lambda^*)}$.
Solving for $\eta'(\lambda)$ from equations (4) & (5) we have

$$\begin{aligned} \eta'(\lambda^*) = \left[\frac{d}{d\lambda}\{Re\xi(\lambda)\}\right]_{\lambda=\lambda^*} &= -\frac{Q(\lambda^*)N(\lambda^*) + P(\lambda^*)M(\lambda^*)}{P^2(\lambda^*) + Q^*(\lambda^*)}\\ &= \frac{a_3'(\lambda^*) - a_1'(\lambda^*)a_2(\lambda^*) - a_1(\lambda^*)a_2'(\lambda^*)}{a_1^2(\lambda^*) + a_2(\lambda^*)} > 0, \end{aligned}$$

$$\text{provided } \ [a_1(\lambda)a_2(\lambda)]'_{\lambda=\lambda^*} < a_3'(\lambda^*).$$

Also $\xi_3(\lambda^*) = -a_1(\lambda^*) < 0$. Thus the transversality conditions hold and hence Hopf bifurcation occurs at $\lambda = \lambda^*$. Hence the theorem.

Remark: If there exists a critical value δ^* (corresponding to handling time) such that $a_1(\delta^*) > 0$, $a_1(\delta^*)a_2(\delta^*) - a_3(\delta^*) = 0$, $[a_1(\delta)a_2(\delta)]'_{\delta=\delta^*} < a'_3(\delta^*)$ then when $\delta > \delta^*$, the steady state E^* is stable; when $\delta = \delta^*$, E^* loses its stability and the Hopf bifurcation occurs at E^*, when $\delta < \delta^*$, E^* becomes unstable and a family of periodic solutions bifurcates from E^*.

1.3. *Mathematical study of system (1) with diffusion*

Let $(\tilde{S}, \tilde{I}, \tilde{Z})$ denotes the general uniform steady state of the spatial model (1). To investigate the stability of the model (1), we use the perturbations of the form (see ref. [7])

$$\begin{aligned} S(t,x) &= \tilde{S} + S^d \cos(lx)\exp(\nu t), \\ I(t,x) &= \tilde{I} + I^d \cos(lx)\exp(\nu t), \\ Z(t,x) &= \tilde{Z} + Z^d \cos(lx)\exp(\nu t). \end{aligned}$$

where $l(> 0)$ is the wave number of the spatial perturbation and $\nu > 0$ is the time evaluation rate. Substituting the above expressions in (1) and using the conditions for steady state $(\tilde{S}, \tilde{I}, \tilde{Z})$ of the system (1) we have the corresponding system of ordinary differential equations. We linearize this system around the steady state $(\tilde{S}, \tilde{I}, \tilde{Z})$ and obtain the variational matrix as

$$V(\tilde{E}) = \begin{pmatrix} -m_{11} - l^2 D_1 & -m_{12} & -m_{13} \\ m_{21} & -m_{22} - l^2 D_2 & -m_{23} \\ m_{31} & m_{32} & -m_{33} - l^2 D_3 \end{pmatrix}.$$

The eigenvalues of the variational matrix for the steady state $E_0(0,0,0)$ in presence of diffusion are $r - l^2 D_1, -\mu - l^2 D_2$ and $-d - l^2 D_3$. Hence in presence of diffusion E_0 is uniformly stable if $D_1 > D_1^* = \frac{r}{l^2}$. This implies that if the diffusion coefficient of susceptible prey exceeds the threshold D_1^*, the system becomes collapse.

Again the eigenvalues of the variational matrix for the steady state $E_1(k,0,0)$ in presence of diffusion are $-r - l^2 D_1$, $\lambda k - \mu - l^2 D_2$ and $\theta \alpha k - d(a+k) - l^2 D_3$. In absence of diffusion E_1 is locally asymptotically stable for $\lambda K < \mu$ and $\theta \alpha k < d(a+k)$. Under the same conditions E_1 (in presence of diffusion) is also spatially stable.

For the diffusive system (1) at zooplankton free steady state E_{12}, the characteristic equation of the variational matrix is

$$\left\{x + d + l^2 D_3 - \theta\left(\frac{\alpha\bar{S}}{a+\bar{S}} + \beta\bar{I}\right)\right\} \times$$
$$\left\{x^2 + \left(\frac{r\bar{S}}{k} + \frac{\lambda^2\delta\bar{S}\bar{I}}{(1+\lambda\delta\bar{I})^2} + l^2(D_1+D_2)\right)x + A\right\} = 0, \qquad (6)$$

where $A = \left(\frac{r\bar{S}}{K} + l^2 D_1\right)\left(\frac{\lambda^2\delta\bar{S}2\bar{I}}{(1+\lambda\delta\bar{I})^2} + l^2 D_2\right) + \frac{\lambda^2\bar{S}\bar{I}}{(1+\lambda\delta\bar{I})^3}$.

Now in absence of diffusion E_{12} is stable if $\theta(\frac{\alpha\bar{S}}{a+\bar{S}} + \beta\bar{I}) < d$ condition is satisfied. Under the same condition E_{12} is also spatially stable because all the eigenvalues of the variational matrix for the diffusive system at E_{12} have negative real parts.

In presence of diffusion one eigenvalue of the variational matrix at E_{13} is $\lambda\tilde{S} - \beta\tilde{Z} - \mu - l^2 D_2$ and other two are the roots of

$$x^2 + \left\{\frac{r\tilde{S}}{k} - \frac{\alpha\tilde{S}\tilde{Z}}{(a+\tilde{S})^2} + l^2(D_1+D_3)\right\}x +$$
$$l^2 D_3\left\{\frac{r\tilde{S}}{k} - \frac{\alpha\tilde{S}\tilde{Z}}{(a+\tilde{S})^2} + l^2 D_1\right\} + \frac{a\theta\alpha^2\tilde{S}\tilde{Z}}{(a+\tilde{S})^3} = 0. \qquad (7)$$

If $r(a+\tilde{S})^2 > \alpha k\tilde{Z}$ and $\lambda\tilde{S} < \lambda + \beta\tilde{Z}$ hold then E_{13} is locally asymptotically stable without diffusion. Under the same condition E_{13} is also spatially stable due to the fact that $\lambda\tilde{S} - \lambda - \beta\tilde{Z} - l^2 D_2 < 0$ and all the coefficients of (7) are positive.

The characteristic equation of the variational matrix at the interior equilibrium (S^*, I^*, Z^*) of the system (1) is

$$\xi^3 + p_1\xi^2 + p_2\xi + p_3 = 0, \qquad (8)$$

where

$$p_1 = m_{11} + m_{22} + l^2(D_1 + D_2 + D_3),$$
$$p_2 = m_{11}m_{22} + m_{13}m_{31} + m_{23}m_{32} + m_{12}m_{21}$$
$$+ l^2(D_1 m_{22} + D_2 m_{11} + D_3 m_{11} + D_3 m_{22}) + l^4(D_1 D_1 + D_2 D_3 + D_3 D_1),$$
$$p_3 = m_{11}m_{23}m_{32} - m_{12}m_{23}m_{31} + m_{13}m_{21}m_{32} + m_{13}m_{22}m_{31}$$
$$+ l^2\{D_1 m_{23}m_{32} + D_2 m_{13}m_{31} + D_3(m_{11}m_{22} + m_{12}m_{21})\}$$
$$+ l^4(m_{11}D_2 D_3 + m_{22}D_1 D_3) + l^6 D_1 D_2 D_3.$$

Now $p_1 > 0$ and $p_2 > 0$, $p_3 > 0$
if (i) $r(a+S^*)^2 > \alpha k Z^*$, (ii) $(1+\lambda\delta I^*)^2 c\beta Z^* > \lambda\alpha S^*$ are satisfied. Also

$$\begin{aligned}
p_1.p_2 - p_3 &= m_{11}(m_{13}m_{31} + m_{11}m_{22} + m_{22}^2) + m_{12}(m_{11}m_{21} + m_{21}m_{22} + m_{23}m_{31}) \\
&+ m_{32}(m_{22}m_{23} - m_{13}m_{21}) \\
&+ l^2\{D_1(2m_{11}m_{22} + m_{22}m_{22} + m_{13}m_{31} + m_{12}m_{21}) \\
&+ D_2(2m_{11}m_{22} + m_{11}m_{11} + m_{23}m_{32} + m_{12}m_{21}) \\
&+ D_3(m_{11}m_{11} + 2m_{11}m_{22} + m_{22}m_{22} + m_{13}m_{31} + m_{23}m_{32})\} \\
&+ l^4\{m_{11}(D_1^2 + D_2^2 + D_3^2 + 2D_1D_3 + D_1D_2 + 2D_2D_3) \\
&+ m_{22}(D_1^2 + D_1^2 + D_3^2 + 2D_1D_3 + D_1D_2 + 2D_2D_3)\} \\
&+ l^6\{D_1(D_1 + D_3)(D_1 + D_2 + D_3) + D_2D_3(D_2 + D_3)\} > 0.
\end{aligned}$$

Thus all conditions of Routh-Hurwitz criterion are satisfied and the system (1) is locally asymptotically stable around the positive equilibrium point E^*. Hence introducing diffusion on the system (2) we can not experience diffusive instability, provided $r(a+S^*)^2 > \alpha k Z^*$ and $(1+\lambda\delta I^*)^2 c\beta Z^* > \lambda\alpha S^*$ holds.

Theorem 1.3.1. *The local stability of the equilibrium point E^* implies its global stability in the interior of* Γ.

Proof: Define $L : \{(S, I, Z) \in \Gamma : S, I, Z > 0\} \to \mathbf{R}$ by

$$\begin{aligned}
L = \int_0^R &\left\{\theta(S - S^* - S^* \log \frac{S}{S^*}) + \theta(I - I^* - I^* \log \frac{I}{I^*}) \right. \\
&\left. + (Z - Z^* - Z^* \log \frac{Z}{Z^*})\right\} dx
\end{aligned}$$

Then L is C^1 in the interior of Γ and L has global minimum at E^* with $L(S^*, I^*, Z^*) = 0$.

The time derivative of L along the solution of (1) is

$$\frac{dL}{dt} = \int_0^R \left[\theta\left(\frac{S-S^*}{S}\right)\frac{\partial S}{\partial t} + \theta\left(\frac{I-I^*}{I}\right)\frac{\partial I}{\partial t} + \left(\frac{Z-Z^*}{Z}\right)\frac{\partial Z}{\partial t}\right] dx$$

$$= \theta\int_0^R (S-S^*)\left\{r\left(1-\frac{S}{K}\right) - \frac{\lambda I}{1+\lambda\delta I} - \frac{\alpha Z}{a+S}\right\} dx$$

$$+ \theta D_1 \int_0^R \left(\frac{S-S^*}{S}\right)\frac{\partial^2 S}{\partial x^2} dx$$

$$+ \theta\int_0^R (I-I^*)\left\{\frac{\lambda S}{1+\lambda\delta I} - \mu - \beta Z\right\} dx + \theta D_2 \int_0^R \left(\frac{I-I^*}{I}\right)\frac{\partial^2 I}{\partial x^2} dx$$

$$+ \int_0^R (Z-Z^*)\left\{\frac{\theta\alpha S}{a+S} + \theta\beta I - d\right\} dx + D_3 \int_0^R \left(\frac{Z-Z^*}{Z}\right)\frac{\partial^2 Z}{\partial x^2} dx$$

Using the conditions of existence and stability of E^* and also the boundary condition of system (1) we have

$$\frac{dL}{dt} \le -\int_0^R \left[\frac{\theta r}{k}(S-S^*)^2 + \theta D_1 \frac{S^*}{S^2}\left(\frac{\partial S}{\partial x}\right)^2 + \theta D_2 \frac{I^*}{I^2}\left(\frac{\partial I}{\partial x}\right)^2 \right.$$

$$\left. + D_3 \frac{Z^*}{Z^2}\left(\frac{\partial Z}{\partial x}\right)^2\right] dx < 0.$$

Again $\frac{dL}{dt} = 0$ if and only if $S = S^*, I = I^*$ and $Z = Z^*$. Therefore the largest invariant compact set in $\{(S, I, Z) \in \Gamma : \frac{dL}{dt} = 0\}$ is the singleton set $\{E^*\}$. Hence by LaSalle's invariant principle [18] E^* is globally asymptotically stable in the interior of Γ.

1.4. *Numerical Simulation*

We investigate the nature of stability of several equilibria of the system (1) and system (2) by using numerical simulation. For this we select our hypothetical parameter values satisfying the existence criteria of the positive equilibrium. The hypothetical parametric values are given in the Table 1.

Table 1. Parameters values and units for the system (2).

Parameter	Value	Unit	Parameter	Value	Unit
r	2.75	$(day)^{-1}$	β	0.055	$(day)^{-1}$
k	20.0	$individual/day$	θ	0.75	–
λ	0.05	ml/day	μ	0.005	$(day)^{-1}$
δ	0.045	ml/day	a	1.0	$(day)^{-1}$
α	0.075	$(day)^{-1}$	d	0.175	$(day)^{-1}$

For the set of parameter values in Table 1 and with initial value $X(0) = (3.5,\ 2.5,\ 5.5)$ we have Fig. 1, which demonstrates that the system (2) exhibit locally asymptotically stable interior equilibrium point. We also draw the phase portrait diagram Fig. 2, which also demonstrates that the interior equilibrium E^* is locally asymptotically stable.

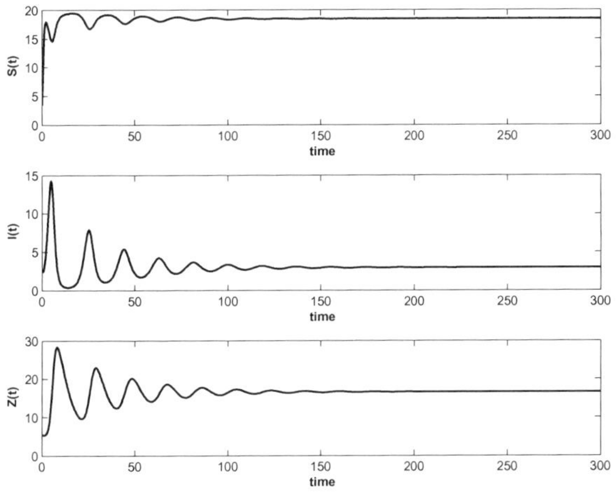

Figure 1. The time series solution depicts E^* is locally asymptotically stable.

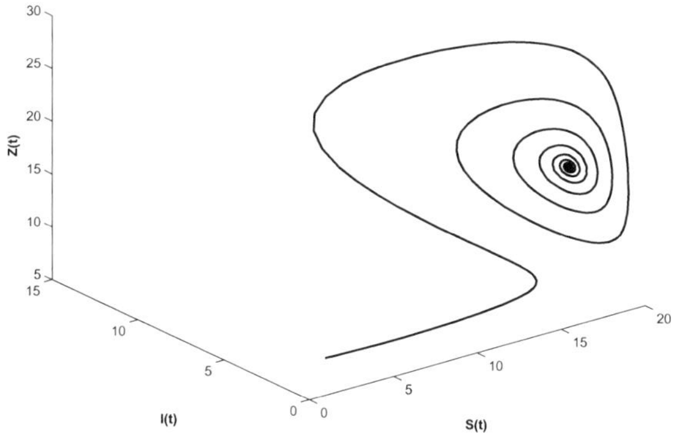

Figure 2. Phase portrait diagram depicts E^* is locally asymptotically stable.

Again if we simply increase the value of the rate of infection λ from 0.05 to 0.6 we observe that the all the species initially oscillates (See Figs. 3 & 4) and ultimately become stable.

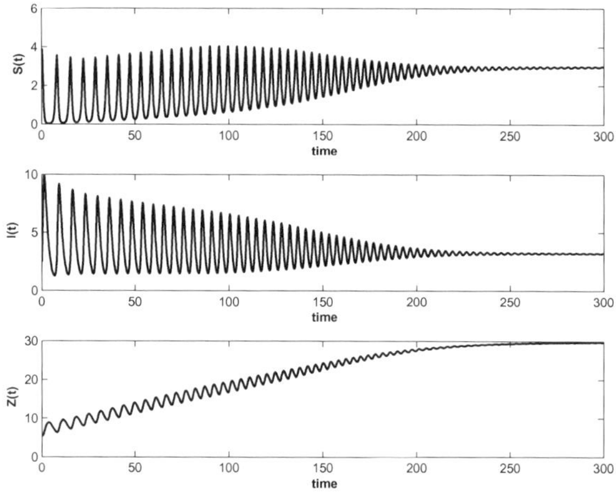

Figure 3. Time series solution depicts E^* is locally asymptotically stable followed by decreasing oscillation.

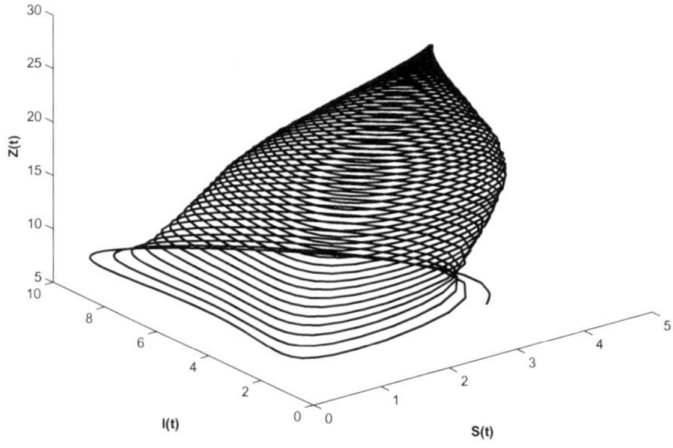

Figure 4. Phage portrait diagram depicts oscillatory behaviour but ultimately stable near E^*.

Now for $\lambda = 0.075$ and all other parameters values as in Table 1 and with diffusion coefficients $D_1 = 5,\ D_2 = 5,\ D_3 = 4$, we have the Figs. 5, 6, 7, which demonstrate that all the species show stable biomass distribution and the interior positive equilibrium point E^* is spatially stable. Now if we increase the value of rate of infection λ from 0.075 to 0.475 and all other parameters values as in Table 1 and with diffusion coefficients $D_1 = 5,\ D_2 = 5,\ D_3 = 4$, we have the Figs. 8, 9, 10, which demonstrate that the biomass distribution is oscillatory with decreasing oscillations. Now for $\lambda = 0.015$ and all other parameters values as in Table 1 and with diffusion coefficients $D_1 = 5,\ D_2 = 5,\ D_3 = 4$, we have the Figs. 11, 12, 13, which depict that the biomass distribution of all species is unstable. Also for $\lambda = 0.75,\ \theta = 0.95$ and and all other parameters values as in Table 1 and with diffusion coefficients $D_1 = 5,\ D_2 = 5,\ D_3 = 4$, we have the Figs. 14, 15, 16, which depict that the biomass distribution of all species is also unstable.

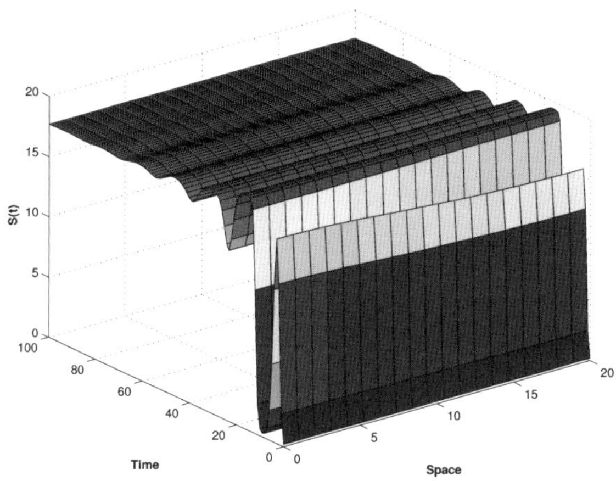

Figure 5. Stable homogeneous biomass distribution of susceptible phytoplankton $S(t)$.

1.5. *Observation and Conclusion*

In this paper, we have studied the stability and bifurcation analysis of an epidemic model in plankton system with diffusion. We have considered that the disease transmitted only on prey population with a nonlinear incidence rate $\dfrac{\lambda I}{1+\lambda\delta I}$ (See Ref. [15]). Also we include diffusion in the system to study the nature of biomass distribution and for occurrence of diffusive instability.

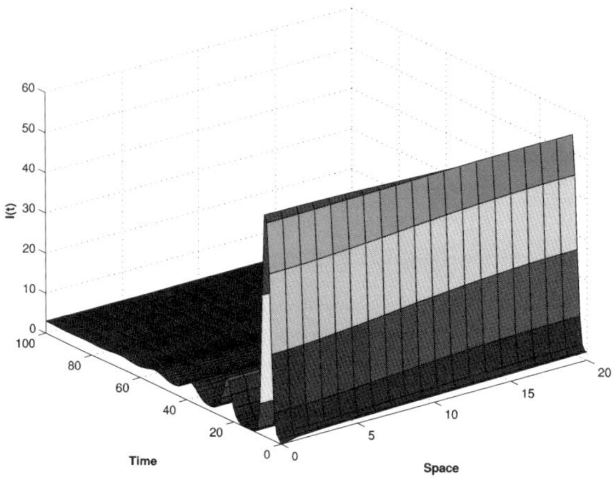

Figure 6. Stable homogeneous biomass distribution of infected phytoplankton $I(t)$.

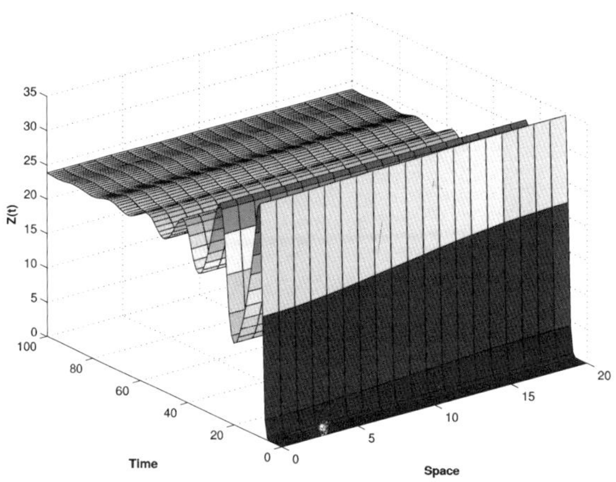

Figure 7. Stable homogeneous biomass distribution of zooplankton $Z(t)$.

We have shown that the dynamics of this model are very rich inside the region Ω.

From theoretical study and numerical observations it is clear that the system (2) around positive equilibrium is locally asymptotically stable (See Figs. 1, 2). Also for the same parameters values as in Table 1 with diffusion coefficients $D_1 = 5,\ D_2 = 5,\ D_3 = 4$, the system (1) is spatially stable (See Figs. 5, 6, 7). Therefore for this choice of system parameters diffusion can not alter the stability of positive equilibrium point E^*. Now if we increase

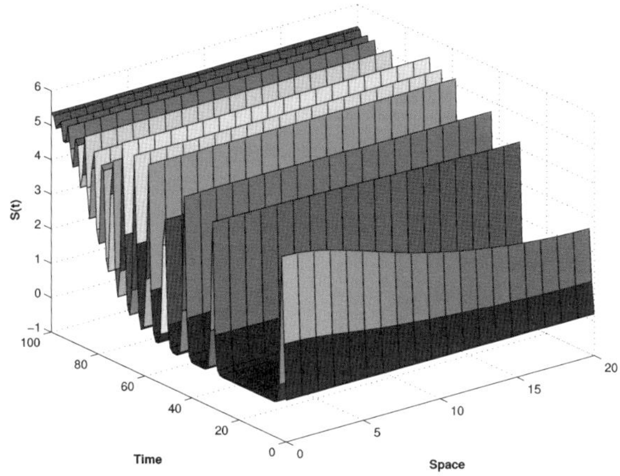

Figure 8. Biomass distribution of susceptible phytoplankton is not uniform.

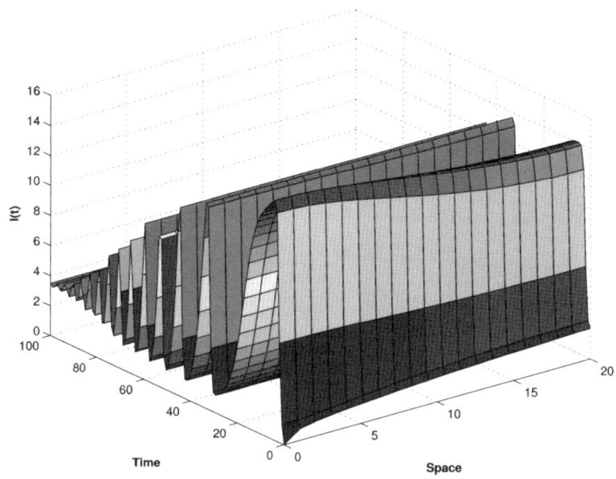

Figure 9. Biomass distribution of infected phytoplankton is not uniform.

the value of rate of infection λ from 0.075 to 0.475 and all other parameters values as in Table 1, we observe the oscillatory biomass distribution with decreasing oscillations (See Figs. 8, 9, 10). Also Figs. 11, 12, 13 depict that the biomass distribution of all species is unstable. Again for $\lambda = 0.75$, $\theta = 0.95$ and all other parameters values as in Table 1, we have the Figs. 14, 15, 16, which depict that the biomass distribution of all species is also unstable. It has been observed that incorporate of diffusion and restriction of rate of infection arises diffusion instability. Also all the species

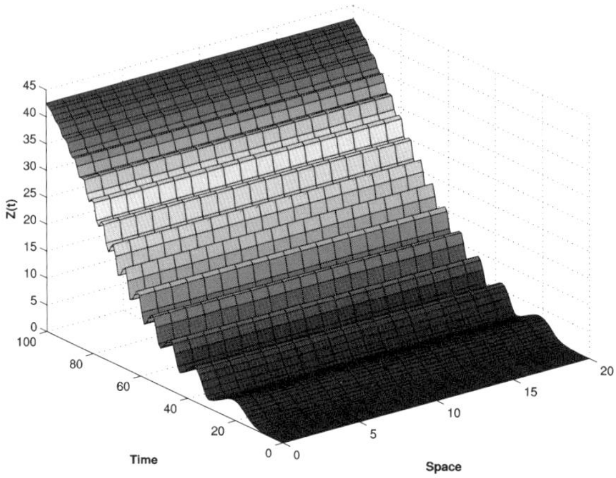

Figure 10. Biomass distribution of zooplankton is not uniform.

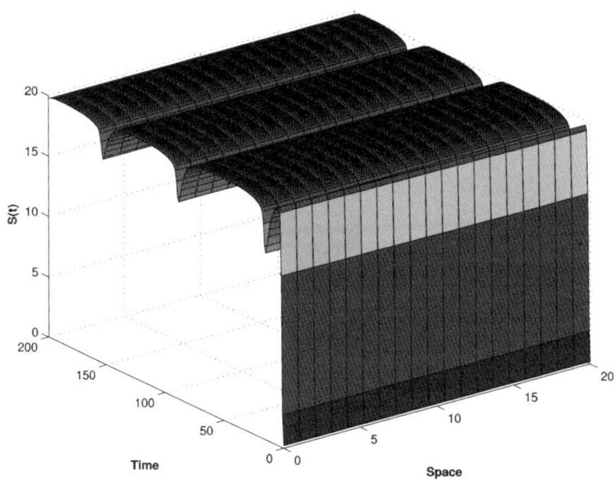

Figure 11. Unstable biomass distribution of susceptible phytoplankton with same amplitude.

coexist with homogeneous biomass distribution for specific choice of system parameters.

Thus, based on the above observations it is clear that:

• The system (2) around positive equilibrium is stable for a specific range of rate of infection λ.

• If the rate of infection and the conversion factor be restricted the all the species coexists and the system (2) is stable.

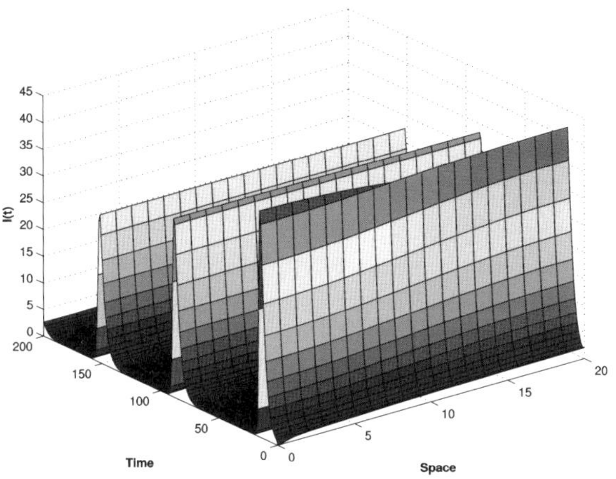

Figure 12. Unstable biomass distribution of infected phytoplankton with same amplitude.

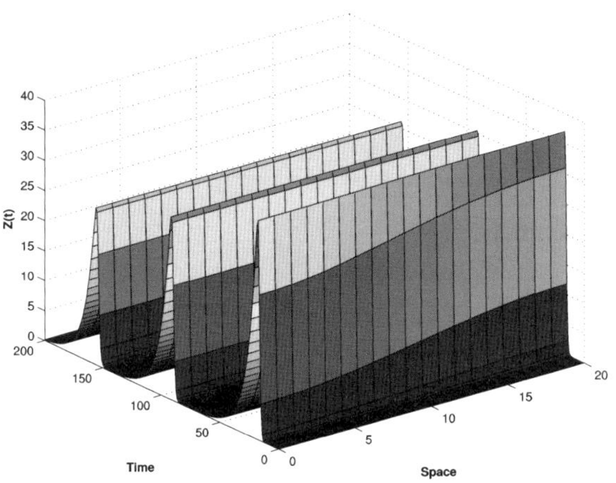

Figure 13. Unstable biomass distribution of zooplankton with same amplitude.

• Also by inclusion of diffusion in the system we observe homogeneous biomass distribution and diffusion instability.

Acknowledgments

I, Dr. Nanda Das acknowledge the funding authority 'University Grants Commission (U G C)', India, for the financial support towards this study (Grant no. PSW-37/12-13(ERO), dated 05.02.2013).

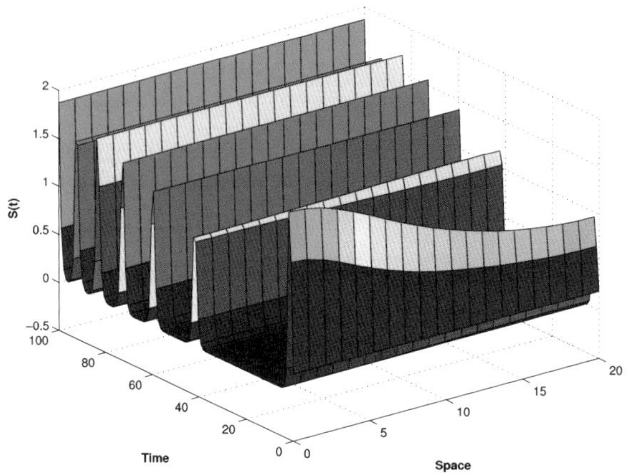

Figure 14. Diffusive instability of susceptible phytoplankton.

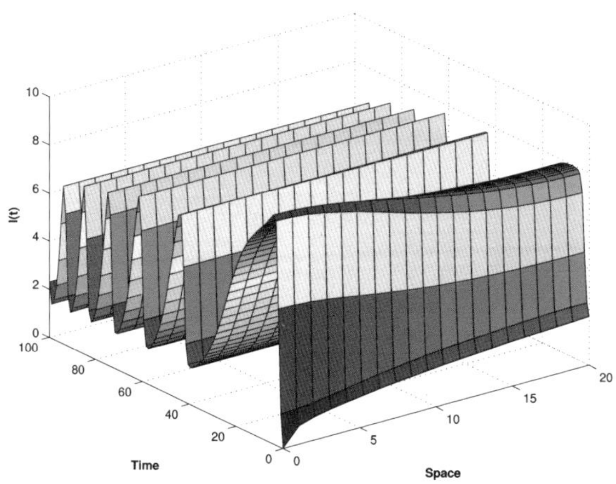

Figure 15. Diffusive instability of infected phytoplankton.

References

1. S. Jacquet,M. Heldal,D. Iglesias-Rodriguez, A. Larsen, W. Wilson, and G. Bratbak, Flow cytometric analysis of an *Emilla huxleyi* bloom terminated by viral infection, *Aquat. Microb. Eclo.*, **27**, 111 (2002).
2. C. J. Rhodes, J. E. Truscoot and A. P. Martin, Viral infection as aregulator of oceanic plankton populations, *Journal of Marine Systems*, **74**, 216 (2008).
3. J. Chattopadhyay and O. Arino, A predator prey model with disease in the prey, *Nonlinear Analysis, Pergamon Press, U.S.A.*, **36**, 747 (1999).

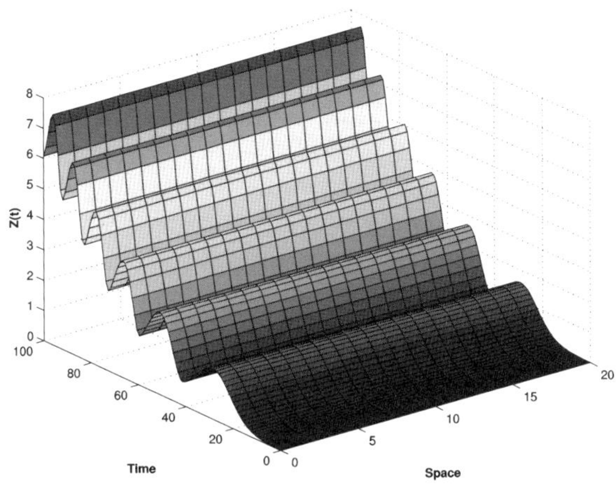

Figure 16. Diffusive instability of zooplankton.

4. J. Chattopadhyay, S. Pal and A. El abdellaoui, Classical predator-prey system with infection on prey population -a mathematical model, *Mathematical Methods in the Applied Sciences*, **26**, 1211 (2003).
5. H. W. Hethcote, W. W. L. Han and Ma. Zhien, A predator-prey model with infected prey, *Theo. pop. biol.*, **66**, 259 (2004).
6. S. Roy and J. Chattopadhyay, Disease-selective predation may lead to prey extinction, *Mathematical Methods in the Applied Sciences*, **28**, 1257 (2005).
7. S. Chaudhuri, J. Chattopadhyay and E. Venturino, Toxic phytoplankton-induced spatiotemporal patterns. *Journal of Biological Physics*, **10 December**, 1 (2011).
8. L. A. Segel and J. L. Jackson, Dissipative structure: an explanation and an ecological example, *J. Theoret. Biol.*, **37**, 545 (1972).
9. S. A. Levin and L. E. Segel, Hypothesis for origin of planktonic patchiness, *Nature*, **259**, 659 (1976).
10. S. A. Levin, A more functional response to predator-prey stability, *Amer. Natur.*, **111**, 381 (1977).
11. G. T. Csanady, Turbulent Diffusion in the Environment, *D. Reidel*, **Boston**, 1973.
12. A. F. Bennett and K. L. Denman, Phytoplankton patchiness: inferences from particle statistics, *J. Mar. Res.*, **43** , 307 (1985).
13. M. X. Wang Stationary patterns caused by cross-diffusion for a three-species prey - predator model, *Comput. Math. Appl.*, **52**, 707 (2006).
14. N. Das, S. Pal and J. Chattopadhyay, Spatial effects on viral disease in plankton system, *Nonlinear Studies*, **20(1)**, 105 (2013).
15. V. Capasso and G. Serio, A generalization of the Kermack-McKendrick deterministic epidemic model, *Math. Biosci.*, **42**, 43 (1978).
16. M. Nagumo, Uber die lage der integralkurven gewohnlicher differentialgle-

ichungen, *Proc. Phys. Math. Soc. Japan*, **24**, 551 (1942).
17. G. Birkhoff and G. C. Rota, Ordinary Differential equations, *Ginn Boston*, (1982).
18. J. P. LaSalle, The Stability of Dynamical Systems. *SIAM,Philadelphia, PA* (1976).

A (1+2)-DIMENSIONAL KELLER-SEGEL MODEL: LIE SYMMETRY AND EXACT SOLUTIONS FOR THE CAUCHY PROBLEM

ROMAN CHERNIHA

School of Mathematical Sciences, University of Nottingham, University Park, Nottingham, NG7 2RD, UK
E-mail: cherniha@gmail.com

MAKSYM DIDOVYCH

Institute of Mathematics, National Academy of Science of Ukraine, 3, Tereshchenkivska Str., Kyiv 01601, Ukraine
E-mail: m.didovych@gmail.com

This research is a natural continuation of our recent paper "Exact solutions of the simplified Keller-Segel model" published in Commun Nonlinear Sci Numer Simulat 2013; 18: 2960-2971. It is shown that a (1+2)-dimensional Keller-Segel type system is invariant with respect to infinity-dimensional Lie algebra. Lie symmetry operators are used for constructing exact solutions of the Cauchy problems based on the Keller-Segel system in question. In particular, we have proved that the Cauchy problem for the (1+1)-dimensional Keller-Segel system can be linearized and solved in an explicit form. Moreover, additional biologically motivated restrictions were established in order to obtain uniqueness of solution.

1. Introduction

In 1970-1971, E.F.Keller and L.A.Segel published a remarkable paper [1,2], in which they constructed the mathematical model for describing the chemotactic interaction of amoebae mediated by the chemical (acrasin) in. Nowadays their model is called the Keller-Segel model and used for modeling a wide range of processes in biology and medicine. The one-dimensional (with respect to the space variable) version of the Keller-Segel model reads as

$$\begin{aligned} N_t &= [D^1(N,P)N_x - \chi(P)NP_x]_x, \\ P_t &= [D^2(P)P_x]_x + \alpha(P)N - \beta(P), \end{aligned} \tag{1}$$

where unknown functions $N(t,x)$ and $P(t,x)$ describe the densities of cells (species) and chemicals, respectively, t and x denote the time and space variables, respectively, $D^1(N,P)$ and $D^2(P)$ are the diffusivities of cells (species) and chemicals, while $\alpha(P)$ and $\beta(P)$ are known non-negative smooth functions. The function $\chi(P)$ (usually a constant χ_0) is called the chemotactic sensitivity. Nowadays a wide range of simplifications of the Keller-Segel model are used for modeling processes in biology and medicine. Here we restrict ourselves on the(1+2)-dimensional Keller-Segel system of the form [3,4,5,6]

$$\begin{aligned} &N_t(t,x,y) = d_1\triangle N(t,x,y) - \chi_0\nabla(N(t,x,y)\nabla P(t,x,y)),\\ &0 = \triangle P(t,x,y) + \alpha N(t,x,y) - \beta P(t,x,y), \end{aligned} \tag{2}$$

where the parameters d_1, χ_0, α and β are non-negative constants, moreover, $\chi_0\alpha \neq 0$ (otherwise the model loses its biological meaning). Nowadays system (2), including the special case $\beta = 0$, is extensively examined by means of different mathematical techniques, in particular, several talks were devoted to this model at a special session within 10th AIMS Conference on Dynamical Systems, Differential Equations and Applications [7].

However, to the best of our knowledge, there are no papers devoted to application of the Lie symmetry method for investigation of system (2), notably for construction of exact solutions. In this paper, we show that this nonlinear system with $\beta = 0$ is invariant with respect infinity-dimensional Lie algebra generated by the operators involving three arbitrary functions, which depend on time variable. Using this algebra we construct a wide range of initial profiles when the corresponding Cauchy problems also admit infinity-dimensional Lie algebra and reduce the problems in question to (1+1)-dimensional problems. In particular, exact solutions in explicit form are constructed for some problems with correctly-specified initial profiles. Finally, a biological interpretation of the results derived is discussed.

This research is a natural continuation of our recent paper [8].

2. Lie symmetry of (1+2) Keller-Segel system

First of all, one notes that all the parameters, excepting β, can be dropped in system (2) if one introduces non-dimensional variables using the standard re-scaling procedure, i.e. this simplified Keller-Segel system is equivalent to

$$\begin{aligned} &\rho_t(t,x,y) = \triangle\rho(t,x,y) - \nabla(\rho(t,x,y)\nabla S(t,x,y)),\\ &0 = \triangle S(t,x,y) + \rho(t,x,y) - \beta^* S(t,x,y). \end{aligned} \tag{3}$$

where $\beta^* = \beta d_1/\alpha$. Obviously, one may set $\beta^* = 0$ provided $\beta d_1/\alpha = \varepsilon <<$ 1 in (2), hence the nonlinear system

$$\begin{aligned} \rho_t(t,x,y) &= \triangle\rho(t,x,y) - \nabla(\rho(t,x,y)\nabla S(t,x,y)), \\ 0 &= \triangle S(t,x,y) + \rho(t,x,y) \end{aligned} \tag{4}$$

is obtained.

Theorem 1. *Maximal algebras of invariance (MAI) of the (1+2) KS system is the infinity-dimensional Lie algebra generated by the operators*

$$\begin{aligned} &G_1^\infty = f_1(t)\tfrac{\partial}{\partial x} + xf_1^{'}(t)\tfrac{\partial}{\partial S},\ G_2^\infty = f_2(t)\tfrac{\partial}{\partial y} + yf_2^{'}(t)\tfrac{\partial}{\partial S}, \\ &X_S^\infty = g(t)\tfrac{\partial}{\partial S},\ P_t = \tfrac{\partial}{\partial t},\ J_{12} = -x\tfrac{\partial}{\partial y} + y\tfrac{\partial}{\partial x}, \\ &D = 2t\tfrac{\partial}{\partial t} + x\tfrac{\partial}{\partial x} + y\tfrac{\partial}{\partial y} - 2\rho\tfrac{\partial}{\partial \rho}, \end{aligned} \tag{5}$$

where $f_1(t)$, $f_2(t)$ and $g(t)$ are arbitrary smooth functions.

Proof of the theorem is obtained by straightforward calculations using the well-known technique created by Sophus Lie in 80s of 19 century. Nowadays this routine can be done using computer algebra packages therefore we used Maple 16.

However, a PDE (system of PDEs) cannot model any real process without additional conditions on unknown functions. Probably the most common additional conditions are the correctly specified initial profiles of unknown functions. Thus, the Cauchy problem is obtained. In the case of system (4) the Cauchy problem can be formulated as follows:

$$\begin{aligned} &\rho_t = \triangle\rho - \nabla(\rho\nabla S), \\ &0 = \triangle S + \rho, \\ &t = 0:\ S = \phi(x), \quad \rho = -\phi_{xx}(x), \end{aligned} \tag{6}$$

where $\phi(x)$ is an arbitrary smooth function (in general case this function may depend on the variable y too). In order to guarantee existence of a classical solution of the Cauchy problem (6), the initial profile for the component ρ should be specified as above.

Obviously, Lie algebra (5) cannot be MAI of the Cauchy problem (6). To determine which of the operators listed in (5) are Lie symmetry operators of Cauchy problem in question we use Bluman's criteria [9] (see also its generalization [10]). According to the criteria one should examine whether initial conditions from (6) are invariant under the operator in question. In order to check this for all the operators listed in (5), we take their linear combination

$$X = a_1 G_1^\infty + a_2 G_2^\infty + a_3 X_S^\infty + a_4 P_t + a_5 J_{12} + a_6 D =$$

$$(a_4 + 2ta_6)\frac{\partial}{\partial t} + (a_1 f_1(t) + a_5 y + a_6 x)\frac{\partial}{\partial x} + (a_2 f_2(t) - a_5 x + a_6 y)\frac{\partial}{\partial y} +$$

$$(a_3 g(t) + a_1 f_1^{'}(t)x + a_2 f_2^{'}(t)y)\frac{\partial}{\partial S} - 2a_6\rho\frac{\partial}{\partial \rho}.$$

Applying the operator X to the manifolds $t = 0$, $S = \phi(x)$ and $\rho = -\phi_{xx}(x)$, one arrives at one restriction and two equations:

1) $a_4 = 0$.

2) $(a_1 f_1(0) + a_5 y + a_6 x)\phi_x(x) = a_3 g(0) + a_1 f_1^{'}(0)x + a_2 f_2^{'}(0)y$.

3) $2a_6\phi_{xx}(x) + (a_1 f_1(0) + a_5 y + a_6 x)\phi_{xxx}(x) = 0$.

Since equation 3) is a differential consequence of 2) we need analyze 2) only (obviously the restriction $a_4 = 0$ means the well-known fact that the Cauchy problem is not invariant w.r.t. time translation).

Equation 2) implies certain limitations on the function $\phi(x)$. This function can only be arbitrary when both sides of equation vanish. When the multiplier in the left-hand-side is non-zero, one obtains ODE to find $\phi(x)$. This ODE can be easily solved and proved that four different cases arise and each of them is described in Tables 2–5.

Let us consider the most general case. Equation 2) is fulfilled for arbitrary smooth function $\phi(x)$ when $a_1 f_1(0) + a_5 y + a_6 x = 0$ and $a_3 g(0) + a_1 f_1^{'}(0)x + a_2 f_2^{'}(0)y = 0$. It means that all different cases can be established by simple analysis of possible values for constants $a_1, \ldots a_6$ and setting $g(0) = 0$, $f_1(0) = 0$, $f_1^{'}(0) = 0$ and $f_2^{'}(0) = 0$ provided the corresponding parameters are non-zero. Thus, this case was completely described and the result is presented in Table 1, which contains the operators along with restrictions they should meet.

Table 1. $\phi(x)$ is an arbitrary smooth function.

	Restrictions	Operators
1	$f_1(0)=0, f_1^{'}(0)=0$	G_1^{∞}
2	$f_2^{'}(0)=0$	G_2^{∞}
3	$g(0)=0$	X_S^{∞}
4	$f_1(0)=0, f_1^{'}(0)=0, f_2^{'}(0)=0$	$G_1^{\infty}, G_2^{\infty}$
5	$g(0)=0, f_1(0)=0, f_1^{'}(0)=0$	$X_S^{\infty}, G_1^{\infty}$
6	$g(0)=0, f_2^{'}(0)=0$	$X_S^{\infty}, G_2^{\infty}$
7	$g(0)=0, f_1(0)=0, f_1^{'}(0)=0, f_2^{'}(0)=0$	$X_S^{\infty}, G_1^{\infty}, G_2^{\infty}$

As we mentioned above, ODE 2) should be solved to find $\phi(x)$ in the case $a_1 f_1(0) + a_5 y + a_6 x \neq 0$. Thus, solving this ODE with $a_6 \neq 0$ one finds $\phi(x) = \frac{a_3 a_6 g(0) - a_1^2 f_1(0) f_1^{'}(0)}{a_6^2} \ln \left| x + \frac{a_1 f_1(0)}{a_6} \right| + \frac{a_1 f_1^{'}(0)}{a_6} x + C$, where C is arbitrary constant. This function can be simplified by the transformation $x + \frac{a_1 f_1(0)}{a_6} \to x$ and the new notations $\frac{a_1}{a_6} \to a_1$ and $\frac{a_3}{a_6} \to a_3$. The final result is presented in Table 2 provided $a_3 a_6 g(0) - a_1^2 f_1(0) f_1^{'}(0) \neq 0$.

The restriction $a_3 a_6 g(0) - a_1^2 f_1(0) f_1^{'}(0) = 0$ immediately leads to a linear function. The relevant symmetries are presented in Tables 3 and 4.

Finally, examination of ODE 2) with $a_6 = 0$ leads to the function $\phi(x)$, which is a parabola, hence without losing a generality we may set $\phi(x) = \lambda_2 x^2 + \lambda_0$, $\lambda_2 \neq 0$. Thus, the Lie symmetry operators obtained are presented Table 5, which contains the operators along with restrictions they should meet.

Table 2. $\phi(x) = \big(a_3 g(0) - a_1^2 f_1(0) f_1'(0)\big)\, \ln|x| + a_1 f_1'(0)\, x + c,\ x \neq 0.$

	Restrictions	Operators
1	$a_1^2 + a_3^2 \neq 0, (f_1(0))^2 + (f_1'(0))^2 \neq 0, g(0) \neq 0,$ $a_1 f_1(0) f_1'(0) \neq a_3 g(0)$	$a_1 G_1^\infty + a_3 X_S^\infty + D$
2	$a_1 f_1(0) f_1'(0) \neq 0, g(0) = 0$	$X_S^\infty, a_1 G_1^\infty + D$
3	$a_3 g(0) \neq 0, f_1(0) = 0, f_1'(0) = 0$	$G_1^\infty, a_3 X_S^\infty + D$
4	$a_1^2 + a_3^2 \neq 0, (f_1(0))^2 + (f_1'(0))^2 \neq 0, g(0) \neq 0,$ $a_1 f_1(0) f_1'(0) \neq a_3 g(0), f_2'(0) = 0$	$a_1 G_1^\infty + a_3 X_S^\infty + D, G_2^\infty$
5	$a_1 f_1(0) f_1'(0) \neq 0, g(0) = 0, f_2'(0) = 0$	$X_S^\infty, a_1 G_1^\infty + D, G_2^\infty$
6	$a_3 g(0) \neq 0, f_1(0) = 0, f_1'(0) = 0, f_2'(0) = 0$	$G_1^\infty, a_3 X_S^\infty + D, G_2^\infty$

Table 3. $\phi(x) = \lambda_0$.

	Restrictions	Operators
1	—	J_{12}, D
2	$f_1'(0) = 0$	G_1^∞, J_{12}, D
3	$f_2'(0) = 0$	G_2^∞, J_{12}, D
4	$g(0) = 0$	X_S^∞, J_{12}, D
5	$f_1'(0) = 0, f_2'(0) = 0$	$G_1^\infty, G_2^\infty, J_{12}, D$
6	$g(0) = 0, f_1'(0) = 0$	$X_S^\infty, G_1^\infty, J_{12}, D$
7	$g(0) = 0, f_2'(0) = 0$	$X_S^\infty, G_2^\infty, J_{12}, D$
8	$g(0) = 0, f_1'(0) = 0, f_2'(0) = 0$	$X_S^\infty, G_1^\infty, G_2^\infty, J_{12}, D$

Table 4. $\phi(x) = \lambda_1 x, \lambda_1 \neq 0$.

	Restrictions	Operators
1	$g(0) = 0, f_1(0) = 0, f_1^{'}(0) = \frac{a_6\lambda_1}{a_1}$, $f_2^{'}(0) = \frac{a_5\lambda_1}{a_2}, a_1 \neq 0, a_2 \neq 0$	$X_S^\infty, a_1 G_1^\infty + a_6 D$, $a_2 G_2^\infty + a_5 J_{12}$
2	$g(0) = 0, f_1(0) = 0, f_1^{'}(0) = \frac{a_6\lambda_1}{a_1}$, $a_1 \neq 0$	$X_S^\infty, a_1 G_1^\infty + a_6 D$
3	$g(0) \neq 0, f_1(0) = 0, f_1^{'}(0) = \frac{a_6\lambda_1}{a_1}$, $a_1 \neq 0$	$a_1 G_1^\infty + a_6 D$
4	$g(0) \neq 0, f_2^{'}(0) = \frac{a_5\lambda_1}{a_2}$, $a_2 \neq 0$	$a_2 G_2^\infty + a_5 J_{12}$
5	$g(0) \neq 0, f_1(0) = 0, f_1^{'}(0) = \frac{a_6\lambda_1}{a_1}$, $f_2^{'}(0) = \frac{a_5\lambda_1}{a_2}, a_1 \neq 0, a_2 \neq 0$	$a_1 G_1^\infty + a_6 D$, $a_2 G_2^\infty + a_5 J_{12}$
6	$g(0) = \frac{a_1\lambda_1 f_1(0)}{a_3}, f_1^{'}(0) = a_6\lambda_1$, $f_1(0) \neq 0, a_3 \neq 0$	$a_1 G_1^\infty + a_3 X_S^\infty + a_1 a_6 D$
7	$g(0) = \frac{a_1\lambda_1 f_1(0)}{a_3}, f_1^{'}(0) = a_6\lambda_1, f_2^{'}(0) = \frac{a_5\lambda_1}{a_2}$, $f_1(0) \neq 0, a_2 \neq 0, a_3 \neq 0$	$a_1 G_1^\infty + a_3 X_S^\infty + a_1 a_6 D$, $a_2 G_2^\infty + a_5 J_{12}$

Table 5. $\phi(x) = \lambda_2 x^2 + \lambda_0,\ \lambda_2 \neq 0$.

	Restrictions	Operators
1	$f_1^{'}(0) = 2\lambda_2 f_1(0)$	G_1^∞
2	$g(0) = 0, f_1^{'}(0) = 2\lambda_2 f_1(0)$	X_S^∞, G_1^∞
3	$f_1^{'}(0) = 2\lambda_2 f_1(0), f_2^{'}(0) = 0$	G_1^∞, G_2^∞
4	$g(0) = 0, f_1^{'}(0) = 2\lambda_2 f_1(0), f_2^{'}(0) = 0$	$X_S^\infty, G_1^\infty, G_2^\infty$

3. Application of Lie Symmetry for Constructing Exact Solutions of Cauchy problems

Cauchy problem (6) can be reduced to a set of (1+1)-dimensional ones by using the Lie symmetry operators and their linear combinations. Nowadays it is a standard routine, provided an algebra of invariance is of lower dimensionality. One notes that all MAIs listed in Tables 1–5 are still infinity-dimensional, hence the reduction process can be highly non-trivial for Cauchy problem (6) with a specified profiles. Here we restrict ourselves on case 2 from Table 1.

Because the operator G_2^{∞} contains an arbitrary function $f_2(t)$ with the property $f_2'(0) = 0$, we may consider the special case $f_2(t) = conts \neq 0$, i.e. $G_2^{\infty} = \partial_y$. Obviously this operator produces Ansatz

$$S(t,x,y) = S(t,x), \quad \rho(t,x,y) = \rho(t,x). \tag{7}$$

Substituting ansatz (7) into (6), the following (1+1)-dimensional Cauchy problem

$$\begin{aligned} &\rho_t = \rho_{xx} - (\rho S_x)_x \\ &0 = S_{xx} + \rho \\ &t = 0 : \rho(0,x) = -\phi_{xx}(x), S(0,x) = \phi(x). \end{aligned} \tag{8}$$

is obtained. Thus, the same problem is derived, however, in the case of one spacial variable. Let us reduce the governing system to a single equation, extracting $\rho = -S_{xx}$ from the second equation of (8) and substituting into the first. Hence the 4th order differential equation

$$S_{txx} = S_{xxxx} - (S_{xx}S_x)_x$$

is obtained, which is equivalent to the 3rd order equation

$$S_{tx} = S_{xxx} - S_{xx}S_x + \theta(t)$$

where $\theta(t)$ is an arbitrary function. The obvious substitution

$$W(t,x) = S_x(t,x)$$

transforms the last equation to the nonlinear 2nd order equation

$$W_t = W_{xx} - W_x W + \theta(t). \tag{9}$$

It can be noted that the substitution

$$W(t,x) = U(t,y) + \int_0^t \theta(\tau)\,\mathrm{d}\tau$$
$$x = z + j(t),$$

where $j(t) = \int_0^t (\int_0^\tau \theta(\tau_1)\,\mathrm{d}\tau_1)\mathrm{d}\tau$, reduces equation (9) to the Burgers equation

$$U_t = U_{zz} - UU_z.$$

The Burgers equation is linearizable via the famous Cole-Hopf substitution[11]

$$U(t,z) = -2\frac{V_z(t,z)}{V(t,z)}$$

to the linear heat equation

$$V_t = V_{zz}. \tag{10}$$

All the substitutions mentioned above could be united as follows

$$\begin{aligned} S_x(t,x) &= -2\tfrac{V_z(t,z)}{V(t,z)} + \int_0^t \theta(\tau)\,\mathrm{d}\tau, \\ x &= z + j(t). \end{aligned} \tag{11}$$

It should be stressed that the substitution (11) reduces the nonlinear Cauchy problem (8) to the linear problem for the heat equation (10), which can be exactly solved. In fact, having the specified initial profiles in (8), we find the initial condition for equation (10) as follows:

$$-2\,\frac{V_z(0,z)}{V(0,z)} = (\phi(z))_z$$

i.e.

$$V(0,z) = a\,e^{-\frac{1}{2}\phi(z)} = a\,g(z),\ a > 0. \tag{12}$$

Obviously the exact solution of Cauchy problem (10) and (12) is

$$V(t,z) = \frac{a}{\sqrt{4\pi t}} \int_{-\infty}^{+\infty} g(\xi)\,e^{-\frac{(z-\xi)^2}{4t}}\,\mathrm{d}\xi. \tag{13}$$

Thus, calculating the derivatives $V_z(t,z)$ and $V_{zz}(t,z)$, and using substitution (11), one construct the solution of the initial Cauchy problem (8)

$$
\begin{aligned}
S(t,x) = \int W(t,x)\,\mathrm{d}x = \int \left(U(t,z) + \int_0^t \theta(\tau)\,\mathrm{d}\tau \right) \mathrm{d}x = \\
-2\ln(V(t,z)) + x \cdot \int_0^t \theta(\tau)\,\mathrm{d}\tau + A(t) = \\
-2\ln\left(\frac{1}{\sqrt{4\pi t}} \int_{-\infty}^{+\infty} g(\xi)\, e^{-\frac{(x-j(t)-\xi)^2}{4t}}\,\mathrm{d}\xi \right) + x \cdot \int_0^t \theta(\tau)\,\mathrm{d}\tau + A(t)
\end{aligned}
$$

$$
\begin{aligned}
\rho(t,x) = -U_z(t,z) = 2\left(\frac{V_{zz}(t,z)}{V(t,z)} - \left(\frac{V_z(t,z)}{V(t,z)} \right)^2 \right) = \\
\frac{1}{2t^2} \cdot \frac{\int_{-\infty}^{+\infty} g(\xi)\,(x-j(t)-\xi)^2\, e^{-\frac{(x-j(t)-\xi)^2}{4t}}\,\mathrm{d}\xi}{\int_{-\infty}^{+\infty} g(\xi)\, e^{-\frac{(x-j(t)-\xi)^2}{4t}}\,\mathrm{d}\xi} - \frac{1}{t} - \\
-\frac{1}{2t^2} \cdot \left(\frac{\int_{-\infty}^{+\infty} g(\xi)\,(x-j(t)-\xi)\, e^{-\frac{(x-j(t)-\xi)^2}{4t}}\,\mathrm{d}\xi}{\int_{-\infty}^{+\infty} g(\xi)\, e^{-\frac{(x-j(t)-\xi)^2}{4t}}\,\mathrm{d}\xi} \right)^2
\end{aligned}
$$

The last expression can be simplified to the form

$$
\begin{aligned}
\rho(t,x) = -\frac{1}{t} \cdot \frac{\int_{-\infty}^{+\infty} g'(\xi)\,(x-j(t)-\xi)\, e^{-\frac{(x-j(t)-\xi)^2}{4t}}\,\mathrm{d}\xi}{\int_{-\infty}^{+\infty} g(\xi)\, e^{-\frac{(x-j(t)-\xi)^2}{4t}}\,\mathrm{d}\xi} - \\
-\frac{1}{2t^2} \cdot \left(\frac{\int_{-\infty}^{+\infty} g(\xi)\,(x-j(t)-\xi)\, e^{-\frac{(x-j(t)-\xi)^2}{4t}}\,\mathrm{d}\xi}{\int_{-\infty}^{+\infty} g(\xi)\, e^{-\frac{(x-j(t)-\xi)^2}{4t}}\,\mathrm{d}\xi} \right)^2.
\end{aligned}
$$

Finally, we need to examine the behavior of the solution as $t \to 0$:

$$
\lim_{t\to 0} S(t,x) = -2\ln\left(\lim_{t\to 0} (V(t,x)) \right) + A(0) = -2\ln(e^{-\frac{1}{2}\phi(x)}) + A(0) = \phi(x) + A(0).
$$

Hence, having chosen the function $A(t)$ so that $A(0) = 0$, one we obtain $\lim_{t\to 0} S(t,x) = S(0,x)$, i.e. the function $S(t,x)$ is continuous at $t = 0$.

In order to prove that $\lim_{t\to 0} \rho(t,x) = \rho(0,x)$ one needs to show that $\lim_{t\to 0} V_z(t,z) = V_z(0,z)$ and $\lim_{t\to 0} V_{zz}(t,z) = V_{zz}(0,z)$. The proof of the equality $\lim_{t\to 0} V_z(t,z) = V_z(0,z)$ can be found at [11]. We have proved that $\lim_{t\to 0} V_{zz}(t,z) = V_{zz}(0,z)$ under the restriction $\phi(x) = o(x^2), x \to \infty$ (here the relevant calculations are omitted). Thus, the following statement can be formulated.

Theorem 2. *The solution of Cauchy problem (8), in case when $\phi(x)$ is differentiable twice and $\phi(x) = o(x^2)$, $x \to \infty$ is continuous when $t \to 0$ and can be presented as*

$$
\begin{aligned}
S(t,x) = &-2\ln\left(\frac{1}{\sqrt{4\pi t}}\int_{-\infty}^{+\infty} g(\xi)\, e^{-\frac{(x-j(t)-\xi)^2}{4t}}\,\mathrm{d}\xi\right) + \\
&+ x\cdot\int_0^t \theta(\tau)\,\mathrm{d}\tau + A(t), \\
\rho(t,x) = &-\frac{1}{t}\cdot\frac{\int_{-\infty}^{+\infty} g'(\xi)\,(x-j(t)-\xi)\,e^{-\frac{(x-j(t)-\xi)^2}{4t}}\,\mathrm{d}\xi}{\int_{-\infty}^{+\infty} g(\xi)\,e^{-\frac{(x-j(t)-\xi)^2}{4t}}\,\mathrm{d}\xi} - \\
&-\frac{1}{2t^2}\cdot\left(\frac{\int_{-\infty}^{+\infty} g(\xi)\,(x-j(t)-\xi)\,e^{-\frac{(x-j(t)-\xi)^2}{4t}}\,\mathrm{d}\xi}{\int_{-\infty}^{+\infty} g(\xi)\,e^{-\frac{(x-j(t)-\xi)^2}{4t}}\,\mathrm{d}\xi}\right)^2,
\end{aligned}
\tag{14}
$$

where $g(\xi) = e^{-\frac{1}{2}\phi(\xi)}$, $A(0) = 0$.

Obviously, the exact solution (14) is not unique because one contains two arbitrary functions $\theta(t)$ and $A(t)$. To specify these function one needs additional biologically motivated restrictions. We remind the reader that function $S(t,x)$ describing the density of chemicals should be bounded in space. So functions $\theta(t)$ must vanish (in this case function $j(t)$ that depends on $\theta(t)$ also vanishes). In order to specify both functions $\theta(t)$ and $A(t)$ one needs to assume that the quantity of the chemical $S(t,x)$ is finite in space and time, i.e. $\int_{-\infty}^{+\infty} |S(t,x)|dx < 0$ for $\forall t > 0$. This assumption immediately leads to the unique solution of Cauchy problem (8) in the form

$$S(t,x) = -2\ln\left(\tfrac{1}{\sqrt{4\pi t}}\int_{-\infty}^{+\infty} g(\xi)\, e^{-\frac{(x-\xi)^2}{4t}}\, \mathrm{d}\xi\right),$$

$$\rho(t,x) = -\frac{1}{t}\cdot\frac{\int_{-\infty}^{+\infty} g'(\xi)\,(x-\xi)\,e^{-\frac{(x-\xi)^2}{4t}}\,\mathrm{d}\xi}{\int_{-\infty}^{+\infty} g(\xi)\,e^{-\frac{(x-\xi)^2}{4t}}\,\mathrm{d}\xi} - \frac{1}{2t^2}\cdot\left(\frac{\int_{-\infty}^{+\infty} g(\xi)\,(x-\xi)\,e^{-\frac{(x-\xi)^2}{4t}}\,\mathrm{d}\xi}{\int_{-\infty}^{+\infty} g(\xi)\,e^{-\frac{(x-\xi)^2}{4t}}\,\mathrm{d}\xi}\right)^2. \tag{15}$$

Now we apply the operator G_2^∞ with a non-constant function $f_2(t)$. In order to reduce system (6) we need to construct ansatz by solving the corresponding system of characteristic equations for this operator. After rather standard calculations the Ansatz

$$S(t,x,y) = \tfrac{f_2'(t)}{f_2(t)}\cdot\tfrac{y^2}{2} + \psi(t,x), \quad \rho(t,x,y) = \Psi(t,x) \tag{16}$$

is obtained, however the function $\Psi(t,x)$ can be immediately specified using the second equation in (6), hence

$$\rho(t,x,y) = -\frac{f_2'(t)}{f_2(t)} - \psi_{xx}(t,x)$$

(hereafter $\psi(t,x)$ is to-be-determined function). Substituting this ansatz into the first equation in (6) one arrives at the fourth order partial differential equation

$$\psi_{txx} = \psi_{xxxx} - (\psi_{xx}\psi_x)_x - \frac{2f_2'(t)}{f_2(t)}\psi_{xx} - \frac{f_2''(t)}{f_2(t)},$$

which can be integrated twice w.r.t. the variable x. Thus, the second order nonlinear equation

$$\psi_t = \psi_{xx} - \frac{1}{2}(\psi_x)^2 - \frac{2f_2'(t)}{f_2(t)}\psi - \frac{f_2''(t)}{2f_2(t)}x^2 + c_1(t)x + c_0(t), \tag{17}$$

is obtained ($c_1(t)$ and $c_0(t)$ are arbitrary smooth functions).

In contrary to nonlinear equation (9), we were unable to linearize equation (17) (we remind the reader that $f_2^{'}(t) \neq 0$). To find particular solutions we used non-Lie ansatz (see ref. 12 for example)

$$\psi(t,x) = \psi_2(t)x^2 + \psi_1(t)x + \psi_0(t). \tag{18}$$

in order to reduce the partial differential equation (17) to a system of ODEs. In fact, substituting (18) into equation (17) the system of three ODEs

$$\begin{aligned} &\psi_2^{'} + 2\psi_2^2 + \tfrac{2f_2^{'}}{f_2}\psi_2 + \tfrac{f_2^{''}}{2f_2} = 0, \\ &\psi_1^{'} + 2\psi_1\psi_2 + \tfrac{2f_2^{'}}{f_2}\psi_1 = c_1(t), \\ &\psi_0^{'} - 2\psi_2 + \tfrac{1}{2}\psi_1^2 + \tfrac{2f_2^{'}}{f_2}\psi_0 = c_0(t) \end{aligned} \tag{19}$$

is obtained to find unknown functions $\psi_0(t), \psi_1(t)$ and $\psi_2(t)$. Since the system (19) has the same structure as eq. (38) from ref. 12 one is integrable. In fact the first equation can be solved for arbitrary function $f_2(t)$. By substituting the known function $\psi_2(t)$ into the second equation one obtains the first order linear ODE to find the function $\psi_1(t)$. Finally, having the known functions $\psi_1(t)$ and $\psi_2(t)$, the third equation of system (19) can be easily solved. As result the general solution of (19) has the form

$$\begin{aligned} &\psi_2(t) = \tfrac{1}{2(t-t_0)} - \tfrac{f_2^{'}}{2f_2},\ t_0 \in \mathbb{R}, \\ &\psi_1(t) = \tfrac{1}{(t-t_0)f_2(t)}\left(\textstyle\int (t-t_0)f_2(t)c_1(t)\,\mathrm{d}t + k_1\right),\ k_1 \in \mathbb{R}, \\ &\psi_0(t) = \frac{k_0}{f_2^2(t)} - \frac{1}{2} + \frac{1}{f_2^2(t)}\int\left(c_0(t)f_2^2(t) + \frac{f_2^2(t)}{t-t_0}\right)\mathrm{d}t - \\ &\qquad \frac{1}{f_2^2(t)}\int\left(\frac{(\int (t-t_0)f_2(t)c_1(t)\,\mathrm{d}t + k_1)^2}{2(t-t_0)^2}\right)\mathrm{d}t,\ k_0 \in \mathbb{R}. \end{aligned} \tag{20}$$

Taking into account that $c_1(t)$ and $c_0(t)$ are arbitrary smooth functions, we may simplify the general solution (20) to the form

$$\begin{aligned} &\psi_2(t) = \tfrac{1}{2(t-t_0)} - \tfrac{f_2^{'}(t)}{2f_2(t)},\ t_0 \in \mathbb{R}, \\ &\psi_1(t) = c_1^*(t), \\ &\psi_0(t) = c_0^*(t), \end{aligned} \tag{21}$$

by introducing new notations $c_1^*(t)$ and $c_0^*(t)$, which are arbitrary smooth functions, while t_0 is an arbitrary parameter.

Thus, using (16), (18) and (21) we construct the exact solution

$$\begin{aligned} S(t,x,y) =& \frac{f_2^{'}(t)}{2f_2(t)}y^2 - \frac{f_2^{'}(t)}{2f_2(t)}x^2 + \frac{1}{2(t-t_0)}x^2 + \\ & c_1^*(t)x + c_0^*(t), \\ \rho(t,x,y) =& \tfrac{1}{t_0 - t} \end{aligned} \tag{22}$$

of Cauchy problem (6) with the correctly-specified initial profile $\phi(x) = -\frac{1}{2t_0}x^2 + c_1 x + c_0$.

Obviously, this solution blows up for the finite time $t_0 > 0$ what makes its immediate biological interpretation unlikely. On the other hand, it is in agreement with ref. 6 (see also the references cited therein) because the global solution of this problem requires the constrain $\int_{\mathbb{R}^2} \rho(0,x,y)\mathrm{d}x\mathrm{d}y < 8\pi$, which does not take place for $\rho(0,x,y) = \frac{1}{t_0}$.

4. Conclusions

In this work we studied a simplified version of (1+2)-dimensional Keller-Segel model. It is well-known that Keller-Segel model is widely used for modeling a wide range of processes in biology and medicine (especially tumour growth) therefore one is extensively examined by means of different mathematical techniques.

It was established that MAI of system (4) is the infinity-dimensional Lie algebra. Moreover we have proved that each Cauchy problem for this system of the form (6) still admits a infinity-dimensional Lie algebra. Moreover, using Bluman's criteria, all inequivalent problems (depending on the form of initial profiles) were found, which admit different MAIs (see Tables 1-5).

In order to construct the exact solutions of Cauchy problems the Lie symmetry operators were applied. In particular, we have proved that the Cauchy problem for the one-dimensional (in space) Keller-Segel system in question can be linearized. As result the exact solution of the Cauchy problem was constructed in explicit form. Moreover, additional biologically motivated restrictions were established in order to obtain uniqueness of solution.

Finally, the exact solution for the Cauchy problem with quadratic initial profile was found and its properties were briefly discussed.

References

1. Keller E. F., Segel L. A., J. Theor. Biol. 26, 399-415 (1970).
2. Keller E. F., Segel L. A., J. Theor. Biol. 30, 235-248 (1971).
3. Horstmann D., Leipzig: Max-Planck-Institute for Mathematics in the Sciences, Preprint 3 (2003).
4. Calvez V., Yasmin Dolak-Strub, Mathematical Modeling of Biological System 2, 323-329 (2007).
5. Nagai T., Adv. Differential Equations 16, no. 9-10, 839-866 (2011).
6. Nagai T., Differential Integral Equations 24 , no. 1-2, 29–68 (2011).
7. Mathematical models of chemotaxis. *http://www.aimsciences.org/AIMS-Conference/conf-reg2014/abstracts/ss115_Abstract.pdf*
8. Cherniha R., Didovych M., Commun. Nonlinear Sci. Numer. Simulat. 18, 2960-2971 (2013).
9. G.W. Bluman, and S. Kumei, Berlin: Springer (1989).
10. R. Cherniha, S. Kovalenko, J. Phys. A: Math. Theor. vol. 44, 485202 (25 pp.) (2011).
11. Burgers J. M., The Nonlinear Diffusion Equation, D. Reidel Publishing Company (1973).
12. Cherniha R., J. Phys. A Math. 31: 8179-8198 (1998).

EXPLORATION OF DIFFERENT WAVE PATTERNS IN A MODEL OF THE BOVINE ESTROUS CYCLE BY FOURIER ANALYSIS

C. STÖTZEL[1,*], R. EHRIG[1], H.M.T. BOER[2], J. PLÖNTZKE[1,*], S. RÖBLITZ[1,*]

[1] *Computational Systems Biology Group,*
Zuse Institute Berlin (ZIB),
Takustraße 7,
14195 Berlin,
Germany

[2] *Animal Breeding and Genomics Centre,*
Wageningen UR Livestock Research,
8200 AB Lelystad,
The Netherlands

Cows typically have different numbers of follicular waves during their hormonal cycle. Understanding the underlying regulations leads to insights into the reasons for declined fertility, a phenomenon that has been observed during the last decades. We present a systematic approach based on Fourier analysis to examine how parameter changes in a model of the bovine estrous cycle lead to different wave patterns. Even without any biological considerations, this allows to detect the responsible model parameters that control the type of periodicity of the solution, thus supporting experimental planning of animal scientists.

Introduction

In the bovine female, ovulation takes place once in every estrous cycle at the end of the follicular growth and maturation process. Each cycle includes several wave-like patterns of follicle development[1] in which follicles from the pool of primordial follicles start to grow and compete to become the dominant follicle. Only the dominant follicle of the last wave continues to

*Work supported by the German Federal Ministry of Education and Research (BMBF project BovSys)

grow and ovulates. The number of anovulatory waves occuring before the ovulatory wave differs between cows. In total, cows have between one to four, but mostly two or three follicular waves per cycle. The motivation to study the number of waves per cow is its relation to fertility that is discussed in literature. Further details will be given in Section 1.

The factors that regulate the number of waves in the bovine are not fully explored, though experimental effort has been made to search for the responsible endocrine mechanisms. In this work, we aim to study the differences in hormonal patterns between two- and three-wave cows based on a mathematical model. We want to identify and study systematically those parameters and mechanisms which are responsible for changing the wave patterns.

An ordinary differential equation model, called BovCycle, has been built for the bovine estrous cycle, which describes the underlying biological mechanisms and their regulation[3,4]. Model simulations reproduce periodic estrous cycles lasting 21 days including the development of the follicles and the corpus luteum and the key reproductive hormones involved. The BovCycle model consists of 15 ODEs and 60 parameters and is available in SBML at the BioModels database[5].

A systematic approach to analyze how parameter changes lead to different wave patterns can be taken with the help of Fourier analysis. This allows to detect the responsible model parameters that control the type of periodicity of the solution. During the analysis, we calculate the fraction of the first Fourier coefficients while varying parameter values. The order of the coefficients indicates the order of the most dominant oscillations and thus the number of waves per cycle. A change in the order leads also to an abrupt change in the period length. In a one-dimensional approach, all parameters are changed individually. In a two-dimensional approach, two parameters are changed at a time. Fourier analysis based on FFT (Fast Fourier Transformation) is performed for varying parameter values. This requires knowledge of the exact period length and initial values which are obtained by solving a boundary value problem in every step.

The paper is organized as follows. In Section 1, we start with a short review of the biological mechanisms behind different wave numbers, followed by a brief presentation of the model BovCycle in Section 2. In Section 3, the application of the Fourier analysis to BovCyle is explained, and the results are described in Section 4. A list of abbreviations that are used in this work is given in Table 1.

Table 1. List of abbreviations for the substances modelled in BovCycle.

Abbreviation	Explanation
$GnRH_{Hyp}$	Gonadotropin Releasing Hormone in the Hypothalamus
$GnRH_{Pit}$	Gonadotropin Releasing Hormone in the Pituitary
FSH_{Pit}	Follicle Stimulating Hormone in the Pituitary
FSH_{Bld}	Follicle Stimulating Hormone in the Blood
LH_{Pit}	Luteinizing Hormone in the Pituitary
LH_{Bld}	Luteinizing Hormone in the Blood
Foll	Follicular Capacity to produce steroids
CL	Corpus Luteum
P4	Progesterone
E2	Estradiol
Inh	Inhibin
Enz	Enzymes
OT	Oxytocin
$PGF_{2\alpha}$	Prostaglandin F2α
IOF	Intra-Ovarian Factors

1. Biological Background

The number of follicular waves per cycle differs between cows. Most cows have two or three waves[2,6], *i.e.* one or two anovulatory waves plus the ovulatory wave in each cycle. Some cows may have only one or even four follicular waves per cycle, and often the number of waves differs from cycle to cycle.

Different productive and reproductive characteristics of cows with different follicular wave patterns have been reported in literature. For example, Bleach *et al.*[2] found that cows with two follicular waves during the cycle produce more milk than those with three waves. Another motivation to study this difference is to investigate the relation between the number of waves and fertility.

Two-wave cows, *i.e.* cows with usually two follicular waves per cycle, have a shorter cycle than three-wave cows. Since in a long time span, a two-wave cow ovulates more often, one can derive that, at a randomly chosen time point, a herd of two-wave cows probably includes more cows that are at the stage around ovulation. Therefore, one could assume that two-wave cows have higher fertility rates compared to three-wave cows. However, reality is more complex. While some studies show no difference regarding fertility rates[2,7], other studies report better fertility in three-wave cycles compared to two-wave cycles[8], and it has been suggested that the older and larger ovulatory follicles in cycles with two waves contain oocytes of less quality than cycles with three waves[9].

The reason for this could be the following. The follicle that is dominant at the moment of CL regression ovulates. Therefore, the number of follicular waves in a cycle is largely affected by the interplay of follicle growth rate and the time point of CL regression. Thus, it is influenced by the timing of two major rhythm drivers of the cycle: follicle growth under control of FSH, and CL regression under control of $PGF_{2\alpha}$. When the CL is regressed at the moment that a prolonged dominant follicle is present, the oocyte could be of inferior quality[9].

The factors that regulate the number of waves in bovine are not fully explored, though experimental effort has been made to search for endocrine mechanisms that could be responsible for controlling these factors. According to Adams *et al.*[12], breed or age do not affect the number of waves per cycle. Also, Wolfenson *et al.*[6] did not find any difference in number of waves between cows and heifers. However, some findings on differences between two- and three-wave cows have been reported. Jaiswal *et al.*[10] observed that CL regression occurs 2.5 days earlier in two-wave compared to three-wave cows. The onset of luteolysis thus might play an important role. Bleach *et al.*[2] found that ovulatory follicles in two-wave cycles have a lower growth rate compared to the ovulatory follicles in three-wave cycles. Parker *et al.*[11] observed that cows with three-wave cycles have lower FSH and Inhibin blood concentrations during non-ovulatory waves compared to two-wave cows. Medan *et al.*[14] found that immunization against Inhibin A increased the number of waves per cycle. Summarizing, three biological mechanisms have been reported to influence the number of waves: the time of CL regression, the growth rate of the follicles, and low Inhibin and FSH concentrations.

2. The model BovCycle

The estrous cycle, *i.e.* the periodic development of multiple substances, is the result of a large feedback loop of regulations. The ODE model BovCycle, presented in Boer *et al.*[3] and extended in Stötzel *et al.*[4], describes the key feedback mechanisms behind the cycle, and is able to generate periodic solutions of length 21 days. The mechanisms are depicted in Figure 1, abbreviations are explained in Table 1. Parameter values and initial values can be found in Stötzel *et al.*[4].

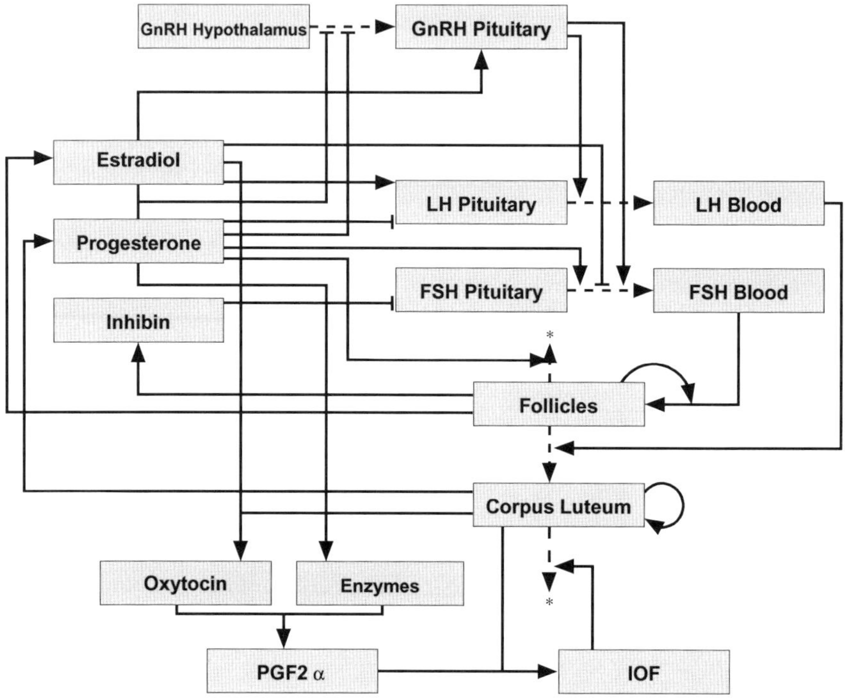

Figure 1. Flowchart for the model BovCycle. A pointed arrow marks a stimulatory effect, a stump arrow an inhibitory influence. A dashed arrow means a transition, and * marks a degraded substance.

In the hypothalamus, the hormone GnRH is synthetized and released into the pituitary. There, it stimulates the release of the hormones FSH and LH into the bloodstream, where they distribute and influence several functions in the body. They regulate processes in the ovaries, where follicles and corpus luteum develop. The ovarian structures produce the steroids E2, P4 and Inh that are released into the blood. From therein, they influence GnRH, FSH and LH in the hypothalamus and pituitary, and stimulate oxytocin and different enzymes that control the action of PGF2α. Together with several intra-ovarian factors, this initiates the decay of the CL. These mechanisms are modelled in BovCycle as a closed system which allows to analyze how the physiological components in different parts of the whole body function together. No external stimuli are needed for the periodic behavior which results only from the developed dynamics and the parameterization of the model.

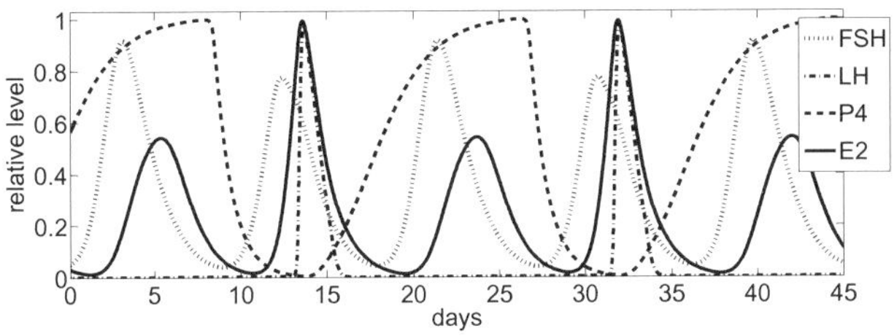

(a) Two-wave cycle

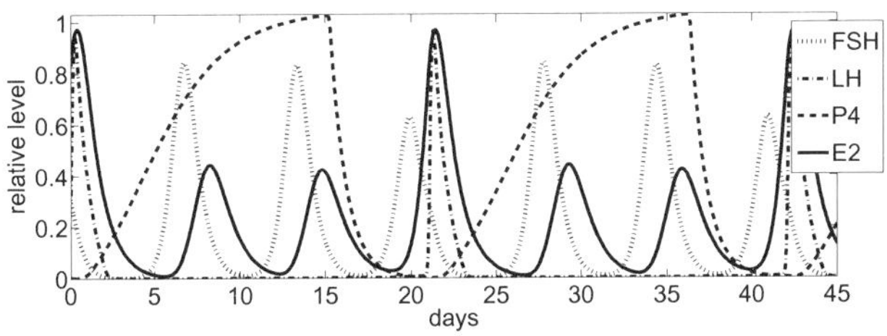

(b) Three-wave cycle

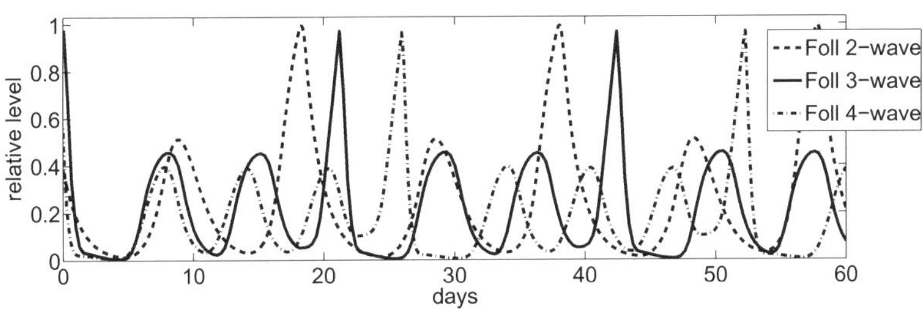

(c) Cycles with different wave numbers

Figure 2. Different parameterizations of the model BovCycle lead to a simulation output with two, three, or four follicular waves per cycle.

In Boer *et al.*[13], 10 parameters of BovCycle, that can directly be associated with the biological mechanisms described in Section 1, were tested for inducing a change in wave number. Their values were varied, sometimes in combination with each other, and for certain parameters, a change in wave number was obtained. In the following, we present a more systematic approach to test the parameter variations.

3. Exploration of Wave Patterns by Fourier Analysis

From the simulations it is clear that mainly two types of components exist in BovCycle. First, there are variables that peak once every cycle. They are the ones associated with luteal development, *i.e.* the corpus luteum itself, progesterone, or the substances involved in luteolysis. Second, other variables have several peaks per cycle. These are the ones associated with follicular development, *i.e.* the follicles and the hormones produced by them. For substances like GnRH, LH, or OT it is not clear which clock is more dominant, but they represent a bridge between the CL and the follicles and play a role in the change of the model behavior regarding the number of waves in the simulation. However, the number of waves per cycle can be counted by comparing the one-peak variables with the several-wave variables.

For the following analysis, we take the variable describing the follicles, y_{Foll}, as representative for the several-wave-variables, and y_{CL} as representative for the one-peak variables. In the following, the *rhythm* of a variable denotes the time interval between two maxima. The rhythm of the one-peak variables is the cycle length, while the rhythm of the several-wave-variables can be calculated through spectral analysis. To compare the two rhythms via the calculation of their Fourier coefficients, first the precise cycle length is needed.

To obtain the precise cycle length T, we have to solve a boundary value problem of the form

$$y' = f(y, p), \quad y(T) - y(0) = 0,$$

whereby p are fixed parameter values and $y(0)$ and T are the unknowns. For this purpose, the code PERIOD[15] is used. This program is an implementation of a multiple shooting method for the computation of periodic solutions of ODEs together with a global underdetermined Gauss-Newton method with adaptive trust region strategies (algorithm NLSCON)[16].

As the algorithm used in PERIOD is sensitive to starting values, a good first approximation of the cycle length is needed. A rough approximation by hand is not good enough. Thus, the Fourier transformation of the simulation of the variable y_{CL} is used for improvement. The maximum coefficient of the Fourier transformation is then taken as initial guess for the length of the period. With this starting value, PERIOD uses NLSCON to iteratively optimize the initial values of the ODE system to obtain a precise periodic solution.

Knowing the true period length T and initial values of the system, a Fourier analysis is now performed on the several-wave-variable y_{Foll},

$$y_{\text{Foll}}(t) = \sum_{k=1}^{10} c_k \exp(i\frac{k\pi}{T}t).$$

This delivers the contribution of the different frequencies to the total time course of y_{Foll}. Ranking of the c_k according to their amplitude allows for the identification of the most prominent frequencies. For y_{Foll}, generally c_2, c_3, or c_4 are the largest factors because this variable peaks 2, 3, or 4 times per cycle. However, the order of c_2, c_3, or c_4 changes depending on parameter values.

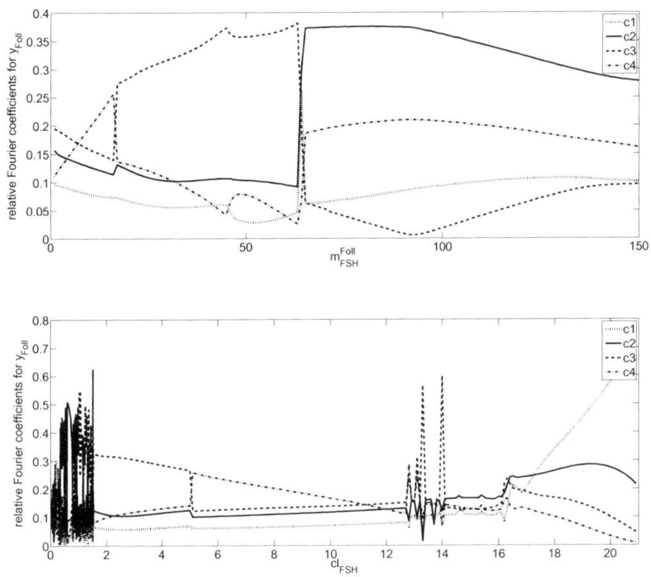

Figure 3. The contribution of the Fourier coefficients $c1$, $c2$, $c3$, and $c4$ (the most dominant oscillations) to y_{Foll} as function of a single parameter. A change in the order of the lines indicates a change in wave numbers in the simulation at this parameter value.

This approach is performed with different parameter values. Possibly, the component that has the most dominant oscillation changes. For example, it changes from the two-wave component having the highest contribution to the three-wave component being the most dominant oscillation of

y_{Foll}, indicating a change from a two-wave to a three-wave cycle pattern of the whole system. Two examples of such changes are depicted in Figure 3. Varying one or two parameters at a time and performing the Fourier analysis simultaneously leads also to a varying period length.

4. Results

The described spectral analysis is performed for all 60 parameters, and the plots have been evaluated visually. For the original parameter value p_{orig}, we here investigate the range of values $[0.1 \cdot p_{orig}, 10 \cdot p_{orig}]$. Within this range, there is a change in the order of the dominant Fourier coefficients for 15 parameters, keeping the other parameter values fixed. This indicates a change in the number of waves per cycle. These 15 parameters, together with the corresponding ranges in which a certain pattern is obtained, are listed in Table 2. For the remaining parameters, there is no change in the order of the oscillatory components in the tested range of values.

Table 2. Parameters and their regions in which the simulation results in a certain order of Fourier coefficients indicating the number of waves per cycle. One parameter is varied at a time.

par	dominance c_1	dominance c_2	dominance c_3	dominance c_4
$T^{\text{FSH}}_{\text{Inh}}$	.	.	0.14-0.4	0.4-0.62
cl_{FSH}	.	.	1.5 - 5.0	5.0 - 10.7
b_{FSH}	.	.	0.55 - 1.7	1.7 - 24.0
$m^{\text{LH}}_{\text{E2}}$	.	5.5 - 30.0	0 - 5.5	.
$m^{\text{Foll}}_{\text{FSH}}$	.	0.92 - 1.8	0.45 - 0.92	0.36 - 0.45
$T^{\text{FSH}}_{\text{Foll}}$	.	.	0-0.077, 0.62-20	0.077-0.57
$m^{\text{FSH}}_{\text{P4}}$	.	2.17-3.22, 4.26-4.72	0.34-1.45, 1.65-2.0, 3.22-3.32	3.32-3.43
$T^{\text{PGF}}_{\text{Enz}}$	.	.	0.91 - 1.8	1.8 - 2.8
$T^{\text{PGF}}_{\text{OT}}$	.	.	0 - 2.4	2.4 - 7.7
cl_{PGF}	.	.	0 - 4.5	4.5 - 24.0
SF	0.58 - 5.0	0.3 - 0.51	0.15 - 0.3	0.045 - 0.15
$m^{\text{P4}}_{\text{CL}}$	.	5.0 - 15.7	1.62 - 5.0	0.72 - 1.62
cl_{Enz}	.	.	1.8 - 3.9	3.9 - 5.7
$m^{\text{OT}}_{\text{E2}}$	.	.	1.0 - 20.0	0.22 - 1.0
cl_{OT}	.	.	0 - 1.4	1.4 - 5.5

One has to keep in mind that a certain order of dominance of the oscillatory components does not guarantee a particular wave pattern. However, the analysis gives a good suggestion about which parameters are able to control the wave pattern of the simulation.

In biology, one expects the number of waves to depend not only on a single parameter, but eventually on a combination of multiple parameters. Extension of the above described analysis technique into higher dimensions, thus varying several parameters at a time and investigating the order of the oscillatory components of the simulation, is not difficult. Visual evaluation of the results, however, needs more careful consideration because evaluation of all graphs even for only two dimensions would be too time-consuming. Thus, a systematic approach has been chosen that takes into account sensitivities with respect to the cycle length.

In the one-dimensional case, it has been observed that a change in the order of the oscillatory components comes along with an abrupt change in the cycle length at a specific value of the parameter, which corresponds to a step-like shape of the period-length-curve, see *e.g.* Figure 4. This is reasonable, since this change of order results from the interplay between the wave variables (e.g. follicles) and the peak-variable (e.g. $PGF_{2\alpha}$).

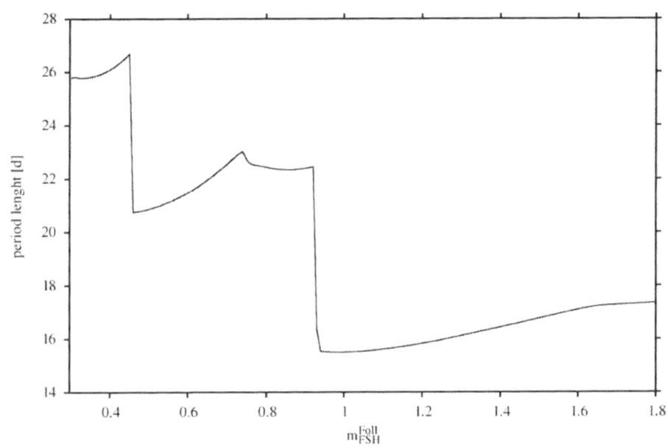

Figure 4. Cycle length during the variation of the parameter $m_{\mathrm{FSH}}^{\mathrm{Foll}}$. At two parameter values, the period length undergoes an immediate large change, which corresponds to a change in the order of the Fourier coefficients, as can be observed in Figure 3.

Thus, a sensitivity analysis of the cycle length L with respect to the parameters, dL/dp, obtained through numerical differentiation, can give a hint to which parameters might lead to a change in the order of the components. Since the period length is quite sensitive to these parameters locally,

Table 3. The 10 most sensitive parameters with respect to the period length, together with their investigated ranges and the therein obtained change in the period length.

par	upper bound	lower bound	change in period length
$m_{\mathrm{CL}}^{\mathrm{CL}}$	0.03530	0.00035	-52.73
$T_{\mathrm{Inh}}^{\mathrm{FSH}}$	0.11800	0.00118	-24.26
$m_{\mathrm{Foll}}^{\mathrm{Foll}}$	0.22000	0.00220	-10.97
$m_{\mathrm{P4}}^{\mathrm{Foll}}$	1.10000	0.01100	-6.79
$m_{\mathrm{FSH}}^{\mathrm{Foll}}$	0.56200	0.00562	6.66
$m_{\mathrm{LH}}^{\mathrm{Ovul}}$	0.20000	0.00200	-5.87
$T_{\mathrm{P4}}^{\mathrm{LH}}$	0.02690	0.00027	-4.21
b_{FSH}	0.94800	0.00948	-2.64
$m_{\mathrm{E2}}^{\mathrm{FSH}}$	0.39600	0.00396	-2.45
$T_{\mathrm{E2}}^{\mathrm{GnRH,2}}$	0.64800	0.00648	2.41

their changes may lead to a rapid and large change in the cycle length at some point. The most sensitive parameters with respect to the period length are given in Table 3. Note that the sensitivity of the cycle-length to the parameters is not a global information, but a hint to which parameters should be examined further. In the one-dimensional case, the sensitivity analysis yields that out of the 15 parameters which can by themselves control the number of waves, 6 are among the 8 most sensitive.

In the two-dimensional case it is therefore convenient, instead of visually checking all $60 \cdot 59/2 = 1741$ possible combinations, to restrict the examination to the most sensitive parameters, in order to find at least some combinations that lead to a change in the number of waves. It is assumed that a parameter that is itself very sensitive is also sensitive in any combination. It can be observed that in the 58 most sensitive combinations, $m_{\mathrm{P4}}^{\mathrm{Foll}}$ is always one of the two parameters in the combination. To also check different parameter combinations, besides the seven most sensitive combinations, also the five most sensitive combinations without $m_{\mathrm{P4}}^{\mathrm{Foll}}$ have been checked. The corresponding sensitivities with respect to the period length are given in Table 4. Two examples of the development of the Fourier fractions are illustrated in Figure 5. Among all 12 of the checked combinations, there is a change in the order of the most dominant fraction, thus a change of waves depending on the values of the parameters.

Summarizing, for the 15 parameters depicted in Table 2, a change in the number of waves per cycle can be obtained by changing the value of a single

Table 4. Parameter combinations and their sensitivities with respect to the period length.

sens. rank	par1	par2	change in period length
1	$m_{\mathrm{FSH}}^{\mathrm{Foll}}$	$m_{\mathrm{P4}}^{\mathrm{Foll}}$	0.01121
2	$T_{\mathrm{Inh}}^{\mathrm{FSH}}$	$m_{\mathrm{P4}}^{\mathrm{Foll}}$	0.01033
3	$m_{\mathrm{P4}}^{\mathrm{Foll}}$	$c_{\mathrm{Foll}}^{\mathrm{Inh}}$	0.01033
4	b_{FSH}	$m_{\mathrm{P4}}^{\mathrm{Foll}}$	0.00997
5	$m_{\mathrm{Foll}}^{\mathrm{Foll}}$	$m_{\mathrm{P4}}^{\mathrm{Foll}}$	0.00988
6	$m_{\mathrm{P4}}^{\mathrm{Foll}}$	$m_{\mathrm{CL}}^{\mathrm{CL}}$	0.00933
7	cl_{FSH}	$m_{\mathrm{P4}}^{\mathrm{Foll}}$	0.00926
59	$T_{\mathrm{Inh}}^{\mathrm{FSH}}$	$m_{\mathrm{FSH}}^{\mathrm{Foll}}$	0.00660749
60	$m_{\mathrm{FSH}}^{\mathrm{Foll}}$	$c_{\mathrm{Foll}}^{\mathrm{Inh}}$	0.00660736
61	b_{FSH}	$m_{\mathrm{FSH}}^{\mathrm{Foll}}$	0.00625
63	$T_{\mathrm{Inh}}^{\mathrm{FSH}}$	$c_{\mathrm{Foll}}^{\mathrm{Inh}}$	0.00572
66	$T_{\mathrm{Inh}}^{\mathrm{FSH}}$	b_{FSH}	0.00536

parameter. These 15 parameters have been identified by evaluating the developments of the contributions of the first four Fourier coefficients. The ranges for the parameter values that lead to a particular wave pattern have been determined. Regarding the simultaneous change of two parameter values, the sensitivity analysis with respect to the cycle length gives good suggestions which parameter combinations can provoke a change in wave patterns.

In addition to Boer *et al.*[13], which used biological knowledge to find parameters that regulate the wave patterns, the here presented method represents a more systematic approach that finds such parameters based on mathematical properties of the simulation output. This extends the previous findings, and it gives a more reliable set of candidates on which experimental effort can be focused on. In addition to the previous work, we calculated ranges of parameter values which lead to a certain wave pattern, see Table 2. The parameters $m_{\mathrm{FSH}}^{\mathrm{Foll}}$ and $T_{\mathrm{OT}}^{\mathrm{PGF}}$ have been found to be decisive in Boer *et al.*[13] and in this work. The other parameters found in Boer *et al.*[13] have not been detected by our new approach. This is probably due to the fact that we here use an extended version of the model which now, amongst others, includes new mechanisms to model luteolysis, and slightly modified mechanisms for ovulation[4].

For the parameter $m_{\mathrm{FSH}}^{\mathrm{Foll}}$, which represents the maximum growth rate of

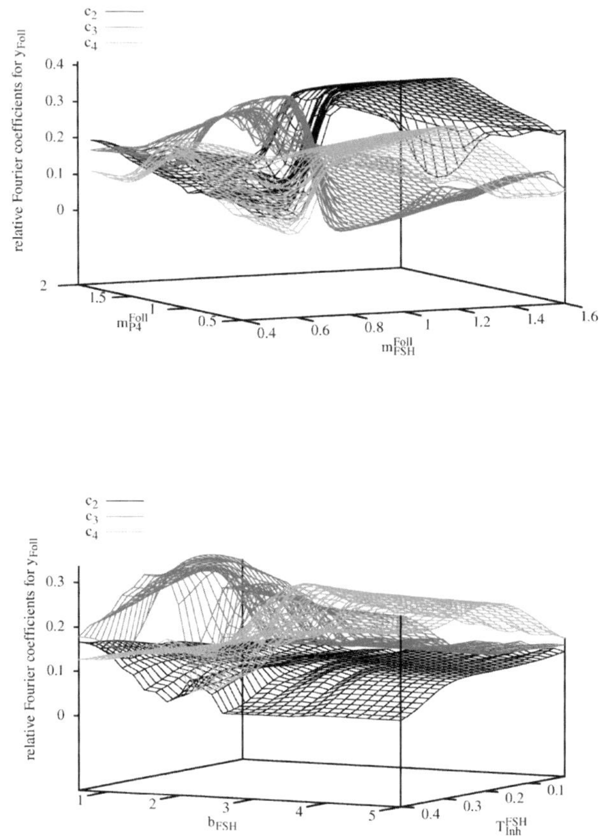

Figure 5. Two examples of the fraction of the first four Fourier coefficients (the most dominant oscillations) to y_{Foll} as function of two parameters. A change in the order of the lines indicates a change in wave numbers in the simulation in this parameter region.

the follicles, one can observe in Table 2 that decreasing its value leads to an increase of waves per cycle. This is likely due to a resulting lower oxytocin growth rate, which results in a later $\mathrm{PGF}_{2\alpha}$ appearance and action. Thus, the CL decays later and the inhibitory effect of P4 on the follicles, LH and GnRH is longer. Therefore, the next ovulation takes place later.

5. Conclusion

The described Fourier analysis has been performed for all model parameters of BovCycle. For 15 parameters there is a change in wave numbers, represented by a change of order of the oscillatory Fourier components in the considered range of values. The ranges for the parameter values that lead to a particular wave pattern have been determined. Furthermore, it has been observed that a change in the order of the oscillatory components comes along with an abrupt change in the cycle length at the specific value of the parameter, which corresponds to a step-like shape of the period-length-curve.

Regarding the simultaneous change of two parameter values, a sensitivity analysis with respect to the cycle length has been performed to obtain an idea which parameter combinations can provoke a change in wave patterns. In the 58 most sensitive combinations it can be observed, that the parameter which stimulates the follicular decay in dependency of the progesterone level, $m_{\mathrm{P4}}^{\mathrm{Foll}}$, is always one of the two parameters in the combination. For all 12 combinations suspected due to sensitivity analysis, there is a change in the order for the most dominant coefficients, thus changing wave patterns depending on the values of the parameters.

Overall, with the Fourier analysis in combination with PERIOD we were able to identify 15 single parameters and at least 12 parameter combinations that are sensitive for changing wave patterns in the bovine estrous cycle model. Prospectively, these findings provide candidate mechanisms for regulating wave patterns in the cow and might be explored experimentally.

References

1. J. E. Fortune, *Biol. Reprod.* **50**, 225–232 (1994).
2. E. C. L. Bleach, R. G. Glencross and P. G. Knight, *Reproduction* **127(5)**, 621–629 (2004).
3. H. M. T. Boer, C. Stötzel, S. Röblitz, P. Deuflhard, R. F. Veerkamp and H. Woelders, *J. Theor. Biol.* **278**, 20–31 (2011).
4. C. Stötzel, J. Plöntzke, W. Heuwieser, and S. Röblitz, *Theriogenology* **78(7)**, 1415–1428 (2012).
5. BovCycle in SBML *http://www.ebi.ac.uk/biomodels-main/BIOMD0000000481.*
6. D. Wolfenson, G. Inbar, Z. Roth, M. Kaim, A. Bloch, and R. Braw-Tal, *Theriogenology* **62(6)**, 1042–1055 (2004).
7. H. A. Celik, I. Aydin, S. Sendag, and D. A. Dinc, *Reprod. Domest. Anim.* **40(2)**, 87–92 (2005).

8. D. H. Townson, P. C. W. Tsang, W. R. Butler, M. Frajblat, L. C. Griel, C. J. Johnson, R. A. Milvae, G. M. Niksic, and J. L. Pate, *J. Anim. Sci.* **80(4)**, 1053–1058 (2002).
9. I. Revah and W. R. Butler, *J. Reprod. Fertil.* **106(1)**, 39–47 (1996).
10. R. S. Jaiswal, J. Singh, L. Marshall, and G. P. Adams, *Theriogenology* **72(1)**, 81–90 (2009).
11. K. I. Parker, D. M. Robertson, N. P. Groome, and K. L. Macmillan, *Biol. Reprod.* **68**, 822–828 (2003).
12. G. P. Adams, R. Jaiswal, J. Singh and P. Malhi, *Theriogenology* **69(1)**, 72–80 (2008).
13. H. M. T. Boer, S. Röblitz, C. Stötzel, R. F. Veerkamp, B. Kemp and H. Woelders, *J. Dairy Sci.* **94(12)**, 5987–6000 (2011).
14. M. S. Medan, T. Takedomi, Y. Aoyagi, M. Konishi, S. Yazawa, G. Watanabe, and K. Taya, *J. Reprod. Dev.* **52**, 107–113 (2006).
15. P. Deuflhard, *BIT Numerical Mathematics* **24(4)**, 456-466 (1984).
16. P. Deuflhard, *Springer Series in Computational Mathematics* **35** (2004).

A SPECTRAL SIMILARITY MEASURE BETWEEN TIME SERIES APPLIED TO THE IDENTIFICATION OF PROTEIN-PROTEIN INTERACTIONS

G. E. SALCEDO, A. M. MONTOYA
Grupo de Investigación y Asesoría en Estadística
Universidad del Quindío
Armenia, Colombia
E-mail: gsalcedo@uniquindio.edu.co

A. F. ARENAS
Grupo de Estudio en Parasitología Molecular
Universidad del Quindío
Armenia, Colombia
E-mail: aylanfarid@yahoo.com

In this work we propose a similarity measure to classify a set of time series based on comparison tests of their spectral densities. Similar to the traditional alignment methods but using the spectral properties of the time series rather than the sequences themselves, our methodology can be used to identify similarities or interactions between proteins. A simulation study applied to stationary time series of different size shows that the approach works satisfactorily well. The approach was applied at searching for some known specific interactions between proteins of Toxoplasma-Human infection model.

1. Introduction

Classification and comparison of time series have been a relevant problem with applications in Biology, Medicine, Sismology, Economics, among others. Recall that a time series is a set of observations $\{x_t\}$, where each x_t is recorded at a specific time t. If the observations are recorded on discrete instants we have a discrete time series and they are the most common type of observed time series in practice.

There are two common approaches to analyze a stationary time series: in the time domain and in the frequency domain. In the time domain, the analysis are basically based on the *autocorrelation function* given by

$$\rho_x(k) = \frac{\gamma_x(k)}{\sigma^2}, \quad k = 0, \pm 1, \pm 2, \ldots,$$

where $\gamma_x(k) = Cov(x_t, x_{t-k}) = E[(x_t - \mu)(x_{t-k} - \mu)]$ is the autocovariance function at lag k, μ is the unconditional expectation and σ^2 the unconditional variance of the process. Function $\rho_x(k)$ measures the linear dependence between pairs of observations separated by a lag k. If $\rho_x(k) = 0$ for all $k \neq 0$ and $\mu = 0$, we say that the process has no memory also known as a white noise. In the frequency domain, the correlation structure is represented by the spectral density function defined as

$$f_x(\lambda) = \sum_{k=-\infty}^{\infty} \gamma_x(k) \exp(-i2\pi\lambda k), \quad \lambda \in [-1/2, 1/2],$$

where the frequency λ is measured in cycles per unit of time. Notice that $f_x(\lambda)$ is the Fourier transform of $\gamma_x(k)$ and describes the properties of the process in terms of periodic components at different frequencies.
For a set of observations $\{x_t, t = 1, 2, \ldots, T\}$, the discrete Fourier transform defined on the discrete Fourier frequencies $\lambda_j = j/T, j = 0, 1, 2, \ldots, [T/2]$, is given by

$$d_x(\lambda_j) = \frac{1}{T} \sum_{t=1}^{T} x_t \exp(-i2\pi\lambda_j t).$$

An unbiased estimator of the spectral density function $f_x(\lambda)$ is the periodogram $I_x(\lambda_j)$ defined as

$$I_x(\lambda_j) = |d_x(\lambda_j)|^2.$$

The value of the periodogram at each frequency λ_j represents the amount of time series variance, or power, related to this frequency.

Accordingly, different techniques both in time and frequency domain have been proposed in order to classify a set of time series based on their second order structures (see [1,6,7,10,11,17], for instance).

In order to classify a set of time series, we need to define a measure of similarity between them; Caiado et al.[4] evaluated different similarity measures to classify a set of time series, most of them based on Euclidean distances between the series, their autocorrelation functions, partial autocorrelation functions, autorregresive parameters, normalized periodograms, as well as other measures based on the Kullback-Leibler distance. Another way to compare two time series is through the test of hypothesis where the p-value can be considered as a dissimilarity measure between the time series since it defines a semi-metric. Maharaj[10,11] proposed a methodology to classify a set of time series based on the p-values from testing hypotheses of equality of their autoregressive coefficients.

On the other hand, in their primary structure, proteins are linear chains of amino acids represented each one by 20 letters of the alphabet, and this alphabetic code sequence can be translated into a numerical sequence using different physicochemical properties for each amino acid such as, levels of Electron-Ion Interaction Potential (EIIP), Hydrophobic Residual Buried (HRB), polarity, polarizability, Vander Walls volumes, Ionization Constant (IC), among others. Furthermore, since the distance between consecutive CA atoms in a sequence protein is $3.8A^o$, the points in some corresponding numerical sequence are considered equidistant and the corresponding numerical sequence can be analyzed as a time series. So different methods for extraction of information from signals, via spectral analysis for instance, can be applied for analyzing a set of proteins. In order to extract common spectral characteristics of sequences having the same or similar biological function, the cross spectral function has been used and to determine the common frequency components for a group of n proteins $X_1, X_2, \ldots, X_n$, is calculated the *multiple cross-spectral function* M in the frequency λ_j defined as follows:

$$|M(\lambda_j)| = |I_{X_1}(\lambda_j)||I_{X_2}(\lambda_j)| \ldots |I_{X_n}(\lambda_j)|, \quad j = 0, 1, 2, \ldots, T/2.$$

Peak frequencies in such a multiple cross-spectral function denote common frequency components for all sequences analyzed. These common peaks are sharing either some common biological function or an interaction. The signal-to-noise ratio (S/N) for each peak is defined as a measure of similarity between sequences analyzed. This kind of analysis is called the Informational Spectrum Method (ISM) and has been successfully applied to characterize protein-protein interactions (see [9,16,18]).

Notice that, despite of its success and usefulness, the ISM is an exploratory technique. To move a little forward, we propos to apply a testing hypothesis of equality of spectral densities and consider the p-value of this testing as a measure of spectral similarity between two time series. Then we propos to classify a set of proteins according to their spectral similarity constructing the dissimilarity matriz from this p-value for clustering them. This procedure consider that the nearness between two time series represents some kind of relationship between them. The procedure is applied for clustering a set of proteins of the Toxoplasma-Human infection model and the nearness between two proteins can represent either a common function or an interaction.

The paper is organized as follows. Section 2 describes the test of equality of spectra, our approach to classify a set of proteins according to their spectral similarity is discussed in Section 3 along with a simulation study of their power to classify a set of ARMA processes, the application is presented in Section 4, and finally, some conclusions are given in Section 5.

2. Test of equality of spectra

Let $f_x(\lambda)$ and $f_y(\lambda)$ be the spectral density functions of time series $\{x_t\}$ and $\{y_t\}$, respectively, and let $I_x(\lambda)$ and $I_y(\lambda)$ be their periodograms. Asymptotically we have that,

$$I_x(\lambda) \sim \frac{f_x(\lambda)\chi_2^2}{2}, \quad \lambda \neq 0, 1/2,$$

and

$$I_y(\lambda) \sim \frac{f_y(\lambda)\chi_2^2}{2}, \quad \lambda \neq 0, 1/2.$$

So defining the spectral functions $J(\lambda)$ and $U(\lambda)$ as

$$J(\lambda) = \frac{I_x(\lambda)}{I_y(\lambda)} \quad \text{and} \quad U(\lambda) = \frac{f_x(\lambda)}{f_y(\lambda)}, \quad 0 < \lambda < 1/2,$$

we have that asymptotically,

$$J(\lambda) = \frac{I_x(\lambda)}{I_y(\lambda)} \sim \frac{\frac{f_x(\lambda)\chi_2^2}{2}}{\frac{f_y(\lambda)\chi_2^2}{2}} \sim U(\lambda)F_{2,2}$$

where F is the Fisher-Snedecor distribution. Furthermore,

$$\ln J(\lambda) \sim \text{logistic}\{\ln U(\lambda), 1\},$$

where logistic$(a, 1)$ denotes the logistic distribution with cumulative distribution function

$$F(x) = \left\{1 + e^{-(x-a)}\right\}^{-1}, \quad -\infty < x < \infty.$$

On the other hand, it can be shown that

$$z_i = \ln(1 + J^{-1}(\lambda_i)) \xrightarrow{d} U(\lambda_i)\exp(1),$$

where $\lambda_i = \frac{i}{T}$, $i = 1, ..., m = \left[\frac{T-1}{2}\right]$, $\xrightarrow{d}$ represents a convergence in distribution and $\exp(1)$ is the exponential distribution with mean 1. Thus, under the null hypothesis

$$H_0 : f_x(\lambda) = f_y(\lambda), \forall\lambda \in (0, 1/2),$$

it can be concluded that $U(\lambda_i) = 1$ and

$$z_{\mathrm{i}} \xrightarrow{d} \exp(1).$$

Consequently, the statistics $c_j = \sum_{i=1}^{j} z_i$ constitute the points of a Poisson process with mean 1, then $\left\{o_j = \frac{c_j}{c_m}\right\}$, $j = 1, ..., m$, is a vector of order statistics from the uniform distribution in (0,1)[5]. Accordingly, we can reject H_0 if statistics $o_j, j = 1, \ldots, m$, does not fit the uniform distribution.

Based on the fact that the $p-$value obtained in a testing of comparison of two time series offers a way to assess their "nearness[10,11], and a disimilarity matrix can be constructed from these p-values, we propose the following clustering procedure to classify a set of time series.

3. Clustering procedure

Let consider a set of stationary time series of different length, our approach follows the next steps.

- Step 1: For each pair of time series perform the hypothesis test described in Section 2 for each alignment beetwen them, and select a global p-value associated with the testing. We tried the mean and the maximum p-value.
- Step 2: The dissimilarity structure for the set of time series is represented by a matriz constructed with the before p-values.
- Step 3: A hierarchical clustering algorithm is performed using the Ward's agglomeration method, entering the dissimilarities matrix from step 2.
- Step 4: The results of the procedure are displayed by means of a dendrogram and the nearness between two time series indicates its level of spectral similarity.

In order to assess the performance of this proposal we classify a set of time series generated from different processes ARMA(p, q). $\{x_t, t \in \tau\}$ is an autoregressive mean average or ARMA(p, q) process of zero mean if $x_t = \phi_1 x_{t-1} + \ldots + \phi_p x_{t-p} + \theta_1 a_{t-1} + \ldots + \theta_q a_{t-q} + a_t$, where $\{a_t\}$ is a white noise process. When $q = 0$ we have an Autoregressive process of order p or AR(p), if $p = 0$ we have an mean average process of order q or MA(q), and we have a mixed process otherwise.

3.1. *Simulation study: assessment of clustering procedure*

The clustering procedure was applied to 100 replications of time series generated from 10 kind of ARMA stationary processes in the following way: two time series from process AR(1) with parameter $\phi_1 = 0.8$ and length 500, two time series from process AR(2) with parameters $\phi_1 = 0.8$, $\phi_2 = -0.3$ and length 400, two time series from process MA(1) with parameter $\theta_1 = -0.6$ and length 300, two time series from process MA(2) with parameters $\theta_1 = -0.6$, $\theta_2 = 0.3$ and length 200, and two time series from process ARMA(1,1) with parameters $\phi_1 = 0.8$, $\theta_1 = -0.6$ and lenght 100. In all cases the process $\{a_t, t \in \tau\}$ was generated from a standard normal distribution. For each dendrogram obtained in each one of the 100 replications, we count the parameter *number of exactly correct clusters.* However, because of series from processes ARMA(1,1) could eventually be mixed in some clusters with series from processes AR(1) and MA(1), or series from processes AR(1) can be mixed with series from processes AR(2), and series from processes MA(1) and MA(2) could eventually be mixed in some clusters, we also count the parameter *number of clusters with series from different processes or clusters with mixed series.* In all cases, we just analyzed the solutions of 4, 5 and 6 clusters. Results appear in Tables 3.1 and 3.2.

Table 3.1. Number of exactly correct clusters

	Exactly correct clusters					
Cluster solution	0	1	2	3	4	5
4	2	6	13	79	0	-
5	2	4	11	46	0	37
6	3	5	30	15	47	0

Under the parameter *exactly correct clusters*, the procedure performs reasonably good because is able to recognize at least 3 correct clusters in 79% of times for the solution of 4 clusters, and in 83% of times for the solution of 5 clusters. The misclassification is due to the mixed series. Under the parameter *clusters with mixed series*, for the solution of 5 clusters in 48% and 12% of times there are 1 or 2 clusters mixing the series, respectively. Notice the solution of 4 clusters, recall that our procedure generates 79 cases with 3 exactly correct clusters and in 79 cases appear just 1 (79 cases) or 2 (17 cases) clusters with mixed series. These mixtures can be expected for the similarity between the ARMA processes with the AR and MA processes, causing the misclassification.

Table 3.2. Number of Clusters with series from different processes

	Clusters of mixed series					
Cluster solution	0	1	2	3	4	5
4	0	79	17	4	0	-
5	37	48	12	3	0	0
6	47	46	5	2	0	0

4. Application

In the Toxoplasma-Human infection model, when a parasite invades the host cells, it begins by previously releasing some proteins called Rhoptries(ROPs) that enables the parasite to control a variety of host cell processes [3,2,8,15,14]. For example, the kinase protein ROP16 can phosphorilate the human trascription factors STAT3 and STAT6. There is also evidence that the interaction ROP16-STAT3 affects the cell signaling pathway JAK/STAT (see [12,14,19,20]). Further, it is well known that protein ROP18 can phosphorilate the transcription factor ATF6β driving it to the proteosome for his degradation (see [20]). So, in order to validate our clustering procedure, we choose two sets of 13 and 42 proteins, respectively, from data base KEGG (*www.genome.jp/kegg/pathway*) and applied our procedure to assess its power to identify the interaction ROP16-STAT3. Figures 1 and 2 show the obtained results, notice that our procedure identified this interaction in a first level for both sets where STAT3 is labeled with the number 7 and ROP16 with the number 13. In the same way, with two sets of 30 and 59 proteins, we validated the interaction ROP18-ATF6β, and as can be seen in Figures 3 and 4 our procedure also identified this interaction in a first level. In both cases all considered proteins were transformated into numerical sequences under the levels of hydrophobicity. Values of hydrophobicity for each aminoacid can be consulted in Qiu et al.[13].

5. Conclusions

Based on multiple comparisons between the spectral densities of several time series we developed a clustering algorithm to identify similarities or interactions between a set of proteins. The approach performs reasonably well based on a simulation study where we assessed its ability to identify

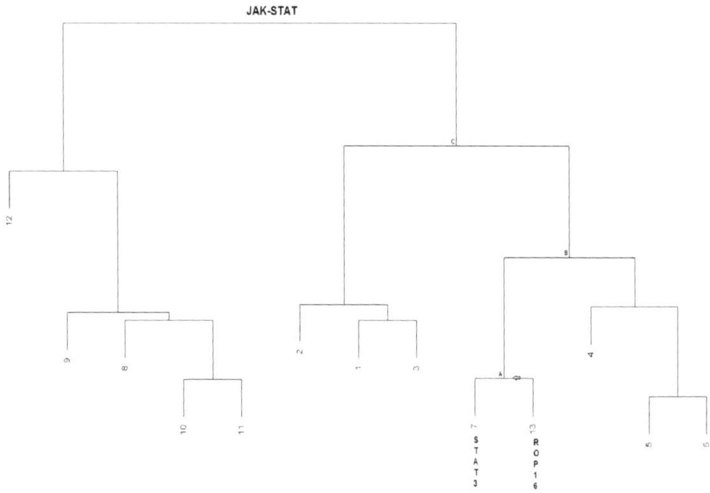

Figure 1. Validation of interaction ROP16-STAT3 for a set of 12 proteins from JAK-STAT pathway and ROP16 protein

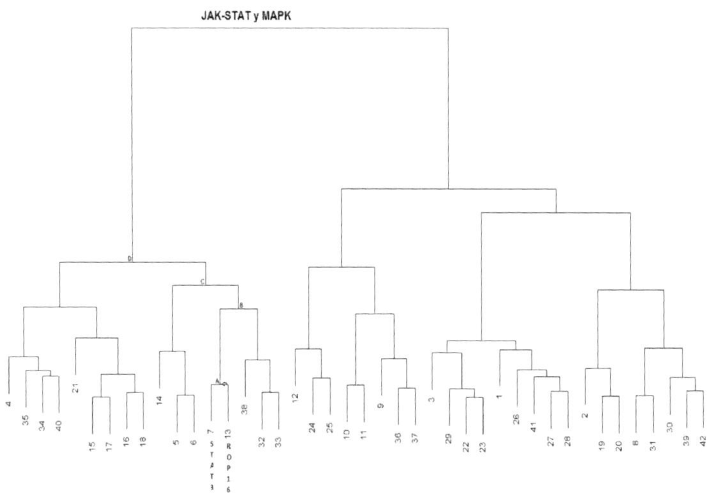

Figure 2. Validation of interaction ROP16-STAT3 for a set of 41 proteins from JAK-STAT, MAPK pathways and ROP16 protein

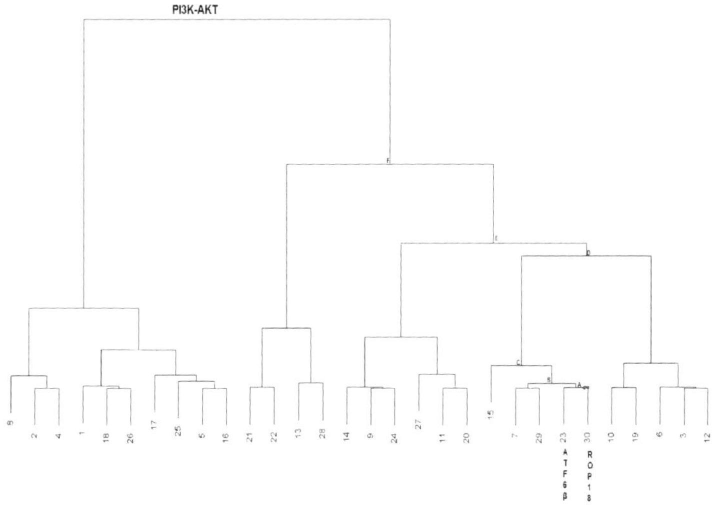

Figure 3. Validation of interaction ROP18-ATF6β for a set of 29 proteins from AKT-PI3K pathway and ROP18 protein

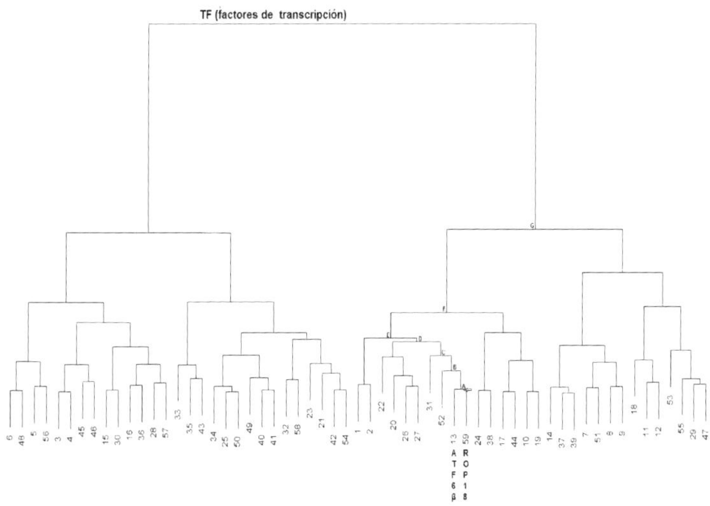

Figure 4. Validation of interaction ROP18-ATF6β for a set of 58 transcription factors and ROP18 protein

spectral similarities between time series from similar ARMA processes. We also could identify some well recognized interactions between proteins from Toxoplasma-Human infection model. The procedure can be applied either to identify or predict some other interactions between proteins from other models of infection.

Acknowledgments

This work was partially supported by Colciencias grant 1113/569/33319.

References

1. Basawa, V., Billard, L., Srinivasan, R., (1984) Large-sample tests of homogeneity for time series models, *Biometrika*, **71**, 203–206.
2. Behnke, M.S. et al. (2011) Virulence differences in Toxoplasma mediated by amplification of a family of polymorphic pseudokinases, *Proc. Natl. Acad. Sci. U.S.A.*, **108**, 9631–9636.
3. Blader, I.J. and Saeij, J.P. (2009) Communication between Toxoplasma gondii and its host: impact on parasite growth, development, immune evasion, and virulence, *APMIS*, **117**, 458–476.
4. Caiado, J., Crato, N. and Peña, D. (2006) A periodogram-based metric for time series classification, *Computational Statistics and Data Analysis*, **50**, 2668–2684.
5. Coates, D.S. and Diggle, P.J. (1986) Test for comparing two estimated spectral densities, *Journal of time series analysis*, **7**, 7–20.
6. Diggle, P.J. and Fisher, N.I. (1991) Non-parametric comparison of cumulative periodograms, *Applied Statistics*, **40**, 423–434.
7. Guo, J.H. (1999) A non-parametric test for the parallelism of two first-order autoregressive processes, *Australian & New Zealand Journal of Statistics*, **41**, 59–65.
8. El Hajj. H, et al. (2007) ROP18 is a rhoptry kinase controlling the intracellular proliferation of Toxoplasma gondii, *PLoS Pathog*, **3**(2): e14.
9. Hu, W. (2010) Highly conserved domains in hemagglutinin of influenza viruses characterizing dual receptor binding, *Natural science*, **2**(9), 1005–1014.
10. Maharaj, E.A. (1999) Comparison and classification of stationary multivariate time series, *Pattern Recognition*, **32**, 1129–1138.
11. Maharaj, E.A. (2000) Clusters of time series, *Journal of Classification*, **17**, 297–314.
12. Ong, Y.C., Reese, M.L., Boothroyd, J.C. (2010) Toxoplasma rhoptry protein 16 (ROP16) subverts host function by direct tyrosine phosphorylation of STAT6, *J. Biol. Chem.*, **285**, 28731–28740.
13. Qiu, J., Liang, R., Zou, X., Mo, J. (2003) Prediction of protein secondary structure based on continuous wavelet transform, *Talanta*, **61**, 285-293.

14. Saeij, J.P.J. et al. (2006) Polymorphic secreted kinases are key virulence factors in toxoplasmosis, *Science*, **314**(5806), 1780–1783.
15. Taylor, S. et al. (2006) A secreted serine-threonine kinase determines virulence in the eukaryotic pathogen Toxoplasma gondii, *Science*, **314**(5806), 1776–1780.
16. De, Trad C. H., Fang Q., Cosic I. (2000) The Resonant Recognition Model (RRM) Predicts Amino Acid Residues in Highly Conservative Regions of the Hormone Prolactin (PRL), *Biophysical Chemistry*, **84**(2), 149–157.
17. Timmer, J., Lauk, M., Vach, W., Lucking, C.H. (1999) A test for the difference between spectral peak frequencies, *Computational Statistics & Data Analysis*, **30** 45–55.
18. Veljkovic, V. (2009) Identification of hemagglutinin structural domain and polymorphisms which may modulate swine H1N1 interactions with human receptor, *BMC Structural Biology*, **9**(62).
19. Yamamoto. M, et al. (2009) A single polymorphic amino acid on Toxoplasma gondii kinase ROP16 determines the direct and strain-specific activation of Stat3, *J. Exp. Med.*, **206**, 2747–2760.
20. Yamamoto, M. et al. (2011) ATF6β is a host cellular target of the Toxoplasma gondii virulence factor ROP18, *J. Exp. Med.*, **208**, 1533–1546.

AN AGENT-BASED MODELING FRAMEWORK TO STUDY THE BURDEN OF PERTUSSIS AND THE IMPACT OF PREVENTATIVE MEASURES

J.-E. POIRRIER, D. CURRAN
GSK Vaccines, Global Health Economics, 1300 Wavre, Belgium

C. PHILEMOTTE
Euranova, 1435 Mont-Saint-Guibert, Belgium

The bacterium *Bordetella pertussis* causes the respiratory disease, commonly known as whooping cough (pertussis). Pertussis may be severe and prolonged, resulting in substantial morbidity and mortality, especially in infants and young children. Despite universal mass vaccination (UMV) programs in many countries, the disease is still persistent worldwide, with recent increases in the number of cases in infants, but also a significant proportion of cases in older children and adults. In the last few years, many countries with a good history of UMV programs have experienced several outbreaks of pertussis, predominantly affecting adolescents and adults. Although the recent increases and outbreaks may be explained by several factors, better understanding of these phenomena are warranted. In this context there is a need for computer-based models and simulation tools that may be of assistance to better understand underlying mechanisms of outbreaks, their determinants and their consequences on public health. One factor that may play an important role in disease outbreaks is the social contexts and consequently the level of person-to-person contact. Therefore we propose here an agent-based framework to model the spread of pertussis in a population of individual agents organized in social contexts. This agent-based model would also potentially allow the study of pertussis outbreaks as well as the impact of vaccination strategies with varying coverage rates on the disease incidence, quality of life of ill persons and economic burden on society. GSK Study Identifier: HO-12-7031.

1. Introduction

1.1. *Pertussis*

Pertussis, also known as whooping cough, is an acute respiratory infection caused by the gram-negative bacillus *Bordetella pertussis* [8,16]. The bacteria infect the human respiratory tract through inhaled airborne respiratory droplets or through direct contact with nasopharyngeal discharges from an infected person [17]. The bacterium overcomes the mucosal immune defenses of the upper respiratory tract and causes the disease in infected individuals. The different clinical stages of pertussis, their durations are summarized in Table 1.

Table 1. Clinical Stages of Pertussis [4, 8]

Clinical stage name	Symptoms	Duration
Catarrhal	Malaise, rhinnorrhea, mild, occasional cough, low grade fever, coryza	1-2 weeks
Paroxysmal	Severe paroxysms of cough ; paroxysms most common at night, cyanosis, vomiting and exhaustion	1-2 months
Convalescent	Persistent cough but of decreasing frequency and severity	1-6 weeks

B. pertussis is a strictly human pathogen with no known reservoir [11]. Therefore, pertussis may be eradicated with an effective control [22].

Before vaccines became widely available, pertussis was one of the most common childhood diseases worldwide. Since the creation of the World Health Organization (WHO)'s Expanded Programme on Immunization, the coverage of 3 doses of pertussis vaccines grew to reach an estimated worldwide coverage of 82% in infants, averting about 687,000 deaths in 2008 [20]. Despite that, the very same year, about 16 million cases of pertussis still occurred [20]. Pertussis historically has been endemic and cyclical, with outbreaks occurring every 3 to 5 years. A gradual increase of pertussis incidence has also been observed over the past few decades in children, adolescents and adults [8].

A number of possible causes of resurgence in pertussis were suggested and considered by experts who met during the Working Group Meeting on Pertussis in 2013 [2]: "short-lived adaptive immunity following immunization; suboptimal balance of immune response; need for additional vaccine antigens for optimal protection; insufficient quantity or incorrect, balance of antigens; antigen mismatch with circulating strains; lack of vaccine protection against transmission; suboptimal schedule or population coverage; differential effectiveness of the vaccines in current use; and increased awareness, better diagnostic tests, and/or improved reporting".

Recently, many countries with historical vaccination programmes have experienced outbreaks of pertussis, predominantly affecting adolescents and adults (United Kingdom [1], United States of America (USA) [13,21], Canada [6] a.o.). In particular, in 2012, increased pertussis activity was observed extending to very young infants, who are at highest risk of serious complications and death.

In this context, there is a need for computer-based modeling and simulation tools that may be of assistance to better understand underlying mechanisms of outbreaks, determinants of outbreaks and consequences of outbreaks on public health and payers' budget.

1.2. *Pertussis outbreak modeling*

Since the 1970s, theoretical approaches were developed to try to build models replicating outbreaks [9,10]. Despite practical approaches developed for various diseases, very few models were specifically published for pertussis. The three most recent papers to our knowledge are from Codeço and Luz [3], Parks et al. [14] and de Vries et al. [5]

(1) Codeço and Luz built a spatially explicit individual-based model developed to simulate the dynamics of pertussis in a population of 2,500 agents in a grid [3]. In this model, they introduce pertussis cases at time zero and transmission occurs given contact between an infectious agent and another agent in the adjacent cells. Transmission could occur during 2 time steps (one time step represents one month) after infection. Immunity wanes with a half-normal curve with a monthly rate of immunity decay of 0.012. With this model, the authors tried to demonstrate different causes for waning of immunity. Therefore, in this model, social contexts are not modeled per se and the one month time step may not be granular enough to reflect the true clinical stages of pertussis [19].
(2) As part of the decision-making model, Parks et al. studied the spread of pertussis in 13,876 agents (census data of Utah, USA) distributed in a two-dimension space [14]. They include a hierarchy of 3 levels: public health, ground truth and top level. The disease mechanism follows a SEIR [a] model. However, their goal was not to study outbreaks per se.
(3) Finally de Vries et al. published a discrete-event simulation model heavily based on the Ordinary Differential Equations-based model philosophy (regarding the disease pathway diagram, the force of infection (FoI), ...) [5]. They use an interesting method to determine

[a]SEIR: Susceptible-Exposed-Infected-Recovered

the FoI. However they assume only one social context. Also, they focus only on the impact of adolescent vaccination.

This paper presents an agent-based model (ABM) of pertussis transmission and prevention based on a population organized into social contexts. Pertussis is spread within the population depending on the strength of links between individuals. The model allows for the characterization of pertussis outbreaks and for the study of the impact of existing vaccination strategies on the disease and its outbreaks.

2. Agent-Based Model structure

2.1. *Agents, characteristics, behavior*

Pertussis is transmitted only between humans and therefore, only one kind of agent is needed, representing human beings (homogeneous population). Their intrinsic properties are: an age (in days), a gender (male or female), a neighborhood membership (see Section 2.2), a health state (see Section 2.3), an age spent in health state (in days), a maximum age in health state (in days), and an immunity level (value within the closed interval $[0, 1]$, 0 being no immunity, 1 being full immunity).

In some situations, the agents are categorized as a function of their age as following: children (aged less than 18 years), adults (aged 18-64 years), and elderly (aged 65 years or more).

A set of N agents forms a population. Its size can vary in the time according to different dynamics such as deaths or births. The probability that an agent would die is based on life tables (adjusted to our time step).

A newborn agent can be added (birth) to the population considering the given Crude Birth Rate (CBR). For each time step, we consider a law of Poisson with an occurrence rate (adjusted to our time step) equals to the CBR. To determine the gender of a newborn, we need to consider the Birth Sex Ratio. The sex of a newborn is then determined thanks to a Bernoulli distribution with a success probability equal to the averaged number of male newborns.

2.2. *Social networks*

The main mechanism of transmission of infectious diseases is usually the direct contact of susceptible individuals with an infectious one. Since this contact is highly localized in space, the majority of ABMs give an important role to space in dynamics of infectious diseases, e.g. [3,15]. However, in the present model, this is not the case. Since what matters for pertussis is the occurrence of daily physical contacts that can support the transfer of the disease, agents are organized in a social network describing which agents are connected and have possible daily physical contacts.

Depending on the social context, the importance of physical contact can vary. A short random contact two agents may have in a street is assumed to be less important than a contact between two agents living in the same house. Therefore the model distinguishes three different kinds of social subnetworks (cluster of connected agents in a given social context determining the importance of the contacts occurring between those agents): household (e.g. a family), local (e.g. a school or a company), and broader (e.g. random contacts in public transports).

In a given social subnetwork, we then define several contact networks, each representing contacts between agents. The global network is the sum of the three main social subnetworks. Technically, any contact network is represented by an hypergraph [b] where the node is an agent and the hyperedge is the neighborhood (approximately a set of m agents that are in contact).

2.2.1. *Household*

The social context "household" describes a place where agents are living together: it is typically a family. The model assumes each child lives in a household (adults can be single).

For the generation of the household context, each agent, depending on his/her age and gender, will be assigned to a household. That assignation is made using a right-centered decision tree [c] (Figures 1 and 2). By doing so, we can easily label each node, and represent each tree as a table provided as input data.

Households are changing over time. The different events that happen to change a household composition are the followings:

(1) Birth of a child (a new agent joins the household if he or she is the newborn of the mother of this household);
(2) Death of any member of the household;

---

[b] A hypergraph is a generalization of a graph where an edge can connect multiple edges. This consideration has an important impact on the implementation and the runtime performance. See Section 3.3 for more details.
[c] A right-centered tree is a tree whose all left branches are leaves.

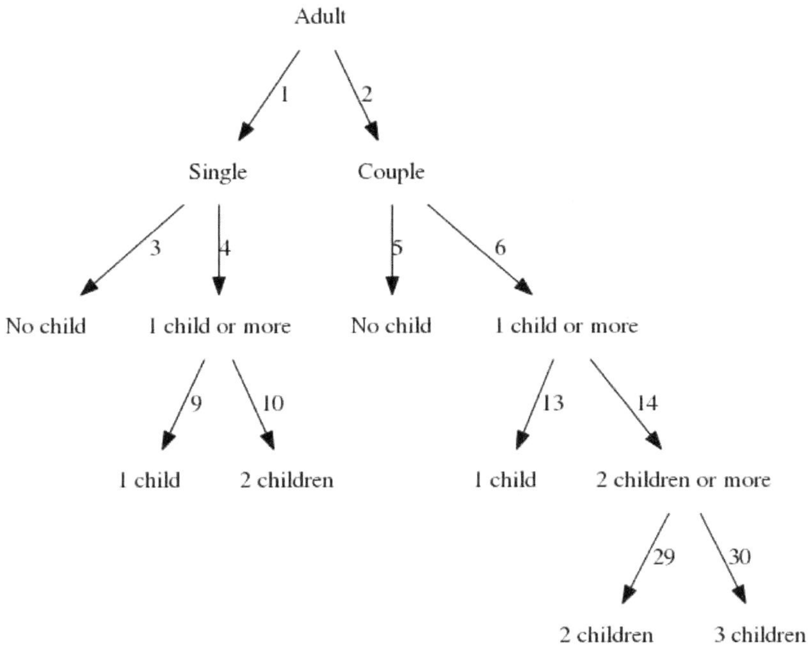

Figure 1. Decision tree for adult agent assignation to household

(3) Child leaving the household to constitute a household of his/her own;
(4) Re-constitution: two families can merge into one unique household provided that they are of type 1, 3, or 4 (in the adult decision tree, on Figure 1). The probability of merging is dependent on the age of both household heads.

In case of death, the agent is simply removed from his/her household. Births will be allocated to families (or households) as per the following procedure:

- Population birth rate will be used to estimate the total number of births during a defined period of time (N_B);
- Eligible families will be identified based on their composition (e.g. only families including a woman aged between 18 and 45 years old);
- Children will be allocated to first N_B families.

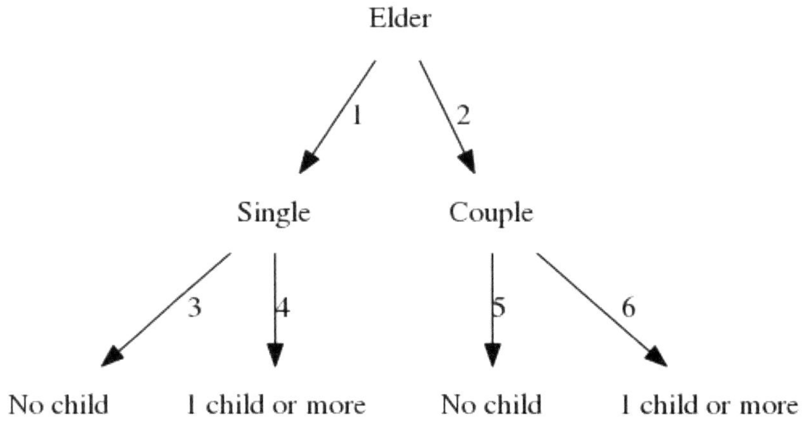

Figure 2. Decision tree for elderly agent assignation to household

2.2.2. *Local*

The social context "local" describes a place where contacts occur outside the family, on a regular basis according to some specific criteria/conditions (such as age, or location). A "local" neighborhood does not change, except due to death (this will evolve in a future version).

In the same population, for the moment, local contexts with overlapping age ranges are not allowed.

Distributions of agents between the different possible local neighborhoods depend on the average size and number of occurrences of their corresponding local context. Distributions are considered as a discrete weighted uniform distribution. The cumulative frequencies distribution (cfd) is used to build the explicit distribution.

2.2.3. *Broader*

The social context "broader" models the public place where two agents can have random contacts. By definition, a broader neighborhood does not last forever (typically it models the contact occurring on a given day). The distribution of neighborhoods in a context follows a power distribution law.

The following algorithm may be used as we assume a low connection degree between agents (< 1.0). The algorithm works as follow:

- To generate a number of contacts per agent with a power law distribution (the mean is the average connection degree).
- To push the index of the agent in a list of agents to meet its number of contacts.
- To shuffle randomly this list of indexes.
- To drop the last element of the list if its size is not even.

This provides a list of indexes of agents, or better, a list of couples of agents. Two agents following each others are put in contact (one whose index is even and the other one whose index is odd in the list).

2.3. *Disease transmission model*

2.3.1. *Health State Diagram*

In order to describe key relevant dynamics of pertussis spread within the different social contexts, we consider a health state diagram inspired by the work of de Vries et al. [5]. The health state of an agent is updated at each new time step $t+1$ thanks to the update function F. The update function F can be defined as a transition state machine as shown on Figure 3.

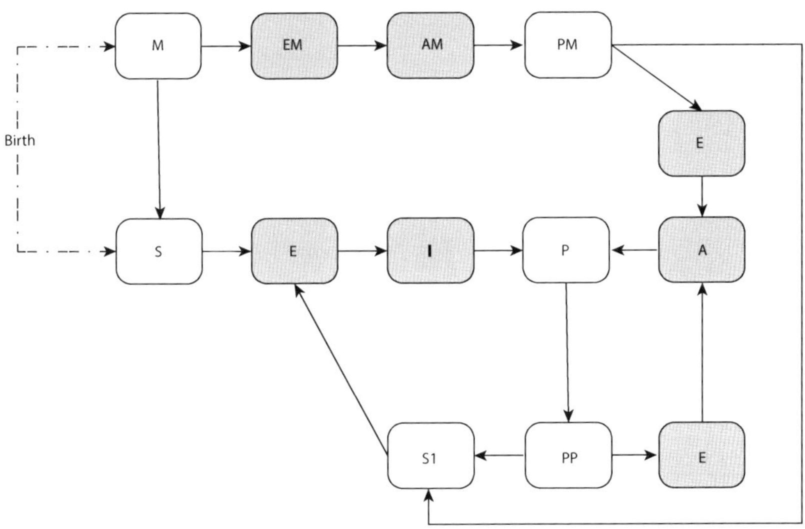

Figure 3. Health state diagram for each agents; greyed boxes represent disease states; legend of states in text.

The different health states are the followings:

- S: susceptible to symptomatic infection without any prior exposure to the disease
- $S1$: susceptible to symptomatic infection after loss of immunity
- E: exposed to infection ("incubating" agents are not infectious)
- EM: exposed to asymptomatic infection in infants ("incubating" agents are not infectious)
- I: infected with symptomatic pertussis (infectious)
- P: fully protected against pertussis
- PP: partially protected against symptomatic infections but not against asymptomatic ones
- PM: protected against symptomatic infection in infants who got maternal antibodies
- A: asymptomatically infected (infectious)
- AM: asymptomatically infected (infectious, in infants who got maternal antibodies)
- M: protected by maternal antibodies

Agents born during a simulation are either fully susceptible to pertussis (S) or protected by maternal antibodies against pertussis (M) if the mother developed the disease before or during her pregnancy or was vaccinated against pertussis during her pregnancy. As immunity conferred by the mother wanes, agents move from M to S.

When susceptible agents meet an infectious agent (I, see below), a probability function determines if pertussis is transmitted (Section 2.3.2). If pertussis is transmitted, the agent becomes first exposed to pertussis (E) and, after a few days, the agent becomes symptomatically infectious (I) after a fixed number of days. At the end of the infectious period, the agent becomes fully protected against pertussis (P) after a fixed number of days too. As immunity wanes the agent becomes partially protected against pertussis (i.e. protected against symptomatic infections but not necessarily against asymptomatic ones, PP) and then susceptible to either another symptomatic infection (if immunity wanes further, $\rightarrow S1$) or exposed to an asymptomatic infection (if they meet an infectious agent and pertussis is transmitted, $\rightarrow E$). Agents in $S1$ can become exposed to infection (E) if they meet an infectious agent and pertussis is transmitted. Agents in E (not coming from S) becomes asymptomatically infectious (A) and then fully protected against infection (P), each time after a fixed number of days.

If a baby protected by maternal antibodies against pertussis (M) meets an infectious agent and pertussis is transmitted (same probability function as above), the baby becomes exposed to infection (EM), asymptomatically infectious (AM) after a fixed number of days and finally protected against symptomatic infection (PM). From PM, the baby can become infected (E) and then asymptomatically infectious (A) or, if immunity wanes, susceptible to symptomatic infection ($S1$). This asymptomatic infection path for infants is created to differentiate vaccination of fully susceptible agents from vaccination of previously exposed agents (see next section).

When the initial population is built, each agent is set to a state depending on a given distribution of health state per age. During a simulation, a newborn agent is always put in state M or $S1$ (the ratio $M/S1$ is determined by the vaccination coverage of pregnant women). The state of the death D is the terminal state (and not represented here).

2.3.2. *Daily individual probability of infection*

For pertussis, the basic reproduction number (R_0) is equal to 17. By definition, it is represented by $R_0 = \frac{\beta}{\gamma}$ (where γ is the removal rate and β the effective contact rate). Since the inverse of a constant removal rate is the duration of infection (here 21 days), $\beta = 0.81$ is the effective contact rate.

In a SIR-type model (Susceptible-Infected-Recovered), the susceptible cohort is depleted following equation: $S(t) = 1 - e^{-\lambda * t}$. For an effective infection (at the end of the infectious period), $S(21) \simeq \beta$ and therefore, $\lambda = -\frac{ln(1-\beta)}{\text{duration}} = 0.079$ or 7.9%.

On day 1, $S(1) = 7.6\%$, i.e. the daily individual probability of infection in the household (for a susceptible agent), at day 1.

For a susceptible agent in the model, we define the daily probability of being infected by an infected agent (P) as the following:

$$
\begin{aligned}
P = 1 - (\ & (1-p_{HI})^{HI} * (1-p_{HA})^{HA} * (1-p_{HAM})^{HAM} * \\
& (1-p_{LI})^{LI} * (1-p_{LA})^{LA} * (1-p_{LAM})^{LAM} * \\
& (1-p_{BI})^{BI} * (1-p_{BA})^{BA} * (1-p_{BAM})^{BAM})
\end{aligned} \tag{1}
$$

With the first letter representing the context:

- H = number of individuals in the household context,
- L = number of individuals in the local context,
- B = number of individuals in the broader context.

And the subsequent letters representing the disease state:

- I = number of individuals who are symptomatically infected (in I),
- A = number of individuals who are asymptomatically infected (in A),
- AM = number of individuals who are asymptomatically infected (in AM).

Therefore:

- p_{HI} is the daily probability for a susceptible of being infected by a symptomatically infectious individual (in I) in the same household context ;
- p_{HA} is the daily probability for a susceptible of being infected by an asymptomatically infectious individual (in A) in the same household context ;
- p_{HAM} is the daily probability for a susceptible of being infected by an asymptomatically infectious individual (in AM) in the same household context ;
- p_{LI} is the daily probability for a susceptible of being infected by a symptomatically infectious individual (in I) in the same local context
- etc.

Several population sets of each social context are generated. The sum square of errors between the linear combination of these matrices and well-known contact matrices [12] are then minimized in order to find weights adjusting for the daily number of contacts with infectious agents in each contexts. At the end of each simulation, the effective contact matrix is computed and compared to the theoretical one by a measure of distance.

2.4. *Vaccination*

The dynamic of vaccination also follows a health state diagram. Currently only the primary and booster vaccinations are taken into account. Therefore, transitions described here are all event-driven.

WHO recommends to give, before the age of 1 year, a primary vaccination consisting of 2 or 3 doses of a pertussis-containing vaccine. Usually the vaccine used also contain other antigens [20]. Usually a booster dose of a pertussis-containing vaccine is also given during the second year of life and another dose is given in some countries to children 4-6 years old [7]. In this paper, only characteristics of acellular pertussis vaccines are considered.

The different states available for the primary vaccination arms are (Figure 4):

- $V1$: protected by one dose of a pertussis-containing vaccine
- $V2$: protected by two doses of a pertussis-containing vaccine
- $V3$: protected by three doses of a pertussis-containing vaccine
- S: susceptible without prior exposure to the disease
- E: exposed to infection ("incubating" agents are not infectious)
- P: fully protected against pertussis
- PM: protected against symptomatic infection in infants who got maternal antibodies
- A: asymptomatically infectious

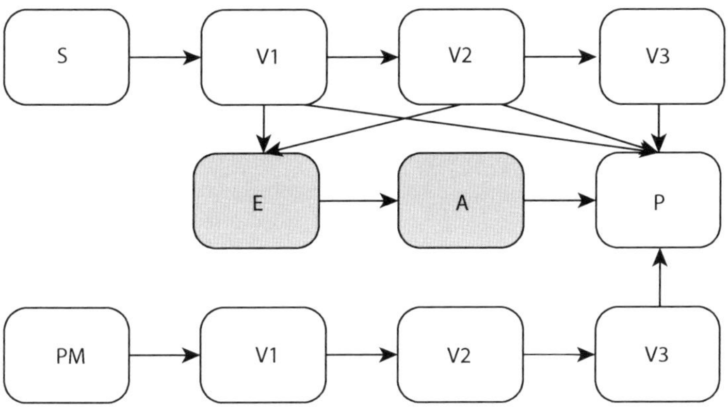

Figure 4. Primary vaccination state diagram for each agents ; greyed boxes represent states with disease ; legend of states in text

Initially, an agent eligible for a primary vaccination is either fully susceptible to pertussis (S) or acquired maternal pertussis antibodies from his mother (PM). The time when a fully susceptible agent receives the first dose of vaccine determines when he or she will receive the second and then the third dose of the vaccine. We consider the protection conferred by the first and second dose as sufficient to protect against symptomatic infections but insufficient to protect against asymptomatic infections. Therefore, if a vaccinated agent (in states $V1$ or $V2$) is in contact with an infectious agent and the infection is transmitted, the vaccinated agent becomes exposed (E), then asymptomatically infectious (A) and finally fully protected against pertussis (P). Baby agents who developed an asymptomatic infection (PM ; Figure 3) before are also eligible for primary vaccination but will not develop the disease if ever in contact with another infectious agent.

The different states in which agents are eligible for the booster vaccination are:

- S: susceptible without prior exposure to the disease
- $S1$: susceptible after loss of immunity
- E: exposed to infection ("incubating" agents are not infectious)
- P: fully protected against pertussis
- PP: partially protected against pertussis (protected against symptomatic infection but not asymptomatic ones)
- PM: protected against symptomatic infection in infants who got maternal antibodies
- A: asymptomatically infectious

Other states are ineligible for a booster vaccine either for age reason (e.g. agents in M are too young to receive a booster) or for medical reasons (e.g. a doctor or a nurse will not administer the vaccine if the agent had a pertussis episode in its recent history, i.e. agent in state I).

A booster dose is considered as an event that brings the agent in state P after which she follows the health state diagram for the disease itself.

On top of the primary and booster vaccination, the model implements a rule to take into account maternal immunization (vaccination of pregnant mothers during the third trimester of pregnancy) [18]. The rule specifies that a percentage of newborns will be protected by maternal antibodies (they enter the simulation in the M state, Figure 3).

2.5. *Time and scheduling*

At each time step (i.e. each day), the simulation discretely resolves the different dynamics. All transition rates and probabilities are converted to, respectively, daily rates and daily probabilities.

At each time step, different dynamics occur as a result of the model ruling the demography and the disease: birth and death, ageing, disease transmission, recovery and vaccination. Those dynamics can then alter the state of each agent (Section 2.1). Those dynamics are synchronously resolved (they only consider the states of all the agents at the previous time step to resolve the possible change of state of the agents at the current time step). It is therefore not necessary to follow a particular order to update the state of each agent. We can then write that the state $x_i(t)$ of an agent i at time step t is given by the following equation:

$$x_i(t) = F(x_i(t-1), ..., x_N(t)) \tag{2}$$

where F is the function resulting of the composition of every dynamics function, and N the total number of agents.

Currently, simulations are typically run over 5 years as it should allow the observation of the dynamics of pertussis to be expressed while not involving changes in the population that become incoherent.

3. Implementation

Several aspects are covered below to give a good overview of the current implementation including the requirements and assumptions that have lead it.

3.1. *Requirements*

The first purpose of the model is to support studies of health economics of pertussis. In particular, it should allow to evaluate the economics of outbreaks and various vaccination policies. This consists in the first and main requirement.

Beyond that, several requirements have been stated from the beginning of the project. The simulation software (later on referred to as the "simulator") should be efficient, flexible, not platform-specific, runnable on a *Unix* grid (or similar operating system) and maintainable.

The simulator should be efficient enough to be run in less than 1 hour on a common computer. First, this allows users to quickly get a grasp on the simulator by quickly getting a result. Secondly, due to its inherent stochasticity, the simulator will generate a set of possible stories. It should then be run several times to get a good idea of the distribution of the stories.

As the model is still in a research and development phase, the implementation has be to be flexible enough to allow, for example, adaptation to other diseases.

As most of the users are expected using the *Microsoft* operating system, the implementation is kept as much as possible platform-unspecific. On the other hand, the simulator should be runnable on a *Unix*-type gridas it is a typical solution when it comes to run the simulation several times.

Maintainability is important not only to guarantee the maintainability of the simulator but also an efficient transfer of knowledge, which is critical when working in a team subject to potential turnover.

3.2. *Chosen technologies*

The used programing language is C++ [d]. It has been chosen because it allows writing a very efficient code that is still very abstract and maintainable. For instance, it is possible to achieve a very low coupling without runtime overhead by using template and defining generic types.

As the aim is an implementation as portable as possible, the code is only depending on the STL (included with C++ and provided with every modern compilator) and Boost libraries (efficient, well tested and commonly used set of libraries where most of them are independent).

To achieve functional quality, *Google* test and Mock are used as test frameworks. Testing pushes us to a low coupling and dependency injection that we can achieve thanks to templates.

Several dynamic analysis tools are used in order to achieve memory and time efficiency, and to detect memory leaks and other runtime defects: *Google* Performance Tools, Valgrind, and gcov. Some static analysis tools such as cppcheck or simian are also used to increase the maintainability.

3.3. *Design overview*

This section presents a quick overview of the current implemented design. The simulation has been designed to be as modular as possible to achieve flexibility and maintainability. Several modules are distinguished here:

- demography model;
- disease model;
- agent and population generators;
- data collection;
- tools; and
- simulation controller and configuration.

[d]ISO, *ISO/IEC 14882:2011* (2011).

3.3.1. *Demography Model*

The Demography Model implements everything that concerns the population of agents: the agents themselves, the neighborhoods composed of agents, the social contexts characterizing the neighborhoods, and the dynamics i.e. deaths, births and sex resolution, allocation of a newborn to a household and daily creation of broader neighborhood.

Each part of the Demography Model is orchestrated by the population itself. Actually, running one step of the simulation corresponds to aging the population by one day and this triggers the following dynamics and state resolution:

(1) for each agent:
   (a) check if he/she dies and list it if appropriate;
   (b) age the agent (successively increment its age, increment its state age, update its health state and decays immunity level);
   (c) if in age, an agent possibly gets a vaccine dose;
   (d) if the agent is female, check her age and list it if the agent could potentially be pregnant;

(2) remove deceased agents from the population;
(3) remove empty household;
(4) update the probability of not being infected in a neighborhood for each neighborhood (see Section 2.2);
(5) generate birth to new agents and allocate them to household of appropriately aged female agents; and
(6) remove the previous broader neighborhoods and recreate new ones.

As coupling is kept as low as possible, the agent does only know in which health state it is. Every dynamics allows to determine if the next health state is implemented or not in the disease model (see next Section 3.3.2).

3.3.2. *Disease Model*

This module implements the health states and their transitions, the vaccinations, and the state transition machines.

To resolve the transitions, it is necessary to know several properties of the agent. To break the circular dependency and reduce the coupling, a context embedding the necessary agent properties is passed to the health state. Therefore the implementation does not require any knowledge of the demography.

A transition can be triggered by three different events: a time step, a death, or a vaccination. The building of a state machine consists then in defining which event allows a transition from one state to another. The setup is done based on passed parameters. That design allows a very flexible definition of a state machine.

3.3.3. *Agent and Population Generators*

Even if several dynamics are stochastic processes, the whole simulation is deterministic and stateless when considering the seed of the pseudo-random generator. That means that if the simulation is stopped after a certain number of time steps and the state of simulation is stored (the state of all agents and the seed), the simulation can be restarted and the same story be obtained as if the simulation had never been interrupted.

For that reason, every processes involved in the building of the initial states have been implemented in *ad hoc* generators. There are four generators: population, household, local, and health state. The population generator creates the N agents (age, sex, and unique ID). The household generator allocates each agent to a household neighborhood (with an unique ID). The local generator does the same but for local neighborhood. The health state generator determines the initial health state of each agent.

3.3.4. *Data Collection*

Data collection is implemented in an independent module, flexible enough to be configured to allow for any data collection possibility. It mainly consists in:

- a notepad where the data are collected,
- a measurement function actually collecting the data, and
- a way to plug them into any data struture.

3.3.5. *Tools*

A few *ad hoc* tools and utilities have been implemented such as reading a CSV [e] file, or calculating a Student's t-test. External libraries are not used for these cases to avoid another library dependency. Of course, if the effort to develop the functionality is too high, an existing library is used.

[e]CSV: Coma separated value

3.3.6. *Simulation Controller and Configuration*

Finally, the simulator or generators can be configured in two ways: by providing a configuration file, or by passing arguments to the simulator. This is implemented separately.

4. Discussion

A model includes a set of assumptions that can have important consequences in terms of modeling of the disease and its implementation. This section is discussing the most critical points.

We started the development of this ABM from an internal, equation-based, compartmental model. For that reason, it seemed natural to choose a discrete finite state-transition machine for this model. However, as the compartmental model has a population perspective, the health state diagram had to be adapted, conditions of infection had to be redesigned and the health state transitions had to be redefined (see Section 3.3.2).

At the implementation level, this had also implications on the health states of agents. Indeed, as health states are a value object [f], it is therefore not necessary to create as many health states as agents in the simulation. One version of each state is kept by the state machine and is assigned to an agent when it is in that state. As a consequence the health state objects can not be mutated, and every data describing the relation between the agent and the health state should be stored in the agent (such as the age and the period of the health state). This is the basic application of the Flyweight design pattern.

Another implication of the switch from a compartmental model to an ABM is the time representation. In an ABM, the discreteness of simulation time allows a natural computation: it corresponds to one iteration of the simulation and does not involve any numerical approximation (as requested in a compartmental model). The chosen time unit also implies a special attention on the time scale of other input data and a possible scaling.

The input data requirements were also bigger in this ABM compared to other types of models. Although the Netherlands (our target country) has a lot of data available from demographic, epidemiological and disease standpoints, input data were often only partially available at the required level of detail. For instance, data about age of individuals leaving the

[f] Value objects are two different objects having the same value, in our case representing the same health state.

household, household creation, size of local neighborhood, dynamicity of this local neighborhood are nonexistent but have an important impact on choices made to design the social network and the concept of social context (see Section 2.2). Complex neighborhood dynamics are not yet considered.

The available input data led us to consider a population of 100,000 agents. Indeed, as most collected data are for the whole country, this population seemed to us small enough to allow for good performances without creating too many issues of downscaling nor too many assumptions on relationships between downscaled parameters. Moreover, we do not know how big the population should be in order to correctly model the pertussis transmission at the country level and be comparable with published epidemiological data.

Simulating a population of 100,000 agents can take a long time if naively implemented. To achieve efficiency, we focused on two specific aspects: the amount of memory to store the agent health states, and the amount of memory to store the graph of social neighborhoods and the time needed to traverse it (especially to calculate the probability of not being infected in a neighborhood). Both aspects are discussed in the next paragraphs.

It is assumed that all agents in a neighborhood are in daily contact with each other (Section 2.2). It is then a complete graph. As consequence, the number of edges in a neighborhood grows as the square of the number of agents. This needs a lot of memory for a population of 100,000 agents, even if the whole graph is not complete. If we use the fact that a neighborhood is a complete graph, we can simplify the representation by considering the neighborhood as a hyperedge (a set of several nodes/agents, not only two nodes/agents), and by doing so, the whole social graph is an hypergraph. This reduces considerably the amount of memory to store the unused edges.

Considering the Eq. 1 and the fact that neighborhoods are complete graphs, we can also reduce the time needed to calculate the probability of an agent being infected. A naive way to proceed would be to calculate this probability each time it is necessary to test if the agent is infected or not. This is time consuming as it requires checking each neighbors' health state. Equation 1 is based on the probabilities of not being infected in each neighborhood of the considered agent. Those probabilities depend on the number of infected agents in the neighborhood, and it only changes when agents become infected or just recover. As those events happen less frequently, we can compute that probability for a given neighborhood and update it only when one of the agents become infected or recover in that specific neighborhood. Of course, contacts are still taken into account as a neighborhood models them.

In order to be fully operational, the model needs to address two remaining issues.

The first issue is the immunity level of agents. Linking health states and agents is very interesting from a resources point of view as counting the number of agents in a specific state will give us a basis to compute the economic burden (e.g. the disease cost for agents in the infectious state). However implementing an immunity level as a half-normal curve, as Codeço and Luz did [3], will allow to discriminate between health states like P (protected against any pertussis infections) and PA (protected against asymptomatic pertussis infections only).

The second issue is about constant local contexts. In reality, humans change jobs, move from a school to another, ... Keeping local contexts constant over 5 years of simulation time might not introduce major issues. However, when the simulation time will increase, constant local contexts might reduce the mixing pattern of agents and therefore reduce the opportunity of pertussis to spread in the population. Currently, data on local context dynamics seldom exist.

Finally, we would like to emphasize the fact that our goal has always been to use this ABM for a health economic analysis of preventative measures to reduce the burden of pertussis outbreaks. When considering inputs to include or exclude from the model, a balance has always been taken into account, keeping that goal in mind. On the other hand, the way this model is built allows to start from the classical dichotomy between epidemiological models and economic/quality of life models that are often juxtaposed in health economics models. The current design of health economics aspect in the model includes costs associated with the disease or the vaccination as well as quality of life gained or lost at the level of the agents themselves.

5. Conclusion

In this paper, we presented the foundation of an ABM that allows to study the spread of pertussis (and its prevention) in a population organized in three social contexts: household, local and broader contexts. Agents characteristics, the different contexts, the transmission models for disease and the primary and booster vaccinations were described. Implementation requirements and an overview of the design were explained.

This model will serve as a basis for the study of pertussis spread and outbreaks that happened in the past and the potential impact vaccination could have had. It will also be used for a health economic evaluation of the impact of different vaccination strategies on pertussis outbreaks.

By its modularity, the way it was implemented and its optimization, the model could also be adapted to other diseases where transmission does not occur via a vector and where social contexts and contacts are suspected to play an important role.

Acknowledgments

This work was supported by GlaxoSmithKline Biologicals SA, Rixensart, Belgium, which was involved in all stages of the study conduct and analysis and also funded preparation of the abstract and all related publications (GSK Study Identifier: HO-12-7031). The authors are grateful to Professor Seyed Moghadas (University of York, ON, Canada), Julie Roiz (formerly at GSK Vaccines, Wavre, Belgium, at the time of the study), Christophe Sauboin (GSK Vaccines, Wavre, Belgium), Augustin Terlinden (Navigha, Brussels, Belgium on behalf of GSK Vaccines, Wavre, Belgium) and Nicolas Van de Velde (GSK Vaccines, Wavre, Belgium) for providing input for this model. Editorial support was provided by Bruno Baudoux (Business & Decision Life Sciences, Brussels, Belgium on behalf of GSK Vaccines, Wavre, Belgium).

Conflict of interest

JEP and DC are employees of the GSK group of companies. DC reports stock ownership in the GSK group of companies. CP reports funding from GSK group of companies to his employer, Euranova, during the conduct of the study.

Contributorship

All authors participated in the design or implementation or analysis, and interpretation of the study; and the development of this manuscript. All authors had full access to the data and gave final approval before submission.

References

1. G. Amirthalingam, S. Gupta and H. Campbell, *Eurosurveillance* **18**, 20587 (2013).

2. D.L. Burns, B.D. Meade and N.E. Meissionnier, *J. Infect. Dis.* **209 Suppl**, S32 (2014).
3. C.T. Codeço and P.M. Luz, *CAD Saude Publica* **17**, 491 (2001).
4. Clinical features of Pertussis (Whooping Cough), retrieved from the CDC website at http://www.cdc.gov/pertussis/clinical/features.html on September 29, 2014
5. R. de Vries, M. Kretzschmar, J.F.P. Schellekens, F.G. Versteegh, T. Westra, J.J. Roord and M.J. Postma, *PloS one* **5**, e13392 (2010).
6. S. Fathima, C. Ferrato, B.E. Lee, K. Simmonds, L. Yan, S.N. Mukhi, V. Li, L. Chui and S.J. Drews, *BMC Infect. Dis.* **14**, 48 (2014).
7. G. Gabutti, C. Trucchi, M. Conversano, G. Zivelonghi and G. Zoppi, *Biomed Res. Int.* **2014**, 541319 (2014).
8. J.D. Hartzell and J.M. Blaylock, *Chest* **146**, 205 (2014).
9. M.J. Keeling and K.T.D. Eames, *J. R. Soc. Interface* **2**, 295 (2005).
10. J.S. Lavine, A.A. King and O.N. Bjornstad, *Proc Natl Acad Sci USA* **108**, 7259 (2011).
11. S. Mattoo and J.D. Cherry, *Clin Microbiol Rev* **18**, 326 (2005).
12. J. Mossong, N. Hens, M. Jit, P. Beutels, K. Auranen, R. Mikolajczyk, M. Massari, S. Salmaso, G.S. Tomba, J. Wallinga, J. Heijne, M. Sadkowska-Todys, M. Rosinska and W.J. Edmunds, *PLoS Med* **5**, e74 (2008).
13. C.D.C., *Morb Mortal Wkly Rep* **61**, 517 (2012).
14. A.L. Parks, B. Walker, W. Pettey, J. Benuzillo, P. Gesteland, J. Grant, J. Koopman, F. Drews and M. Samore, *AMIA. Annu. Symp. Proc.* **2009**, 504 (2009).
15. L. Perez and S. Dragicevic, *Int. J. Health. Geogr.* **8**, 50 (2009).
16. S.A. Plotkin, W. Orenstein and P.A. Offit, *Vaccines – 6th edition*, Saunders (2012).
17. J. Schellekens, C.W. von König and P. Gardner, *Pediatr. Infect. Dis. J.* **5 Suppl**, S19 (2005).
18. M.C. Steinhoff, *Vaccine* **31 Suppl 4**, D27 (2013).
19. C.W. von König, S. Halperin, M. Riffelmann and N. Guiso, *Lancet Infect. Dis.* **2**, 744 (2002).
20. W.H.O., *Weekly epidemiological record* **40**, 385 (2010).
21. K. Winter, K. Harriman, J. Zipprich, R. Schechter, J. Talarico, J. Watt and G. Chavez, *J. Pediatr.* **161**, 1091 (2012).
22. S.H. Yeh, *Expert. Rev. Vaccines* **2**, 113 (2003).

ZOONOTIC VISCERAL LEISHMANIASIS: A NOVEL MODEL INVOLVING DYNAMIC INTERACTIONS OF HUMANS, DOGS AND SANDFLIES

HELIO JUNJI SHIMOZAKO

University of São Paulo,School of Medicine, Oscar Freire Institute, 01246-000, Brazil
and York University, Ross Building, Department of Mathematics and Statistics,Toronto - ON, M3J1P3, Canada

JIANHONG WU

York University
Ross Building, Department of Mathematics and Statistics, Toronto - ON, M3J1P3, Canada

EDUARDO MASSAD

University of São Paulo, School of Medicine and LIM 01, HCFMUSP 05405-000, São Paulo, Brazil and London School of Hygiene and Tropical Medicine, London University, UK

Brazil is one of the high endemic countries for visceral leishmaniasis (VL): according to the Brazilian Ministry of Health, the annual number of new human cases and deaths due to VL has been increasing for the last 20 years. In this work, we developed a novel optimization model for VL control with five potential preventive control methods: (1) insecticide-impregnated dog collar, (2) dog vaccination, (3) infected dog culling, (4) infected dog treatment and (5) sandfly population control. Using a set of differential equations we simulated the dynamic of this disease and the optimal control strategy is examined considering the best cost-benefit relationship for these five control methods. Although this model is based on the previous model from Burattini et al. (1998) and Ribas et al. (2013), here we simulated with most of parameters updated and we included new categories that represent vaccinated dogs and dogs with insecticide impregnated collar. As conclusion, we obtained a good representation of the zoonotic visceral leishmaniasis dynamics and we found the sandfly population control as the best individual strategy to reduce the human incidence of this disease.

Keywords: visceral leishmaniasis, mathematical modeling, disease control, epidemiology.

1. Introduction

Visceral leishmaniasis is one of the world deadliest and neglected infectious disease, according to World Health Organization. This disease is endemic in 80 countries worldwide, in which 90% of all visceral leishmaniasis cases occur in Bangladesh, Brazil, India, Nepal and Sudan. Thus, about 360 million of people are exposed to risk of infection in the world [1,2,3,4]. The visceral leishmaniasis is a disease of major human and veterinary medical significance that involves a complex interplay between a protozoal pathogen, arthropod vectors, environmental influence on vector distribution, a small companion animal (dog) reservoir of infection and susceptible human populations. The leishmaniasis is emerging within non-endemic areas mostly because of transportation of dogs from endemic areas and climatic changes with the expansion of the geographical range of the sandfly vector. Then, the efective control will essentially involve interdisciplinary teams of microbiologists, parasitologists, entomologists, ecologists, epidemiologists, immunologists, veterinarians, public health officers and human physicians [5].

Considering only the preventive activities, the currently strategies focuses the reservoirs (dog population) and sandflies population. In order to exemplify these strategies focusing the dog population, we can mention the elimination [6], application of insecticide-impregnated collars [7], vaccination and treatment [9] of infected dogs. On the other hand, focusing the sandflies population, we can mention the environmental intervention by the application of chemical control and environmental sanitation.

Besides the publication of guidelines of visceral leishmaniasis control and the investments made in general surveillance activities, the sandfly and the reservoir in urban areas remains among the major challenges for the control activities. These challenges are due to (1) the necessity to better understand the vector behavior in urban environment; (2) the operational and logistic dificulties to carry out activities in sufficient time to obtain good results; and (3) the high costs involved in these activities [2,10]. In addition, regarding the Americas, the specific relationship between canine and human visceral leishmaniasis is not well understood. Thus, the control of the animal reservoir is complex and frequently involves combined interventions. In particular, the Brazilian Control Program recomends a strategy based on canine culling and vector control with insecticide spraying. Insecticide-impregnated collars for dogs and canine vaccination are not currently recomended as public health control measures [5,11].

In this work we propose a new model for zoonotic visceral leishmaniasis, based on the models previously published by Burattini et al. [12] and Ribas et al. and we included the effect of five control methods (sandfly population control, culling dogs, vaccination, treatment of infected dogs and use of insecticide impregnated dog collar). We simulated this model without the controls and with the controls individually. Then, we compare the efficacy of each one on the human reported case rate, in terms of rate value and cost-beneffit.

2. The Model

The basic idea of our model is based on the models published by Burattini et al. [12] and Ribas et al. [13]. However, in our model we focused on the Zoonotic Visceral Leishmaniasis and we analysed the the effect of the preventive control methods with more details.

The mathematical model used in this study consists of a set of differential equations, where some terms are delayed terms. Those equations represents the human population (with the index h), the dog population (index d). In both human and dog population we have susceptible (indicated by x_i), latent (which means infected but not clinally ill) (l_i), clinically ill (y_i), and recovered (z_i) individuals. In the case of dog population, we also have compartments for dogs that use insecticide impregnated collars (indicated by the superscript C) and vaccinated dogs (indicated by v_d).

In addition to human and dog population, we also included sandfly population, where s_1 is the non infected, s_2 is the infected but not infective, and s_3 is the infected and infective sandfly. Thus, the sandflies in s_3 category is able to infect susceptible human and dogs.

The Figure 1 represents the compartments and the flowchart of our model.

The set of differential equations for human population is represented by Eq. (1) to (4).

$$\begin{aligned} \dot{x_h}(t) &= (\mu_h + r_h)l_h(t) + (\mu_h + \alpha_h)y_h(t) + (\mu_h + \gamma_h)z_h(t) \\ &\quad -(b_h a_h m_h s_3(t))x_h(t) \end{aligned} \tag{1}$$

$$\dot{l_h}(t) = (b_h a_h m_h s_3(t))x_h(t) - (\mu_h + r_h + \delta_h + \varphi_h)l_h(t) \tag{2}$$

$$\dot{y_h}(t) = (\varphi_h)l_h(t) - (\mu_h + \alpha_h + \sigma_h)y_h(t) \tag{3}$$

$$\dot{z_h}(t) = \delta_h l_h(t) + \sigma_h y_h(t) - (\mu_h + \gamma_h)z_h(t) \tag{4}$$

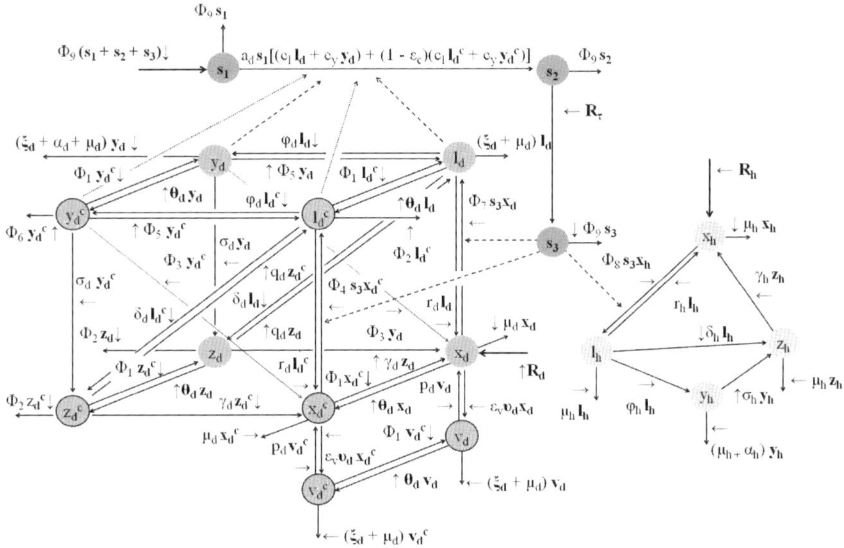

$$\mathbf{R}_\tau = a_d\mathbf{s}_1(\mathbf{t}-\boldsymbol{\tau})\,[(c_l\mathbf{l}_d(\mathbf{t}-\boldsymbol{\tau}) + c_y\mathbf{y}_d(\mathbf{t}-\boldsymbol{\tau})) + (1-\varepsilon_c)(c_l\mathbf{l}_d^c(\mathbf{t}-\boldsymbol{\tau}) + c_y\mathbf{y}_d^c(\mathbf{t}-\boldsymbol{\tau}))]\,e^{-(\mu_s+\xi_s)\tau}$$

$$\mathbf{R}_d = (\xi_d + \alpha_d + \mu_d)(\mathbf{y}_d + \mathbf{y}_d^c) + (\xi_d + \mu_d)(\mathbf{l}_d + \mathbf{l}_d^c + \mathbf{z}_d + \mathbf{z}_d^c + \mathbf{v}_d + \mathbf{v}_d^c) + \mu_d(\mathbf{x}_d + \mathbf{x}_d^c)$$

$$\mathbf{R}_h = (\mu_h + \alpha_h)\mathbf{y}_h + \mu_h(\mathbf{x}_h + \mathbf{l}_h + \mathbf{z}_h)$$

$\Phi_1 = (u_c + \zeta_c)$	$\Phi_6 = (\xi_d + \alpha_d + \mu_d)$
$\Phi_2 = (\xi_d + \mu_d)$	$\Phi_7 = b_d a_d m_d$
$\Phi_3 = (1 - c_k)\psi_d\omega_d$	$\Phi_8 = b_h a_h m_h$
$\Phi_4 = b_d(1 - \varepsilon_c)a_d m_d$	$\Phi_9 = (\xi_s + \mu_s)$
$\Phi_5 = f_d + c_k\psi_d\omega_d$	

Figure 1. Model compartments and the flow between them.

Eq. (5) to (14) represent the dog population.

$$\begin{aligned}
\dot{x_d}(t) = {} & -(\varepsilon_v\nu_d + \theta_d(t) + b_d a_d m_d s_3(t))x_d + (\mu_d + \zeta_d + u_c)x_d^C \\
& +(\xi_d(t) + \mu_d + r_d)l_d(t) + (\xi_d(t) + \mu_d)l_d^C(t) \\
& +(\xi_d(t) + \mu_d + p_d)v_d(t) + (\xi_d(t) + \mu_d)v_d^C(t) \\
& +(\xi_d(t) + \mu_d + \gamma_d)z_d(t) + (\xi_d(t) + \mu_d)z_d^C(t) \\
& +(\xi_d(t) + \alpha_d + \mu_d + (1 - c_k)\psi_d\omega_d(t))y_d(t) \\
& +(\xi_d(t) + \alpha_d + \mu_d)y_d^C(t)
\end{aligned} \tag{5}$$

$$\dot{x_d^C}(t) = r_d l_d^C(t) + (1-c_k)\psi_d\omega_d(t)y_d^C(t) + \gamma_d z_d^C + p_d v_d^C + \theta(t)x_d(t) \\ -((1-\varepsilon_c)b_d a_d m_d s_3(t) + \varepsilon_v\nu_d + \mu_d + \zeta_c + u_c)x_d^C(t) \quad (6)$$

$$\dot{l_d}(t) = b_d a_d m_d s_3(t)x_d(t) - (\xi_d(t) + \theta_d(t) + r_d + \delta_d + \mu_d + \varphi_d)l_d \\ (\zeta_c + u_c)l_d^C(t) + (f_d + c_k\psi_d\omega_d(t))y_d + q_d z_d \quad (7)$$

$$\dot{l_d^C}(t) = (1-\varepsilon_c)b_d a_d m_d s_3(t)x_d^C(t) + \theta_d l_d(t) + c_k\psi_d\omega_d(t)y_d(t) \\ +q_d z_d^C(t) - (\xi_d(t) + \zeta_c + u_c + \mu_d + r_d + \delta_d + \varphi_d)l_d^C(t) \quad (8)$$

$$\dot{y_d}(t) = -(f_d + \sigma_d + \mu_d + \alpha_d + \mu_d + \theta_d(t) + \xi_d(t) + \psi_d\omega_d(t))y_d(t) \\ +\varphi_d l_d(t) + (\zeta_c + u_c)y_d^C(t) \quad (9)$$

$$\dot{y_d^C}(t) = -(f_d + \sigma_d + \mu_d + \alpha_d + \mu_d + \zeta_c + u_c + \xi_d(t) + \psi_d\omega_d(t))y_d^C(t) \\ +\varphi_d l_d^C(t) + \theta_d(t)y_d(t) \quad (10)$$

$$\dot{z_d}(t) = \delta_d l_d(t) + \sigma_d y_d(t) - (q_d + \gamma_d + \mu_d + \theta_d(t) + \xi_d(t))z_d(t) \\ +(\zeta_c + u_c)z_d^C(t) \quad (11)$$

$$\dot{z_d^C}(t) = \delta_d l_d^C(t) + \sigma_d y_d^C(t) + \theta_d(t)z_d(t) + (\zeta_c + u_c)z_d^C \\ -(q_d + \gamma_d + \mu_d + \zeta_c + u_c + \xi_d(t))z_d(t) \quad (12)$$

$$\dot{v_d}(t) = \varepsilon_v\nu_d(t)x_d(t) + (\zeta_c + u_c)v_d^C \\ -(p_d + \mu_d + \theta_d(t) + \xi_d(t))v_d(t) \quad (13)$$

$$\dot{v_d^C}(t) = \varepsilon_v\nu_d(t)x_d^C(t) + \theta_d(t)v_d(t) \\ -(p_d + \mu_d + \zeta_c + u_c + \xi_d(t))v_d^C(t) \quad (14)$$

Eq. (15) to (17) represent sandfly population.

$$\dot{s_1}(t) = (\mu_s + \xi_d(t))(s_2(t) + s_3(t)) - a_d(c_l l_d(t) + c_y y_d(t) + \quad (15) \\ +(1-\varepsilon_c)(c_l l_d^C(t) + c_y y_d^C(t)))s_1(t)$$

$$\dot{s_2}(t) = a_d(c_l l_d(t) + c_y y_d(t) + (1-\varepsilon_c)(c_l l_d^C(t) + c_y y_d^C(t)))s_1(t) \quad (16) \\ -a_d(c_l l_d(t-\tau) + c_y y_d(t-\tau) + (1-\varepsilon_c)(c_l l_d^C(t-\tau) \\ +c_y y_d^C(t-\tau)))e^{-(\mu_s+\xi_s(t))\tau}s_1(t-\tau) - (\mu_s + \xi_d(t))s_2(t)$$

$$\dot{s_3}(t) = a_d(c_l l_d(t-\tau) + c_y y_d(t-\tau) + (1-\varepsilon_c)(c_l l_d^C(t-\tau) \quad (17) \\ +c_y y_d^C(t-\tau)))e^{-(\mu_s+\xi_s(t))\tau}s_1(t-\tau) - (\mu_s + \xi_d(t))s_3(t)$$

where the definition, the biological meaning and the values of each of the parameters used are described in Table 1. Some of those parameters were not found in the literature. Then they were just fitted in order to reproduce the disease dynamics with more realistic approach. Others were just estimated according to sources related to Brazilian Ministry of Health (as a_h, φ_h, σ_h, a_d, μ_s, τ) or according to personal communication (c_l and c_y were estimated by C. N. Rossi, 2013).

A brief description of the set of Eq. (1) to (17) should clarify their meaning.

Let S be the total number of sandflies. The number of bites inflicted in the human host population in an infinitesimal time interval dt is $a_h Sdt$, where is the biting rate on humans. The number of bites inflicted by infected flies is $S = a_h Sdts_3$, where S_3 is the number of infected flies.

Let now X_h be the total number of susceptible individuals in the human population. In an infinitesimal time interval dt, X_h varies as follows:

- a fraction of the infected bites are on uninfected individuals: $a_h Sdts_3 x_h$. A fraction b_h of these becomes latent, so X_h diminishes by $b_h a_h Sdts_3 x_h$;
- in the same time interval $r_h L_h dt + \gamma_h Z_h$ individuals, latent and immune, revert to the susceptible condition, and $\mu_h X_h dt$ die by natural causes other than the disease;
- we must add an entrance term, due to natality, which we choose to be $\alpha_h Y_h dt + \mu_h N_h dt$, where α_h is the disease-induced mortality rate, Y_h is the number of infected humans, and N_h is the total number of humans needed to maintain a constant population.

Therefore we have:

$$\begin{aligned} dX_h = {} & \alpha_h Y_h(t)dt + \mu_h N_h dt - b_h a_h Sdts_3(t)x_h(t) - \mu_h X_h(t)dt \\ & + (r_h L_h(t)dt + \gamma_h Z_h(t)dt) \end{aligned} \tag{18}$$

Dividing this equation by Nhdt and calling $S/N_h = m_h$, we get the first Eq. (1) of our model.

Equations indexed with d in the model are related to the dog's subpopulation. They differ from the original model by Burattini et al. [12] by including control terms. Therefore, the term ω_d refers to the treatment rate, according to which infected dogs are removed from the infectious compartment and are transferred to the susceptible and latent one; the term θ_d refers to the insecticide-impregnated collars, which by repelling the vector reduce the biting rate to the dogs; and finally the term ξ_d refers to the culling of dogs diagnosed as positive by serology test (by hipotesis, all dog that had contact with the parasite or its antigen, is considered positive by serology test). We also included the term ν_d, that means dogs vaccinated with transmission-blocking vaccine, in which are removed from the susceptible category to the vaccinated one.

The last three equations of our model refer to the flies. When infected, a fly remains in a latent stage for a period of time τ. This time corresponds to the extrinsic incubation period of the parasite inside the vector fly. Numerically it lasts for about half the life expectancy of the flies.

Let S_1 be the number of susceptible flies. In an infinitesimal period of time dt, $(a_d((Y_d + L_d + (1-\varepsilon_c)(L_d^C + Y_d^C))/N_d)dt)S_1$ bites due to uninfected flies occur on latent and infected dogs (humans are not considered to be infective for flies; see Tesh [14]). If the latent and infected dogs are using the insecticide impregnated collar, the probability of being bitten is reduced by ε_c. A fraction c_l (if biting a latent dog) and c_y (if bitting a clinically ill dog) of the flies becomes latently infected as a result. Therefore, we have:

$$dS_1 = (S_2(t) + S_3(t))dt - a_d(c_l l_d(t) + c_y y_d(t) \\ +(1-\varepsilon_c)(c_l l_d(t)^C + c_y y_d^C(t)))S_1(t)dt \qquad (19)$$

Dividing by $S = S_1 + S_2 + S_3$ and by dt we get the first of the last three equations of the model. The term $e^{-(\mu_s+\xi_s(t))\tau}$ accounts for sandflies infected at time t, but that die before producing infective sandflies.

3. The Basic Reproduction Number (R_0)

According to van den Driessche and Watmough [27], R_0 is the expected number of secondary cases produced, in a completely susceptible population, by a typical infective individual. Thus, in order to evaluate the R_0 of our model, we did not only base our calculation method on the work published by van den Driessche and Watmough [27], but also considered the assumptions adopted by them. We suggest to see their work for more details [27].

Here, we remark that although this work published by van den Driessche and Watmough [27] does not consider the presence of delayed terms, we proceeded an adaptation on our model in order to calculate the R_0. According to J. Wu and P. van den Driessche (2014, in printing), the delayed terms in our model can be considered as non-deleyed terms (that is, we can apply $\tau = 0$), but we need to keep the term $e^{-\mu_s\tau}$ (that is, for this term, we keep $\tau \neq 0$).

Table 1. Parameters used in the model, their biological meaning and numerical values. In the table, h stands for humans, d for dogs and s for sandflies.

Parameter	Biological meaning	Source	Numerical Value
μ_h	Natural mortality rate	IBGE [15]	$3.67 \times 10^{-5}\ days^{-1}$
α_h	Kalazar specific lethality	World Health Organization[4]	$8.40 \times 10^{-3}\ days^{-1}$
a_h	Average daily biting rate	Estimated	$2.00 \times 10^{-1}\ days^{-1}$
m_h	Vector density per host	Fitted	4.00
r_h	Spontaneous recovery rate	Badaro [16]	$5.48 \times 10^{-4}\ days^{-1}$
γ_h	Loss of immunity rate	Badaro et al. [17]	$2.76 \times 10^{-4}\ days^{-1}$
δ_h	Latent recovery rate	Badaro [16]	$1.20 \times 10^{-2}\ days^{-1}$
φ_h	Inverse of incubation period	Estimated	$4.2 \times 10^{-3}\ days^{-1}$
σ_h	Recovery rate to immunes	Estimated	$5.00 \times 10^{-2}\ days^{-1}$
b_h	Proportion of infective bites	Savani et al. [18]	5.00×10^{-3}
η_h	Proportion of unreported cases	Maia-elkhoury et al. [39]	7.05×10^{-1}
μ_d	Natural mortality rate	Selman et al. [19]	$2.28 \times 10^{-4}\ days^{-1}$
α_d	Kalazar specific lethality	Lanotte [21]	$1.81 \times 10^{-3}\ days^{-1}$
a_d	Average daily biting rate	Estimated	$2.00 \times 10^{-1}\ days^{-1}$
m_d	Vector density per host	Fitted	4.00×10^{1}
p_d	Loss of immunity rate (Leishmune ®vaccination)	Zoetis Brasil ®[22]	2.74×10^{-3}
φ_d	Inverse of incubation period	Fernandez-Cotrina et al. [20]	$1.10 \times 10^{-2}\ days^{-1}$
b_d	Proportion of infective bites	Savani et al. [18]	5.00×10^{-3}
r_d	Spontaneous recovery rate	Lanotte [21]	$5.00 \times 10^{-4}\ days^{-1}$

In general, the disease free equilibrium (DFE) is the scenary where the population remains in the absence of disease. If we consider epidemiological models, they usually have a threshold parameter, known as basic reproduction number, R_0, such that [27]:

- $R_0 < 1$, then the DFE is locally asymptotically stable and the disease cannot invade the population;
- $R_0 > 1$, then the DFE is unstable and invasion is always possible.

Table 2. Parameters used in the model, their biological meaning and numerical values. In the table, h stands for humans, d for dogs and s for sandflies. Continuation of Table 1

Parameter	Biological meaning	Source	Numerical Value
γ_d	Loss of immunity rate (recovery to susceptible)	Kault and March [23]	2.74×10^{-3} $days^{-1}$
σ_d	Recovery rate from clinically ill to immunes	Lanotte [21]	9.04×10^{-3} $days^{-1}$
δ_d	Latent recovery rate	Lanotte [21]	8.22×10^{-3} $days^{-1}$
f_d	Recovery rate from clinically ill to latent	Assumed	0
q_d	Loss of immunity rate (recovery to latent)	Assumed	0
ψ_d	Proportion of dogs who do not give up the treatment	Assumed	1.0
u_c	Inverse of activity period of collar	Scalibor ®[24]	6.00×10^{-3} $days^{-1}$
ζ_c	Loss of insecticide impregnated collar	Reithinger et al. [36]	6.00×10^{-3} $days^{-1}$
μ_s	Natural mortality rate	Estimated	5.00×10^{-2} $days^{-1}$
ε_v	Efficacy of vaccination	Saraiva et al. [25]	7.93 $days^{-1}$
ε_c	Decrease of biting rate due to insecticide impregnated collar	Halbig et al. [26]	8.00 $days^{-1}$
τ	Extrinsic Incubation Period	Estimated	1.40×10^{1} $days$
c_l	Fraction of flies which acquire the infection after biting an latent dog	Estimated	9.30×10^{-1}
c_y	Fraction of flies which acquire the infection after biting an clinically ill dog	Estimated	6.70×10^{-1}
c_k	Proportion of treated dogs that do not become parasite free	Miro et al. [34]	1.54×10^{-1}

In order to compute R_0, it is important to distinguish the new infections from all other changes in population. Let's consider $F_i(x)$ as the rate of appearance of new infections in a compartment of this model, $V_i(x)^+$ as the rate of transfer of individuals (but no including infected) into compartment i by all other means, and $V_i(x)^-$ as the rate of transfer of individuals (including infected) out of compartments, where x is the vector of variables of our model. The index i means to a category of one of the three populations. Thus, $1 \leq i \leq 4$, stands for humans, $5 \leq i \leq 14$, stands for dogs, and $15 \leq i \leq 17$, stands for sandflies. Thus, for each equation of our model, we can organize them as:

$$\frac{d}{dt}x = F_i(x) - V_i(x) \tag{20}$$

where $V_i(x) = V_i(x)^- - V_i(x)^+$.

From this point, we will not describe all the steps of the method. However, all the details are well explained in van der Driessche and Watmough [27]. Thus, before continue, let's define:

$$F(x_0) = \left[\frac{\partial F_i}{\partial x_j}\right] \tag{21}$$

$$V(x_0) = \left[\frac{\partial (V^- - V^+)_i}{\partial x_j}\right] \tag{22}$$

where x_0 is the vector that represents the disease free equilibrium state, i is the total of variables of the model and j represents the compartiments where there are infected individuals. In our model we have $j = 8$ compartments with infected individuals: l_h, l_d, l_d^C, s_2, y_h, y_d, y_d^C, s_3. Therefore, we have to calculate the partial derivative of each expression of our model in respect of each infected compartment.

From Eq. (21) and (22) we can apply them into the expression presented by van der Driessche and Watmough [27]:

$$R_0 = \rho(FV^{-1}) \tag{23}$$

where $\rho(FV^{-1})$ is the spectral radius of matrix FV^{-1}.

Considering our model and if the control methods are not considered ($\theta_d(t) = \xi_d(t) = \nu_d(t) = \omega_d(t) = \xi_d(t) = 0$), we obtain (calculation is not showed):

$$R_0 = b_d a_d^2 m_d e^{-(\mu_s + \xi_s(t))\tau} C \tag{24}$$

where:

$$C = \frac{(\varphi_d c_y + (f_d + \sigma_d + \alpha_d + \mu_d)c_l))}{(f_d + \sigma_d + \alpha_d + \mu_d)c_l)(r_d + \delta_d + \mu_d + \varphi_d) - f_d\varphi_d)\mu_s} \tag{25}$$

From Eq. (24) we can observe a contribution from clinically ill dogs and latent dogs on the R_0 value. Substituting the parameters by their respective values, we obtain the value of 1.65 from clinically ill dog contribution and 5.41 from latent dog contribution (thus, the total is 7.06). This difference of values between those two contributions can be explained by the integrity of the skin of those dogs. The clinically ill dog presents lesions on skin in which it becomes more difficult for a sandfly to be infected. On the other hand, the latent dog has no lesions on skin, in which allows the sandfly to be infected with a higher probability (C. N. Rossi, 2013, in printing).

4. The Linear Chain Trick

The entry of a leishmania protozoan into sandfly organism initiates a cascade of events that lead to the production of new parasites (the promastigota form of leishmania protozoan) (see Greene [29] for more details about the physiopathology). Thus, the set of Eq. (1) to (17) considers that this process occurs instantaneously: as soon as the sandfly is infected, the physiopathological process start and new infective sandflies will appear in τ days. However, in some biological models, this hypothesis is not a good representation (this is not biologically sensible) [28].

In the literature review published by Iggidr et al. [28] we find that fixed delays (the discrete constant delays) are not biologically realistic, and in the context of compartmental systems continuous probability functions of lags (the continuously distributed delays) are far more important than discrete lags. Usually the probability density function (pdf) of the lag is not well known and approximations are used. A widely used family of functions is given by the family of gamma distributions.

In order to take into account the effect of this countinuous distributed delay, we applied the linear chain trick. The linear chain trick allows replacing system with distributed delays whose distributions are linear combinations of gamma functions, by a system of ordinary differential equations. Also, we recall for the fact that the linear chain trick does not provide us an extra or new compartmental model, since it is not a mass-balance system [28]. For details about this method and applications, we suggest to see Iggidr et al. [28] and Azzan and Kane[30].

Thus, we substitute the Eq. (16) and (17) by the following Eq. (26) to (35).

$$\dot{s_2}(t) = a_d(c_l l_d(t) + c_y y_d(t) + (1-\varepsilon_c)(c_l l_d^C(t) + c_y y_d^C(t)))s_1(t) \quad (26)$$
$$- \left(\frac{2}{2+(\mu_s+\xi_d(t))\tau}\right)^2 a_d(c_l E_2^{ld}(t) + c_y E_2^{yd}(t)$$
$$+(1-\varepsilon_c)(c_l E_2^{ldC}(t) + c_y E_2^{ydC}(t))) - (\mu_s + \xi_d(t))s_2(t)$$

$$\dot{s_3}(t) = \left(\frac{2}{2+(\mu_s+\xi_d(t))\tau}\right)^2 a_d(c_l E_2^{ld}(t) + c_y E_2^{yd}(t) \quad (27)$$
$$+(1-\varepsilon_c)(c_l E_2^{ldC}(t) + c_y E_2^{ydC}(t))) - (\mu_s + \xi_d(t))s_3(t)$$

$$\dot{E_1^{ld}}(t) = \left(\frac{2}{\tau} + \mu_s + \xi_s(t)\right)(l_d(t)s_1(t) - E_1^{ld}(t)) \quad (28)$$

$$\dot{E_1^{yd}}(t) = \left(\frac{2}{\tau} + \mu_s + \xi_s(t)\right)(y_d(t)s_1(t) - E_1^{yd}(t)) \quad (29)$$

$$\dot{E_1^{ldc}}(t) = \left(\frac{2}{\tau} + \mu_s + \xi_s(t)\right)(l_d^C(t)s_1(t) - E_1^{ldc}(t)) \quad (30)$$

$$\dot{E_1^{ydc}}(t) = \left(\frac{2}{\tau} + \mu_s + \xi_s(t)\right)(y_d^C(t)s_1(t) - E_1^{ydc}(t)) \quad (31)$$

$$\dot{E_2^{ld}}(t) = \left(\frac{2}{\tau} + \mu_s + \xi_s(t)\right)(E_1^{ld}(t) - E_2^{ld}(t)) \quad (32)$$

$$\dot{E_2^{yd}}(t) = \left(\frac{2}{\tau} + \mu_s + \xi_s(t)\right)(E_1^{yd}(t) - E_2^{yd}(t)) \quad (33)$$

$$\dot{E_2^{ldc}}(t) = \left(\frac{2}{\tau} + \mu_s + \xi_s(t)\right)(E_1^{ldc}(t) - E_2^{ldc}(t)) \quad (34)$$

$$\dot{E_2^{ydc}}(t) = \left(\frac{2}{\tau} + \mu_s + \xi_s(t)\right)(E_1^{ydc}(t) - E_2^{ydc}(t)) \quad (35)$$

5. The Disease Dynamics

In order to analyze the dynamics of the model, we numerically simulated the complete system of 25 equations. However, here we are interested to understand how the disease dynamic is. Thus, we considered this initial simulation with no control methods ($\xi_s(t) = 0$, $\theta_d(t) = 0$, $\omega_d(t) = 0$, $\nu_d(t) = 0$, $\xi_d(t) = 0$). As consequence, our system of equations become simpler, composed by 15 equations. The set of parameters used is shown in Table

1. Figure 2 is the simplified compartment model and flowchart, without control methods. Since we are not considering dog vaccination neither the application of insecticide impregnated collar, there are no compartments for dogs with collar and vaccinated dogs.

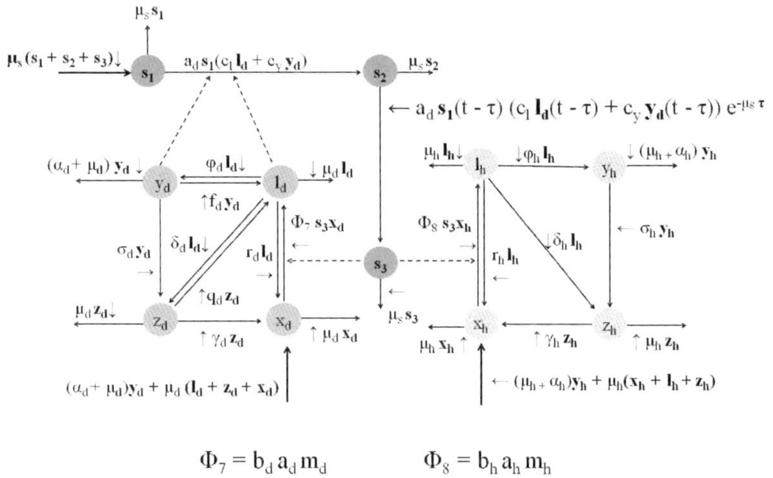

Figure 2. Model compartments and the flow between them after simplification.

Figure 3 ilustrates the dynamic of reported cases by Brazilian Ministry of Health. This dynamic was estimated by the equation:

$$R(t) = (1 - \eta_h)\phi_h l_h(t) \tag{36}$$

where $R(t)$ are the human reported cases, and η_h is the proportion of unreported cases.

The equilibrium state of human reported cases rate is around 1.83×10^{-5}/day. According to data from Brazilian Ministry of Health, the rate of human reported cases was 4.72×10^{-8}/day for 2012 year, considering the whole country. In Brazil, the distribution of zoonotic visceral leishmaniasis is not homogeneus. Thus, we can find states with no human reported cases for 2012 year (as Rio Grande do Sul state). This non-homogeneus distribution is related to geographic featuring and climate changes, in which probably has influenced on the sandfly population dynamics. Therefore, this is one of the arguments to explain the discrepancy regarding to our result and the real data. In addition, although our model used updated

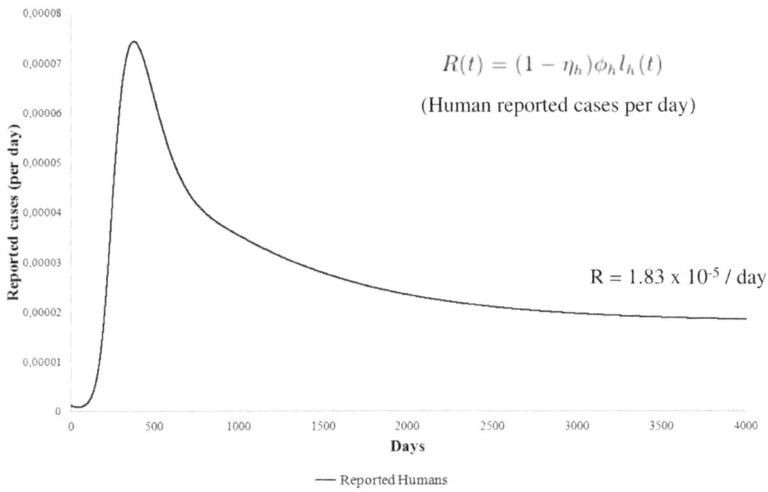

Figure 3. Human reported cases per day ($m_h = 4$, $m_d = 40$). The rate of 1.83×10^{-5} day^{-1} is obtained for the equilibrium state.

parameters, some of them were not able to be updated, and this can also be a reason for this unconformity. Finally, we could reach a close result if we adopt $m_h = 0.511$ ($m_d = 5.11$), however, the analysis of the disease dynamic would be difficult because of extreme low values of output. Since we are firstly interested to present this new model here, we simulated our model using the fitted value of $m_h = 4$ ($m_d = 40$), which we considered very real.

The Figures 4, 5 and 6 ilustrate the dynamic for human, dogs and sandflies, respectively.

In order to compare our results, we were able to find some descriptions about some demographic characterization of those populations regarding to zoonotic visceral leishmaniasis. For instance, our sum of latent and recovery dog was around 0.55, which is in a good agreement with that one found by Cabral et al. [31] (0.50). However, on the other hand, the sum of latent and clinically ill dogs was around 0.157, which is different with the results from Quinnell et al. [32] (0.579).

6. The Optimal Control Strategy

In this model we included five preventive control methods (elimination of positive dogs, sandfly population control, use of insecticide impregnated dog collar, dog vaccination and dog treatment). However, considering the

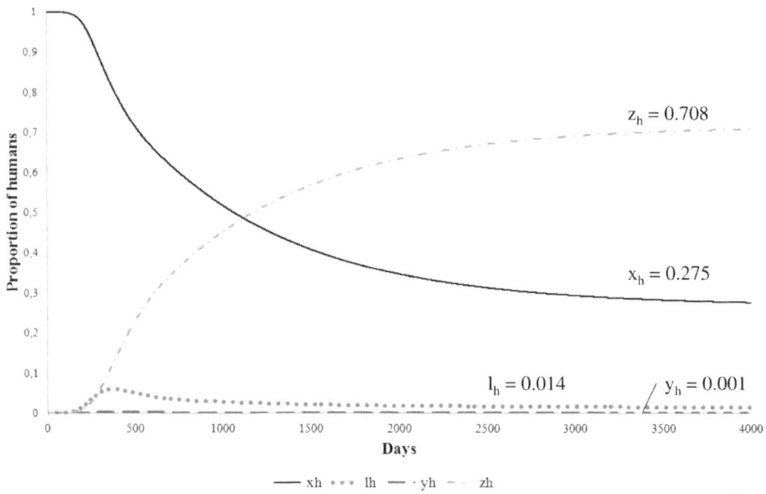

Figure 4. Numerical simulation of the model for human population. The values $x_h = 2.76 \times 10^{-1}$, $l_h = 1.48 \times 10^{-2}$, $y_h = 1.06 \times 10^{-3}$ and $z_h = 7.08 \times 10^{-1}$ are the equilibrium densities.

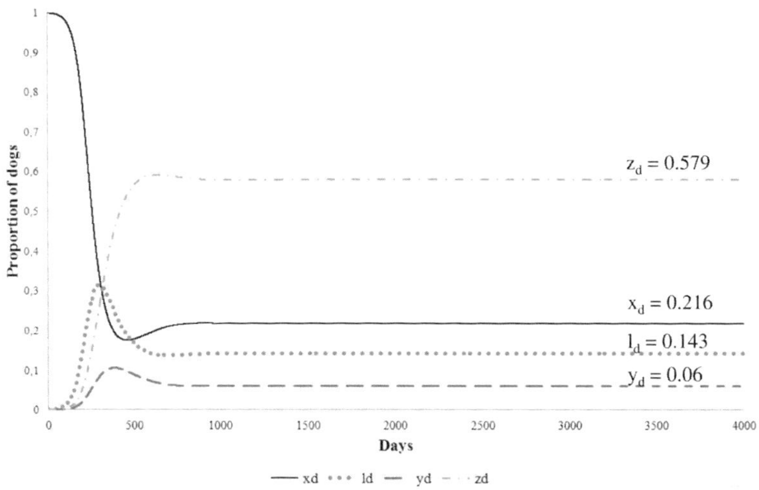

Figure 5. Numerical simulation of the model for dog population. The values $x_d = 2.17 \times 10^{-1}$, $l_d = 1.42 \times 10^{-1}$, $y_d = 6.00 \times 10^{-2}$ and $z_d = 5.79 \times 10^{-1}$ are the equilibrium densities.

official control strategy in Brazil, the Brazilian Ministry of Health adopts the elimination of positive dogs as control method and does not preconize

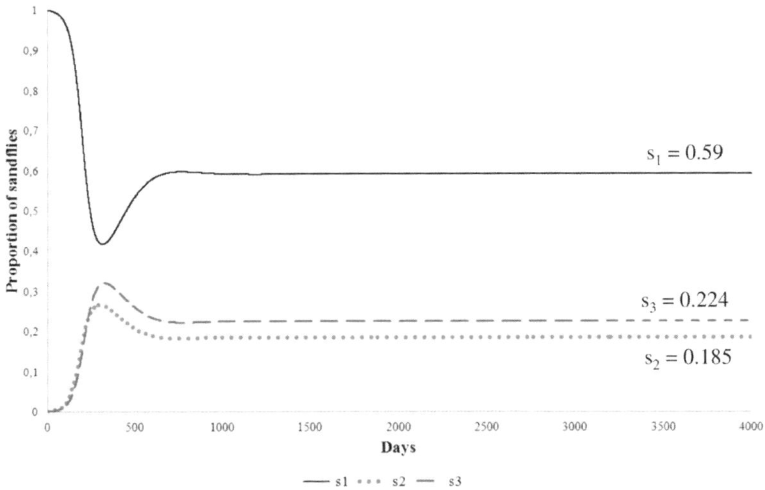

Figure 6. Numerical simulation of the model for sandfly population. The values $s_1 = 5.90 \times 10^{-1}$, $s_2 = 1.85 \times 10^{-1}$, $s_3 = 2.25 \times 10^{-1}$ are the equilibrium densities.

the other four methods mentioned here [39]. However, it has been usual the indiscriminate elimination of dogs, instead of the selective elimination.

In order to check the efficacy of alternative control methods, we numerically simulated the model considering individually each method and all five control methods were evaluated for the same range. As reference of comparison, we observed the effect on the human reported cases rate regarding to the rate values (Figure 7) and the cost per rate unit (Figure 8). For those simulations, we performed with the same set of parameters in Table 1.

Ribas et al. [13] also published a mathematical model for Zoonotic Visceral Leishmaniasis and this model was based on the previous model published by Burattini et al. [12]. In this model published in 2013, Ribas et al. discussed about the optimal control method among those same five methods we presented here. However, differently from Ribas et al., our model allow us to observe the dynamic and movement of dog population if we consider the use of insecticide impregnated dog collar or vaccination, since we included specific categories for them. Futhermore, if we consider the treatment, we included a proportion of dogs that were totally cleared from the parasite $(1 - c_k)$. Finally, since the serology is the official diagnostic method adopted by Brazilian Ministry of Health and considering that this diagnostic method is not able to distinguish an recovered of vaccinated dog

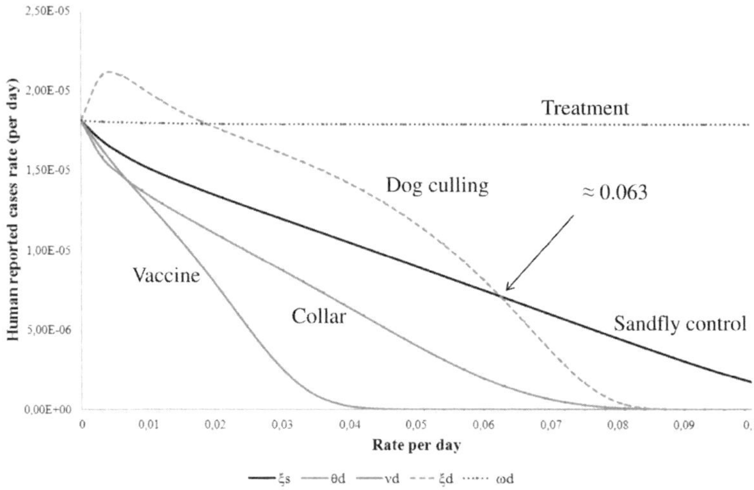

Figure 7. Effect of each control methods (if constant) on reported human cases rate after one year.

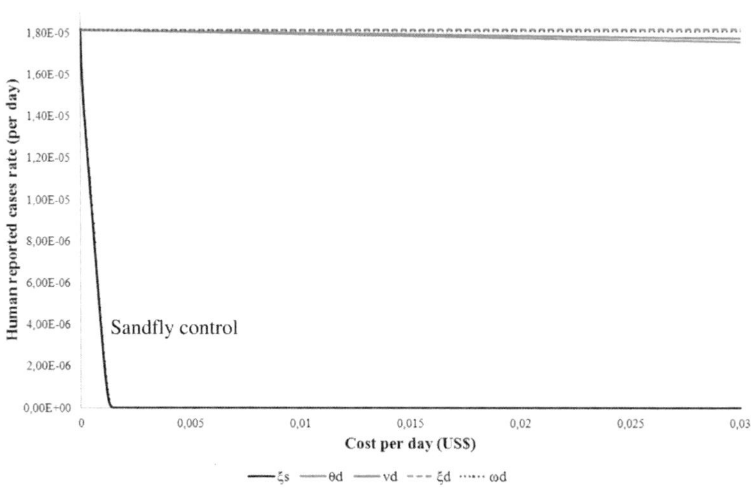

Figure 8. Impact of investment (if constant) on each of strategy control on reported human cases rate after one year.

from the latent or clinically ill dog, we assumed that any dog who had previous contact with the parasite is eligible to be eliminated [40]. Regarding the sandfly population, we assume that this population is r-strategist and, thus, the permanency of sandflies in such area depends on the increment of the reproduction power. Classically, the individuals that are r-strategists are small size individuals, with short life expectancy, maturity quickly reached, high mortality rate and short reproduction cycles [33]. Thus, we simulated the sandfly population control by adding an extra term for mortality rate $\xi_s(t)$.

From Figure 7 and 8 we observed two different results. If we consider only the range of rates and ignore the costs, the Figure 7 shows that the vaccination and the use of insecticide impregnated dog collar were the most efficient methods. They were followed by sandfly population control and dog elimination. The treatment of infected dogs has almost no effect on the human reported cases rate. Those results are different from the conclusions previously presented in Burattini et al.[12] and Ribas et al. [13]. Although they made conclusions on their results on human prevalence, sandfly population control was considered the most efficient strategy of control. However, we recall here that our model has been modified and some parameter values were updated. Furthermore, we observed in our model that the vaccination of dogs causes direct decreasing on the susceptible dog category, in which consequently decreases the size of latent and clinically ill dogs categories. Also, we can observe that the application of insecticide impregnated dog collar not only remove latent and clinically ill dogs from "exposed to sandfly" status to a "non-exposed sandfly" status, but also protect the susceptible and recovered dogs to be bitten by flies. In those both control strategies, since we assumed the density of 40 sandflies per dog, for each dog protected from sandflies bites, we have less 40 sandflies that potentially can infect a susceptible dog or be infected by a infected dog. This is the reason about the sandfly control population was not more efficient than vaccination and insecticide impregnated dog collar.

In our model, all populations are assumed constant. Thus, the elimination of one individual in those population, we have the replacement of a new susceptible individual [12]. Thus, when we compare the elimination of one sandfly to one dog, we observed that elimination of sandflies is more efficient than elimination of dogs until the rate of 0.063/day. This means that in order to obtain the elimination of infected dog as efficient (or more) than sandfly control population, we have to apply a rate of elimination of dogs higher than 0.063 dogs/day.

Finally, we have the treatment of dogs, which is the less effective strategy. Although we assumed an treatment protocol that results in the elimination of the parasites from dog organism [34], our assumption is that the dog becomes to susceptible category. Thus, this not avoid the dog to became infected again and, thus, source of infection.

Now, if we observe Figure 8, we have a different point of view about our previous conclusions. Here we analyzed the effect of the control methods on human reported cases but considering the cost per rate of each method. In this new approach, we assumed the values per rate unit of Table 3 and they were obtained from a sample of costs available in different sources (from Brazilian municipality source to company websites). We will not show the calculation of the costs, but we had to keep some featuring as assumptions and used the exchange rate of Brazilian real to American dollar of 0.41 (verified on 2014 Oct 20), according to Brazilian Central Bank (http://www.bcb.gov.br/). Thus, vaccination, dog culling and insecticide impregnated dog collar application method were budgeted considering two assistents, one veterinary and one driver. The sandfly control considered two assistents and one driver. The treatment, we considered that the medicine is provided to private veterinarians, so we supposed that the Brazilian Ministry of Health does not offer treatment but could provide the medicine to those owners who are interested to treat their dog.

According to Figure 8, by far, the sandfly control strategy is the most efficient in terms of cost effectiveness. As the methods applied on dog population showed low (or none) effect on the human reported cases rate, the sandfly control strategy provides an intense decrease of this rate when we observe the same cost range. However, although the sandfly control is the most cost-effective method, there are logistic implications. In this case, if the insecticide application protocol is not well followed, then it is possible to observe an inverse effect on sandfly population with the appearance of resistant sandflies.

Table 3. Cost of control methods per rate unit (calculations not shown). This estimation were based on material costs and salaries informed in Brazilian municipality source and company websites.

Description	Cost
Total cost for sandfly control	USD 0.01 $\times$ $sandfly^{-1}$
Total cost for culling	USD 485.90 $\times$ dog^{-1}
Total cost for vaccine	USD 45.13 $\times$ dog^{-1}
Total cost for insecticide impregnated collar	USD 41.58 $\times$ dog^{-1}
Total cost for treatment	USD 20.84 $\times$ dog^{-1}

7. Discussion

The model presented in this article provided a new and interesting view about visceral leishmaniasis transmission dynamics among humans, dogs and sandflies population. Also, we developed a mathematical analysis and computer simulation of strategies control application on this system. Finally, we evaluated the optimal strategy when they are compared on the same rate value ranges and in terms of cost benefit.

The researches about control strategies rarely propose the optimization of the strategy in a general approach. That is, most of the results are more empirical than analytical. Thus, it is dificult to proceed to any extrapolation of these practical conclusions to other sceneries. The increasing trend of VL cases observed in endemic countries (like Brazil) and the spread of transmission to previously not affected areas raise doubts about the impact of ongoing control measures [11]. In addition, the studies about the eficacy of control strategies usually analyze them individually and discuss about the comparison with other strategies based on the results of other researches. Again, this does not allow a practical analysis, since the discussion that these articles present are about different experimental fields and/or designs.

The review written by Romero and Boelaert [11] related to visceral leishmaniasis control in Latin America revealed that a lot of work remains to be done in order to clarify the dynamics of Leishmania transmission in human, canine and vector populations. Herein we was able to apply a simulation that predicts what could happen if the control interventions occur. Thus, our theoretical study presents a clear practical application and does not conflict with ethical procedures (since it is a theoretical study). In other hand, although it is possible to develop this optimization by field study, we would probably need to organize dozen of surveillance staffs, each one being responsible for a particular control strategy. Also, we would need a long time to obtain these results. However, this practical way is not simple to proceed, since it would be necessary to submit the study to a bureaucratic assessment by the local public health administration. In addition, we could observe a clear waste of resources and time, because it is hard to evaluate all the possible control strategies.

In our model we could estimate the proportion of the individuals in each stage of disease dynamic. This evaluation is very interesting, since the latent individuals (humans and dogs) are dificult to detect. Furthermore, here we estimated the reported human cases rate for our model

$(1.83 \times 10^{-5}/\text{day})$. Killick-Kendrick [2] highlighted that the key to reach the visceral leishmaniasis control is education. Education was not considered in our model but we can see that public health education and epidemiological surveillance system are very close and work together. As example, the fact of infected people neglected the visceral leishmaniasis provides unreported cases of this disease. As consequence, few cases are reported and the organization of control programs is poorly organized. Thus, if we know the proportion of unreported cases, we could evaluate the eficacy of the surveillance service. Finally, this can indirectly influence the control programs, since the strengthening of the surveillance system capacity is essential to avoid the underreporting of human cases and to follow-up the infection behavior in canine and human population. Strong surveillance will certainly contribute to improve data quality for decision-makers in this complex scenario [11,39].

According to the results presented here, the simulation purposed excellent results even though the culling dog strategy is not applied. We can find some authors that stress the importance of development and use of canine vaccines in order to reduce the canine reservoir of infection. Consequently, this could interrupt the transmission of Leishmania parasites to humans and could eliminate the need for euthanasia [3,5]. Others highlight the effectiveness of insceticide-impregnated collars [7,35]. Thus, we can understand this fact as evidence that the control of visceral leishmaniasis can be reached under ethical constraints. In addition, there are some researches that highlight the effectiveness of control of sandflies population when compared with the culling of infected dog [11,12,36]. From our results we suggested that the control of sandflies is the most effective strategy control among the five strategies considered here. The evidence of a more human and ethical control should attract the attention of most surveillance services of endemic areas that applies the culling of domestic dogs.

If a human and ethical control could be applied (that is, reducing the canine culling), we can suppose that the treatment of infected dog is also possible. Although we can find some articles that do not recommend the treatment of canine visceral leishmaniasis [9,41,38] since parasitological cure has not been achieved, our model ponted that it is also possible to reach the elimination of the infection without culling infected dogs but treating them. Supposing an infected dog has an owner, the other preventive control methods have been applied and that, once the dog is infected, it would become infected for all life, the owner could be free to decide about treating or not his own dog.

As conclusion, according to our model we obtained a good representation of the zoonotic visceral leishmaniasis dynamics and we found the sandfly population control as the best individual strategy to reduce the human incidence of this disease. However, as upcoming work, the non-updated parameters should be reviewed, and it should be interesting to evaluate the optimization of the control strategy, but when all methods are concomitantly applying.

References

1. P.N. Acha, B. Szyfres, *Zoonoses and Communicable Diseases Common to Man and Animals*, Pan American Health Organization, Washington, D.C. (2001).
2. R. Killick-Kendrick, *Bull. World Health Organ.*, **88**, 11 (2010).
3. M.S. Duthie, V.S. Raman, F.M. Piazza, S.G. Reed, *Vaccine*, **30**, 134 (2012).
4. World Health Organization,*http : //http : //www.who.int/en/* (2013).
5. C.B. Palatnik de Souza, M..J. Day, *Parasites & Vectors.*, **4(1)**, 197 (2011).
6. C. M. Caris-Nunes, V.M. F. de Lima, H.B. de Paula, S.H. Perri, A.M. Andrade,F.E. Dias,M. N. Burattini, *Veterinary Parasitology*,**153**,19 (2008).
7. A.S. Mazloumi Gavgani, M.H. Hodjati, H. Mohite, C.R. Davies, *THE LANCET*, **360**, 374 (2002).
8. C. B. Palatnik de Souza, *Vaccine*, **26**, 1709 (2008).
9. L. V. Athanasiou, M. N. Saridomichelakis, V. I. Kontos, G. Spanakos and T. S. Rallis, *Veterinary Parasitology*, **192**, 91 (2013).
10. A. N. S. Maia-Elkhoury, W. A. Alves, M. L. Souza-Gomes, J. M. Sena and E. A. Luna, *Cad. Saude Publica*, **24(12)**, 2941 (2008).
11. G. A. S. Romero, M. Boelart, *PLoS Negl. Trop. Dis.*, **4(1)**, e584 (2010).
12. M. N. Burattini, F. A. B. Coutinho, L. F. Lopez and E. Massad, *J. Biol. Syst.*, **6(4)**, 337 (1998).
13. L. M. Ribas, V. L. Zaher, H. J. Shimozako, E. Massad, *The Scientific World Journal*, **2013**, Article ID 810380,
14. R. B. Tesh, *Am. J. Trop. Med. Hyg.*, **3(52)**, 42 (1995).
15. Brazilian Institute of Geography and Statistics (IBGE), *Life expectancy at birth was 74.6 years in 2012* (2013) (in Portuguese).
16. R. Bardaro, T.C. Jones, E.M. Carvalho, D. Sampaio, S.G. Reed, A. Barral, R. Teixeira, W. R. Johnson *J. Infect. Dis.*, **154(6)**, 1003 (1986).
17. R. Bardaro, E. M. Carvalho, M. G. O. Orge, R. S. Teixeira, H. Rocha *Revista da Sociedade Brasileira de Medicina Tropical*, **18**, 77 (1985).
18. E. S. M. M. Savani, V. L. B. Nunes, E. A. B. Galati, T. M. Castilho, R. A. Zampieri, L. M. Floeter-Winter *Veterinary Parasitology*, **160**, 18 (2009).
19. C. Selman, D.H. Nussey, P. Monaghan *Current Biology*, **10**, R451 (2013).
20. J. Fernandez-Cotrina, V. Iniesta, S. Belinchon-Lorenzo, R. Munoz-Madrid, F. Serrano, J. C. Parejo, L. Gomez-Gordo, M. Soto, C. Alonso, L. C. Gomez-Nieto, *Veterinary Parasitology* **192**, 118 (2013).

21. G. Lanotte, J. A. Rioux, J. Perieres and Y. Vollhardt, *Ann. Parasitol.*, **3(54)**, 7 (1979).
22. Zoetis Brasil website. Accessed on 2015 Jan 15, Avaiable in https://www.zoetis.com.br/node/19883
23. D. A. Kault and K. M. Marsh, *Math. Biosc.*, **103**, 103 (1991).
24. Scalibor website. https://www.medicanimal.com/medicanimal/images/products/PDF/ Scalibor% 20Client% 20DL% 20Leaflet.pdf
25. E.M. Saraiva, A.F. Barbosa, F.N. Santos, G.P. Borja-Cabrera, D. Nico, L.O.P. Souza, C.O. Mendes-Aguiar, E.P. de Souza, P. Fampa, L.E. Parra, I. Menz, J. G. Dias Jr., S. M. de Oliveira, C. B. Palatnik-de-Sousa, *Vaccine*, **13**, 2423 (2006).
26. P. Halbig, H. M. Hodjati, A. S. Mazloumi-Gavgani, H. Mohite, C. R. Davies, *Med Vet Entomol.*, **14(2)**, 223 (2000).
27. P. van den Driessche, J. Watmough, *Mathematical Biosciences* **180**, 29 (2013).
28. A. Iggidr, J Mbang, S Gauthier, *Mathematical Biosciences* **209, 1**, 51 (2013).DOI: 10.1016/j.mbs.2007.01.008
29. C. Greene, *Infectious Diseases of the Dog and Cat*, Saunders, (2011).
30. C. Azzam, J. N. Kane, *Fundamental Mathematics of the Delay Differential Equations in the HIV-1 Virus Model*
31. M. Cabral, J.E. O'Grady, S. Gomes, J.C. Sousa, H. Thompson, J. Alexander, *Veterinary Parasitology* **76**, 173. 1998.
32. R. J. Quinnell, O. Courtenay, S. Davidson, L. Garcez, B. Lambson, P. Ramos, J. J. Shaw, M.-A. Shaw, C. Dye, *Parasitology* **122**, 122, 2000.
33. O. P. Forattini, *Ecology, Epidemiology and Society*, Artes Medicas, Sao Paulo (2011) (in Portuguese).
34. G. Miro, R. Galvez, C. Fraile, M. A. Descalzo, R. Molina, *Parasit Vectors* **4(52)**, (2011) doi: 10.1186/1756-3305-4-52
35. M. Maroli, V. Mizzoni, C. Siragusa, A. D'Orazi and L. Gradoni, *Medical and Veterinary Entomology*, **15**, 358 (2001).
36. R. Reithinger, P. G. Coleman, B. Alexander, E. P. Vieira, G. Assis and C. R. Davies, *International Journal for Parasitology*, **34**, 55 (2004).
37. H.M. Andrade, V.P.C.P. Toledo, M.B. Pinheiro, T.M.P.D. Guimares, N.C. Oliveira, J.A. Castro, R.N. Silva, A.C. Amorim, R.M. S. S. Brando, M. Yoko, A. S. Silva, K. Dumont, M.L. Ribeiro Jr, W. Barthewsky, S.J.H. Monte, *Veterinary Parasitology*, **181**, 83 (2011).
38. S. Rougier, L. Hasseine, P. Delaunay, G. Michel and P. Marty, *Veterinary Parasitology*, **17(1)**, 42 (1974).
39. A. N. Maia-Elkhoury, E. H. Carmo, M. L. Souza-Gomes and E. Mota, *Rev. Saude Publica*, **6(41)**, 913 (2007).
40. M. Marcondes, V. M. F. de Lima, M. de F. L. de Araujo, R. M. Hiramoto, J. E. Tolezano, R. F. C. Vieira, A. W. Biondo, *Veterinary Parasitology*, **197**, 649 (2013).
41. R. M. Anderson and R. M. May, *Infectious Diseases of Humans*, Oxford University Press Inc., New York (1991).

MATHEMATICAL MODELS FOR VACCINATION, WANING IMMUNITY AND IMMUNE SYSTEM BOOSTING: A GENERAL FRAMEWORK

M. V. BARBAROSSA AND G. RÖST

Bolyai Institute, University of Szeged
Aradi vértanúk tere, 1
H-6720 Szeged, Hungary,
E-mail: barbaros@math.u-szeged.hu, rost@math.u-szeged.hu

When the body gets infected by a pathogen or receives a vaccine dose, the immune system develops pathogen-specific immunity. Induced immunity decays in time and years after recovery/vaccination the host might become susceptible again. Exposure to the pathogen in the environment boosts the immune system thus prolonging the duration of the protection. Such an interplay of within host and population level dynamics poses significant challenges in rigorous mathematical modeling of immuno-epidemiology. The aim of this paper is twofold. First, we provide an overview of existing models for waning of disease/vaccine-induced immunity and immune system boosting. Then a new modeling approach is proposed for SIRVS dynamics, monitoring the immune status of individuals and including both waning immunity and immune system boosting. We show that some previous models can be considered as special cases or approximations of our framework.

1. Introduction

Models of SIRS type are a traditional topic in mathematical epidemiology. Classical approaches present a population divided into susceptibles (S), infectives (I) and recovered (R), and consider interactions and transitions among these compartments[9]. Susceptibles are those hosts who either did not contract the disease in the past or lost immunity against the disease-causing pathogen. When a susceptible host gets in contact with an infective one, the pathogen can be transmitted from the infective to the susceptible and with a certain probability the susceptible host becomes infective himself. After pathogen clearance the infective host recovers and becomes immune for some time, afterward he possibly becomes susceptible again (in certain cases one can talk of life-long immunity). The model can be extended by adding vaccination. Vaccinees (V) are protected from infection for some time, usually shorter than naturally infected hosts.

From the in-host point of view, immunity to a pathogen is the result of either active or passive immunization. The latter is a transient protection due to the transmission of antibodies from the mother to the fetus through the placenta. The newborn is thus immune for several months after birth[26]. Active immunization is either induced by natural infection or can be achieved by vaccine administration[35,15].

Let us first consider the case of natural infection. A susceptible host, also called *naive host*, has a very low level of specific immune cells for a pathogen (mostly a virus or a bacterium). The first response to a pathogen is nonspecific, as the innate immune system cannot recognize the physical structure of the pathogen. The innate immune response slows down the initial growth of the pathogen, while the adaptive (pathogen-specific) immune response is activated. Clonal expansion of specific immune cells (mostly antibodies or CTL cells) and pathogen clearance follow. The population of pathogen-specific immune cells is maintained for long time at a level that is much higher than in a naive host. These are the so-called *memory cells* and are activated in case of secondary infection (see Figure 1, adapted from Barbarossa and Röst[6]). Memory cells rapidly activate the immune response and the host mostly shows mild or no symptoms[2].

Each exposure to the pathogen might have a boosting effect on the population of specific memory cells. Indeed, the immune system reacts to a new exposure as it did during primary infection, thus yielding an increased level of memory cells. Though persisting for long time after pathogen clearance, the memory cell population slowly decays and in the long run the host might lose his pathogen-specific immunity[37].

Vaccine-induced immunity works in a similar way as immunity induced by the natural infection. Agents contained in vaccines resemble, in a weaker form, the disease-causing pathogen and force a specific immune reaction without leading to the disease. If the vaccine is successful, the host is immunized for some time. Vaccinees experience immune system boosting and waning immunity, just as hosts recovered from natural infection do. In general, however, disease-induced immunity induces a much longer lasting protection than vaccine-induced immunity does[35].

Waning immunity might be one of the factors which cause, also in highly developed regions, recurrent outbreaks of infectious diseases such as measles, chickenpox and pertussis. On the other side, immune system boosting due to contact with infectives prolongs the protection duration. In

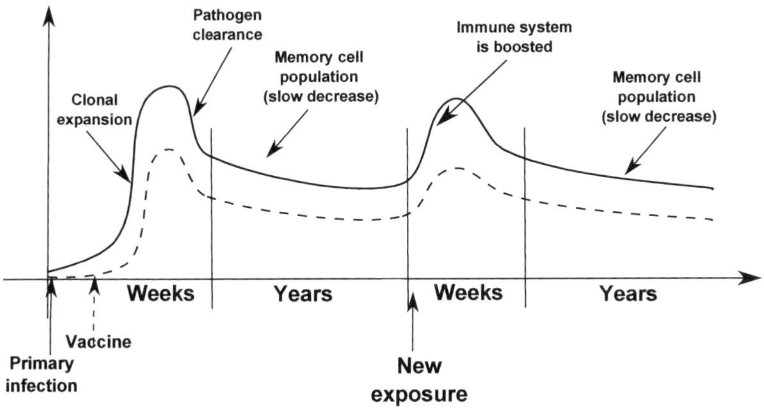

Figure 1. Level of pathogen-specific immune cells with respect to the time. The solid line represents the case of natural infection, the dotted line represents the immune status of a vaccinated host. Generation of memory cells takes a few weeks: once primary infection (respectively, vaccination) occurred, the adaptive immune system produces a high number of specific immune cells (clonal expansion). After pathogen clearance, specific immune cells (memory cells) are maintained for years at a level that is much higher than in a naive host. Memory cells are activated in case of secondary infection.

a highly vaccinated population there are a lot of individuals with vaccine-induced immunity and few infection cases, as well as many individuals with low level of immunity. In other words, if a high portion of the population gets the vaccine, there are very few chances for exposure to the pathogen and consequently for immune system boosting in protected individuals.

In order to understand the role played by waning immunity and immune system boosting in epidemic outbreaks, in the recent past several mathematical models were proposed. Few of these models describe only in-host processes during and after the infection[37,16]. Many more models, formulated in terms of ordinary differential equations (ODEs), consider the problem only at population level, defining compartments for individuals with different levels of immunity and introducing transitions between these compartments[10,17]. Vaccinated hosts or newborns with passive immunity are often included in the model equations and waning of vaccine-induced or passive immunity are observed[33,31,11,3,22,5,30].

To describe the sole waning immunity process, authors have sometimes chosen delay differential equation (DDE) models with constant or distributed delays[21,36,8,7,38]. The delay represents the average duration of the disease-induced immunity. However, neither a constant nor a distributed delay allows for the description of immune system boosting.

Models which include partial differential equations (PDEs) mostly describe an age-structured population[27,20,33] and consider pathogen transmission among the different age groups (newborns, children, pupils, adults, ...). Rare examples suggest a physiologically structured approach with populations structured by the level of immunity, coupling within-host and between-hosts dynamics[25,6].

The goal of the present book chapter is twofold. On the one side, we found necessary to provide a comprehensive overview of previously published models for waning of disease/vaccine-induced immunity and immune system boosting (Sect. 2). On the other side, in Sect. 3 we propose a new modeling framework for SIRVS dynamics, monitoring the immune status of individuals and including both waning immunity and immune system boosting.

2. Mathematical Models for Waning Immunity and Immune System Boosting

In the following we provide an overview on previous mathematical models for waning immunity and immune system boosting. We shall classify these models according to their mathematical structure (systems of ODEs, PDEs or DDEs).

2.1. *Systems of ODEs*

Mossong and coauthors were among the first to suggest the inclusion of individuals with waning immunity in classical SIRS systems[31]. Motivated by the observation that measles epidemics can occur even in highly vaccinated populations, the authors set up a model to study the waning of vaccine-induced immunity and failure of seroconversion as possible causes for recurrent outbreaks. Their compartmental model includes hosts with the so-called "vaccine-modified measles infection" (VMMI) which can occur in people with some degree of passive immunity to the virus, including those previously vaccinated. Assuming that not all vaccinees are protected from developing VMMI, the authors classify vaccinees into three groups: immediately susceptible to VMMI (weak response), temporarily protected who become susceptible to VMMI due to waning of vaccine-induced immunity (intermediate response), and permanently protected from VMMI (strong response). Infection occurs due to contact with infectious individuals (both regular measles infection and VMMI). The resulting compartmental model includes waning of vaccine-induced immunity but not of disease-induced immunity, nor immune system boosting. Similar to McLean and Blower[28], Mossong et al. define a parameter ϕ to describe the impact of the vaccine: if $\phi < 1$, then vaccine failure is possible. Analytical results by Mossong and coauthors[31] show that the main effect of VMMI is to increase the overall reproduction number of the infection.

Inspired by Mossong's work, in 2003-2004 Glass, Grenfell and coauthors[14,13,12] proposed modifications and extensions of the system by Mossong et al.[31]. The basic model is similar to the ODE system by Mossong et al.[31], with a group of subclinical cases which carry the pathogen without showing symptoms[11]. In addition, the distribution of antibody levels in immune hosts (included in the ODEs coefficients) and immune system boosting are introduced: the average antibody level in an immune host increases due to contact with infective or subclinical hosts. This model was used to fit measles data in England[12]. Further, Grenfell and coauthors extend the basic model to consider measles transmission in a meta-population with N patches[13].

Immune system boosting in vaccinees was further studied by Arinaminpathy et al.[3]. The authors introduce two models. In the first one vaccinees are separated from non-vaccinated hosts. Both groups of individuals are classified into susceptible, infective and immune, but in contrast to the models by Glass, Grenfell, Mossong et al.[14,13,12,31], there is no compartment for subclinical cases. Non-vaccinated hosts do not undergo immune system boosting. For vaccinated hosts the authors include a so called "self-boosting" of vaccine, so that contact with infectives moves susceptible vaccinees to the immune vaccinated compartment. The second model extends the first one with a new compartment for hosts with waning immunity (W). These can receive immune system boosting due to contact with infectives or move back to the susceptible compartment due to immunity loss. Numerical simulations show possible sustained oscillations. The SIRWS system was partially analyzed by Dafilis et al.[10].

Heffernan and Keeling[16] proposed an in-host model to understand the behavior of the immune system during and after an infection. Activation of immune system effectors and production of memory cells depend on the virus load. When not stimulated by the virus, the number of activated cells decays (waning immunity). Vaccination is simulated by changing the initial conditions for the virus load. Numerical simulations show that the number of infected immune system cells in a vaccinated patient reaches approximately half of what is reached in a patient who undergoes natural infection. In turn, the level of immunity gained after one dose of vaccine is the same as the level observed in a measles patient 4 years after natural infection. The in-host model by Heffernan and Keeling[16] was extended by the same authors to a population model (SEIRS) with waning immunity and immune system boosting[17]. In contrast to classical SEIRS models, the class R refers here to individuals protected by short-term immune memory, while the class S refers to those individuals who have lost this short-term protection and may experience immune system boosting. Each compartment is classified according to the level of immunity, which can be related to the number of memory cells. Newborns are recruited into the susceptible class S_0 (lowest level of immunity). During exposure and infection the host does not change his level of immunity, that is, transition occurs from S_j to E_j to I_j for each $j \in \mathbb{N}$. Hosts in S and R experience waning immunity and transit from S_j to S_{j-1} (respectively from R_j to R_{j-1}). Immune system boosting is due to recovery from infection and is incorporated into the equations with transition terms from I_j to R_k, with $k \geq j$. The resulting large system of ODEs, with a very high number of parameters, is quite hard to approach from an analytical point of view, hence the authors make use of numerical simulations to investigate the long term behavior. A somehow simplified version of the ODE system by Heffernan and Keeling[17] was proposed by Reluga et al.[32]. A similar large system of ODEs was introduced by Lavine et al.[22], extending the SIRWS model by Mossong, Glass and Grenfell[31,11], by including several levels of immunity for immune hosts (R) and hosts with waning immunity (W), as well as age classes for all compartments. The authors claim that the model can explain several observed features of pertussis in US, in particular a shift in the age-specific incidence and the re-emergence of the disease in a highly vaccinated population.

2.2. *System of DDEs*

Delay models with constant or distributed delay have been introduced to describe waning of disease-induced or vaccine-induced immunity. A simple SIRS system with constant delay is given by

$$\begin{aligned}
\dot{S}(t) &= \mu(1 - S(t)) - \phi S(t) f(I(t)) + \gamma I(t-\tau) e^{-\mu\tau} \\
\dot{I}(t) &= \phi S(t) f(I(t)) - (\mu + \gamma) I(t) \\
\dot{R}(t) &= \gamma I(t) - \mu R(t) - \gamma I(t-\tau) e^{-\mu\tau}.
\end{aligned} \tag{1}$$

This model was studied by Kyrychko and Blyuss[21], who provided results on existence, uniqueness and non-negativity of solutions, linear and global stability of the disease-free equilibrium, as well as global stability of the unique endemic equilibrium. A special case of (1) was considered some years later by Taylor and Carr[36]. An extension of system (1) with distributed delay was proposed by Kyrychko and Blyuss[8] and shortly after by Bhattacharya and Adler[7].

A more general model with distributed delay and vaccination was proposed by Arino et al.[4]. Their system includes three compartments (susceptible, infective and vaccinated hosts) in a population which remains constant in time. Vaccine-induced immunity might be only partial, resulting in vaccinated individuals becoming infective. Systems of ODEs or DDEs can be obtained from the general model by a proper choice of the kernel (see also Hethcote et al.[19,18]).

Recently, Yuan and Bélair proposed a SEIRS model with integro-differential equations which resembles the systems by Arino, Hethcote et al.[4,19]. The probability that an individual stays in the exposed class (E) for t units of time is $P(t)$, hence,

$$E(t) = \int_0^t \beta \frac{S(u)I(u)}{N} e^{-b(t-u)} P(t-u)\, du.$$

Similarly, $Q(t)$ is the probability that an individual is immune t units of time after recovery, thus

$$R(t) = \int_0^t \gamma I(u) e^{-b(t-u)} Q(t-u)\, du.$$

For a certain choice of the probabilities P and Q, the problem can be reduced to a system with one or two constant delays. The authors show existence of an endemic equilibrium and boundedness of solutions in a positive simplex. For the system with one constant delay, results for existence of a global attractor as well as the proof of persistence of the disease in case $R_0 > 1$ are provided.

2.3. *Systems of PDEs*

Structured populations in the context of waning immunity and immune system boosting have been motivated in different ways. Often the structure can be found in the biological age[27,26,20,33], and is used to observe disease transmission among babies, children, adults and seniors. Only few works suggest models for physiologically structured populations[25,6].

McLean and Anderson[27,26] proposed a model for measles transmission which includes a compartment for babies protected by maternal antibodies. Indeed, mothers who have had measles or have been vaccinated transfer measles immunity to the baby through the placenta. For several months after birth (ca. 2 months if the mother was vaccinated, ca. 4 months if she had the disease[26]) the baby is still protected by maternal antibodies and should not be vaccinated. The model by McLean and Anderson[27] considers only waning of maternally induced immunity in the context of measles infection. Few years before McLean, Katzmann and Dietz[20] proposed a bit more general model, which includes also waning of vaccine-induced immunity. In both cases, the age structure was used to determine the optimal age for vaccination. A compartment for adult hosts with waning immunity who can also receive immune system boosting was introduced only years later by Rouderfer et al.[33]. A further deterministic system of ODEs for maternally induced immunity in measles was proposed by Moghadas et al.[29].

Different is the approach when physiologically structured populations are considered. Martcheva and Pilyugin[25] suggest an SIRS model in which infective and recovered hosts are structured by their immune status. In infective hosts the immune status increases over the course of infection, while in recovered hosts the immune status decays at some non-constant rate. When the immune status has reached a critical level, recovered hosts transit from the immune to the susceptible compartment.

A general framework for SIRS systems, modeling waning immunity and immune system boosting, and combining the in-host perspective with the population dynamics, was proposed by Barbarossa and Röst[6].

3. A General Modeling Framework

In this section we extend the model by Barbarossa and Röst[6] to include vaccine-induced immunity. As in the works by Martcheva and Pilyugin, Barbarossa and Röst[25,6], we couple the in-host with the between-hosts dynamics, focusing on the effects of waning immunity and immune system boosting on the population dynamics. In contrast to the models proposed by Heffernan and Keeling, Lavine et al.[17,22], we shall maintain the number of equations as low as possible. The resulting model (V1) is a system of ODEs coupled with two PDEs. The ODE systems by Mossong, Arino, Glass, Grenfell et al.[31,11,3,5,30], as well as extensions of the DDEs systems in the works of Talyor and Carr, Yuan and Bélair [36,38], can be recovered from our modeling framework.

Setting up our model we do not restrict ourselves to a particular pathogen. The model (V1) can be adapted to several epidemic outbreaks (e.g. measles, chickenpox, rubella, pertussis) by ad-hoc estimating coefficients from available experimental data[24,1,23].

3.1. *Model ingredients*

3.1.1. *Originally susceptible and infectives hosts*

Let $S(t)$ denote the total population of *originally susceptible* hosts. These are susceptible individuals which have neither received vaccination nor have been infected before. Newborns enter the susceptible population at rate $b(N)$, dependent on the total population size N. For simplicity we assume that the natural death rate $d > 0$ does not depend on N. Assume that $b : [0, \infty) \to [0, b_+]$, $N \mapsto b(N)$, with $0 < b_+ < \infty$, is a nonnegative function, with $b(0) = 0$. Finally, assume that in absence of disease-induced death there exists an equilibrium N^* such that $b(N^*) = d\,N^*$.

Let $I(t)$ denote the total infective population at time t. Infection of susceptible individuals occurs by contact, at rate $\beta I/N$. Infected hosts recover at rate $\gamma > 0$. When we include disease-induced death at rate $d_I > 0$, the equilibrium N^* satisfies

$$b(N^*) = d\,N^* + d_I I^*.$$

3.1.2. *Immune individuals*

Let us denote by $r(t, z)$ the density of recovered individuals with disease-induced immunity level $z \in [z_{min}, z_{max}]$ at time t. The total population of recovered hosts is given by

$$R(t) = \int_{z_{min}}^{z_{max}} r(t, z)\, dz.$$

The parameter z describes the immune status and can be related to the number of specific immune cells of the host. The value z_{max} corresponds to maximal immunity, whereas z_{min} corresponds to low level of immunity. Individuals who recover at time t enter the immune compartment with maximal level of immunity z_{max}. The level of immunity tends to decay in time and when it reaches the minimal value z_{min}, the host becomes susceptible again. However, exposure to the pathogen can boost the immune system from $z \in [z_{min}, z_{max}]$ to any higher status. It is not straightforward to determine how this kind of immune system boosting works, as no experimental data are available. Nevertheless, laboratory analysis on vaccines tested on animals or humans suggest that the boosting efficacy might depend on several factors, among which the current immune status of the recovered host and the amount of pathogen he receives[1,24]. Possibly, exposure to the pathogen can restore the maximal level of immunity, just as natural infection does[6].

Let $p(z, \tilde{z})$, $z \geq \tilde{z}$, $z, \tilde{z} \in \mathbb{R}$ denote the probability that an individual with immunity level $\tilde{z}$ moves to immunity level z, when exposed to the pathogen. Due to the definition of $p(z, \tilde{z})$, we have $p(z, \tilde{z}) \in [0, 1]$, $z \geq \tilde{z}$ and

$$p(z, \tilde{z}) = 0, \quad \text{for all} \quad z < \tilde{z}.$$

As we effectively consider only immunity levels in the interval $[z_{min}, z_{max}]$, we set

$$p(z, \tilde{z}) = 0, \quad \text{for all} \quad \tilde{z} \in (-\infty, z_{min}) \cup (z_{max}, \infty).$$

Then we have

$$\int_{-\infty}^{\infty} p(z, \tilde{z})\, dz = \int_{\tilde{z}}^{z_{max}} p(z, \tilde{z})\, dz = 1, \quad \text{for all} \quad \tilde{z} \in [z_{min}, z_{max}].$$

Exposure to the pathogen might restore exactly the immunity level induced by the disease (z_{max}). In order to capture this particular aspect of immune system boosting, we write the probability $p(z, \tilde{z})$ as the combination of a continuous (p_0) and atomic measures (Dirac delta):

$$p(z, \tilde{z}) = c_{max}(\tilde{z})\delta(z_{max} - \tilde{z}) + c_0(\tilde{z})p_0(z, \tilde{z}) + c_1(\tilde{z})\delta(z - \tilde{z}),$$

where

- $c_{max} : [z_{min}, z_{max}] \to [0, 1]$, $y \mapsto c_{max}(y)$, is a continuously differentiable function and describes the probability that, due to contact with infectives, a host with immunity level y boosts to the maximal level of immunity z_{max}.
- $c_0 : [z_{min}, z_{max}] \to [0, 1]$, $y \mapsto c_0(y)$, is a continuously differentiable function and describes the probability that, due to contact with infectives, a host with immunity level y boosts to any other level $z \in (y, z_{max})$, according to the continuous probability $p_0(z, y)$.
- $c_1(y) = 1 - c_{max}(y) - c_0(y)$ describes the probability that getting in contact with infectives, the host with immunity level $y \in [z_{min}, z_{max}]$ does not experience immune system boosting.

The immunity level decays in time at some rate $g(z)$ which is the same for all recovered individuals with immunity level z. In other words, the immunity level z follows

$$\frac{d}{dt}z(t) = g(z),$$

with $g : [z_{min}, z_{max}] \to (0, K_g]$, $K_g < \infty$ continuously differentiable. The positivity of $g(z)$ is required from the biological motivation. Indeed, if $g(\tilde{z}) = 0$ for some value $\tilde{z} \in [z_{min}, z_{max}]$, there would be no change of the immunity level at $\tilde{z}$, contradicting the hypothesis of natural decay of immune status. In absence of immune system boosting, we have that

$$\int_{z_{min}}^{z_{max}} \frac{1}{g(x)}\, dx$$

is the time a recovered host remains immune (see Barbarossa and Röst[6]).

3.1.3. *Vaccination*

We structure the vaccinated population by the level of immunity as well. Let $v(t, z)$ be the density of vaccinees with immunity level $z \in [z_{min}, z_{max}]$ at time t. The total population of vaccinated hosts is given by

$$V(t) = \int_{z_{min}}^{z_{max}} v(t, z)\, dz.$$

Vaccination infers a level of immunity z_{vax}, which is lower than the level of immunity after natural infection: $z_{max} > z_{vax} > z_{min}$[35]. As in recovered individuals, the level of immunity of a vaccinated host tends to decay in time and when it reaches the minimal value z_{min}, the host becomes susceptible again. However, also in vaccinated hosts, exposure to the pathogen can boost the immunity level $z \in [z_{min}, z_{vax}]$ to any higher value in $[z_{min}, z_{max}]$. Immune system boosting is described by the probability $p(z, \tilde{z})$, as in recovered hosts. We consider the possibility that exposure to the pathogen boosts the immune system of a vaccinated individual to $z \in (z_{vax}, z_{max}]$. Vaccinated hosts with $z \in (z_{vax}, z_{max}]$ have an immune status which can be compared to the one of hosts who recovered from natural infection.

It is reasonable to assume that in vaccinated individuals the immunity level decays in time at the same rate g, as in hosts who underwent natural infection. In absence of exposure to the pathogen (hence in absence of immune system boosting), the time that a vaccinee remains immune is shorter than the time a recovered host does:

$$\int_{z_{min}}^{z_{vax}} \frac{1}{g(x)}\,dx < \int_{z_{min}}^{z_{max}} \frac{1}{g(x)}\,dx.$$

Let us define the vaccination rate at birth $\alpha > 0$. We assume that originally susceptible (adult) individuals get vaccinated at rate $\phi \geq 0$.

3.1.4. *Becoming susceptible again*

In absence of immune system boosting both disease-induced and vaccine-induced immunity fade away. Individuals who lose immunity either after recovery from infection or after vaccination, enter the class S_2 of susceptible individuals who shall not get a new dose of vaccine. A host who had the disease or got vaccination relies indeed on the induced-immunity and is not aware of the fact that his level of immunity might have dropped below the critical immunity threshold.

We denote by $S_2(t)$ the population at time t of susceptible hosts who are not going to receive vaccination.

3.2. *Model equations*

In view of all what we have mentioned above, we can easily write down the equations for the compartments S, I and S_2. Let initial values $S(0) = S^0 \geq 0$, $I(0) = I^0 \geq 0$ and $S_2(0) = S_2^0 \geq 0$ be given. The population of originally susceptible individuals is governed by

$$\dot{S}(t) = \underbrace{b(N(t))(1-\alpha)}_{\text{birth}} - \underbrace{\phi S(t)}_{\text{vaccination}} - \underbrace{\beta\frac{S(t)I(t)}{N(t)}}_{\text{infection}} - \underbrace{dS(t)}_{\text{death}}, \tag{2}$$

whereas hosts who become susceptible due to immunity loss follow

$$\dot{S}_2(t) = -\underbrace{\beta\frac{S_2(t)I(t)}{N(t)}}_{\text{infection}} - \underbrace{dS_2(t)}_{\text{death}} + \underbrace{\Lambda_R}_{\substack{\text{immunity loss}\\ \text{after recovery}}} + \underbrace{\Lambda_V}_{\substack{\text{immunity loss}\\ \text{after vaccination}}}.$$

The term Λ_R (respectively Λ_V), which represents transitions from the immune (respectively, the vaccinated) compartment to the susceptible one, will be specified below together with the dynamics of the recovered (respectively, vaccinated) population.

Both kinds of susceptible hosts can become infective due to contact with infective hosts:

$$\dot{I}(t) = \underbrace{\beta\frac{S(t)I(t)}{N(t)}}_{\text{infection of } S} + \underbrace{\beta\frac{S_2(t)I(t)}{N(t)}}_{\text{infection of } S_2} - \underbrace{\gamma I(t)}_{\text{recovery}} - \underbrace{dI(t)}_{\substack{\text{natural}\\\text{death}}} - \underbrace{d_I I(t)}_{\substack{\text{disease-induced}\\\text{death}}}. \tag{3}$$

To obtain an equation for the recovered individuals, structured by their levels of immunity, one can proceed similarly to size structured models or as it was done for the immune population by Barbarossa and Röst[6]. The result is the following PDE. Let a nonnegative initial distribution $r(0,z) = \psi(z)$, $z \in [z_{min}, z_{max}]$ be given. For $t > 0, z \in [z_{min}, z_{max}]$ we have

$$\begin{aligned} \frac{\partial}{\partial t}r(t,z) - \frac{\partial}{\partial z}\left(g(z)r(t,z)\right) = -dr(t,z) + \beta\frac{I(t)}{N(t)}\int_{z_{min}}^{z} p(z,x)r(t,x)\,dx \\ - r(t,z)\beta\frac{I(t)}{N(t)}, \end{aligned} \tag{4}$$

with the boundary condition

$$g(z_{max})r(t,z_{max}) = \gamma I(t) + \beta\frac{I(t)}{N(t)}\int_{z_{min}}^{z_{max}} p(z_{max},x)r(t,x)\,dx. \tag{5}$$

Equation (4) expresses the rate of change in the density of recovered individuals according to immune level due to natural waning, mortality, and boosting. The boundary condition (5) includes newly recovered individuals as well as those recovered individuals, who just received a boost which elevated their immune system to maximal level.

Next we shall consider the vaccinated population. Again, by structuring this group according to immunity level, one has the PDE

$$\begin{aligned} \frac{\partial}{\partial t}v(t,z) = \frac{\partial}{\partial z}\left(g(z)v(t,z)\right) - dv(t,z) + \beta\frac{I(t)}{N(t)}\int_{z_{min}}^{z} p(z,x)v(t,x)\,dx \\ - v(t,z)\beta\frac{I(t)}{N(t)} + \delta(z - z_{vax})\left(\phi S(t) + \alpha b(N(t))\right), \end{aligned} \tag{6}$$

and

$$g(z_{max})v(t,z_{max}) = \beta\frac{I(t)}{N(t)}\int_{z_{min}}^{z_{max}} p(z_{max},x)v(t,x)\,dx, \tag{7}$$

provided with a nonnegative initial distribution $v(0,z) = \psi_v(z)$, $z \in [z_{min}, z_{max}]$. Observe that newly vaccinated hosts do not enter the vaccinated population at z_{max}, but at the lower value z_{vax}, which is expressed in equation (6) as an impulse at $z = z_{vax}$ by the term with the Dirac delta $\delta(z - z_{vax})$.

It becomes evident that the quantity Λ_R, initially introduced in the S_2 equation to represent the number of hosts who experienced immunity loss, is given by the number $g(z_{min})r(t, z_{min})$ of immune hosts who reached the minimal level of immunity after recovery from natural infection. Similarly, Λ_V is the number $g(z_{min})v(t, z_{min})$ of vaccinated hosts who reached the minimal level of immunity. Hence we have

$$\dot{S}_2(t) = -\underbrace{\beta\frac{S_2(t)I(t)}{N(t)}}_{\text{infection}} - \underbrace{dS_2(t)}_{\text{death}} + \underbrace{g(z_{min})r(t, z_{min})}_{\Lambda_R} + \underbrace{g(z_{min})v(t, z_{min})}_{\Lambda_V}. \tag{8}$$

In the following we refer to the complete system (2) – (8) as to **model (V1)**.

4. Connection to Other Mathematical Models

4.1. *Connection to ODE models*

As it was shown by Barbarossa and Röst[6] for a simpler problem, model (V1) can be reduced to a system of ODEs analogous to those proposed by several authors[31,11,3,22,17,30]. The connection between model (V1) and the ODE system is given by the *method of lines*, a technique in which all but one dimensions are discretized[34]. In our case, we shall discretize the immunity level (z) and obtain a system of ODEs in the time variable.

Let us define a sequence $\{z_j\}_{j \in \mathbb{N}}$, with $h_j := z_{j+1} - z_j > 0$, for all $j \in \mathbb{N}$. To keep the demonstration as simple as possible, we choose a grid with only a few points, $z_1 := z_{min} < z_W := z_{vax} < z_F < z_{max}$ and for simplicity (or possibly after a rescaling) assume that $h_j = 1$ for all j. We define the following subclasses of the immune/vaccinated population:

- $R_F(t) := r(t, z_F)$, immune hosts with high level of immunity at time t. As their immunity level is quite high, these individuals do not experience immune system boosting. Immunity level decays at rate $\mu := g(z_F) > 0$.

- $R_W(t) := r(t, z_W)$, immune hosts with intermediate level of immunity at time t. These individuals can get immune system boosting and move to R_F. Immunity level decays at rate $\nu := g(z_W) > 0$.

- $R_C(t) := r(t, z_{min})$, immune hosts with critically low level of immunity at time t. With probability θ boosting moves R_C individuals to R_W (respectively, with probability $(1-\theta)$ to R_F). Immunity level decays at rate $\sigma := g(z_{min}) > 0$. If they do not get immune system boosting, these hosts move to the class S_2 (become susceptible again).

- $V_R(t) := v(t, z_F)$, vaccinated hosts who thanks to immune system boosting gained a very high level of immunity at time t. These individuals do not experience immune system boosting. Immunity level decays at rate μ.

- $V_0(t) := v(t, z_W)$, vaccinated individuals at time t with maximal vaccine-induced immunity. This class includes new vaccinees. If their immune system gets boosted hosts move to V_R. Immunity level decays at rate ν.

- $V_C(t) := v(t, z_{min})$, vaccinees with critically low level of immunity at time t. With probability ξ boosting moves V_C hosts to V_0 and with probability $(1-\xi)$ to V_R. Immunity level decays at rate σ. If they do not receive immune system boosting, V_C hosts move to S_2.

To show how the PDE system can be reduced to a system of ODEs by means of the method of lines, we consider a simple example. Let us neglect immune system boosting for a moment. Then the PDE for $r(t,z)$ in model (V1) becomes

$$\frac{\partial}{\partial t} r(t,z) = \frac{\partial}{\partial z}\big(g(z) r(t,z)\big) - d r(t,z), \qquad z \in [z_{min}, z_{max}], \tag{9}$$

with boundary condition $R_{z_{max}}(t) := r(t, z_{max}) = \gamma I(t)/g(z_{max})$. Using forward approximation for the z-derivative in (9), we obtain, e.g., for $R_F(t)$

the following differential equation:

$$\begin{aligned}\dot{R}_F(t) &= \frac{\partial}{\partial t} r(t, z_F) \\ &= \frac{\partial}{\partial z}\big(g(z_F) r(t, z_F)\big) - dr(t, z_F) \\ &\approx \frac{g(z_{max}) r(t, z_{max}) - g(z_F) r(t, z_F)}{\underbrace{z_{max} - z_F}_{=1}} - dr(t, z_F) \\ &= g(z_{max}) R_{z_{max}}(t) - \mu R_F(t) - dR_F(t) \\ &= \gamma I(t) - (\mu + d) R_F(t).\end{aligned}$$

Analogously one can find equations for R_W, R_C, V_R, V_0 and V_C. Altogether we obtain a system of ordinary differential equations in which a linear chain of ODEs replaces the PDEs for the immune and the vaccinated class:

$$\begin{aligned}\dot{S}(t) &= (1-\alpha) b(N(t)) - \phi S(t) - \beta \frac{S(t) I(t)}{N(t)} - dS(t) \\ \dot{I}(t) &= \beta \frac{I(t)}{N(t)} (S(t) + S_2(t)) - (\gamma + d + d_I) I(t) \\ \dot{R}_F(t) &= \gamma I(t) - \mu R_F(t) - dR_F(t) \\ \dot{R}_W(t) &= \mu R_F(t) - \nu R_W(t) - dR_W(t) \\ \dot{R}_C(t) &= \nu R_W(t) - \sigma R_C(t) - dR_C(t) \\ \dot{V}_R(t) &= -\mu V_R(t) - dV_R(t) \\ \dot{V}_0(t) &= \phi S(t) + \alpha b(N(t)) + \mu V_R(t) - \nu V_0(t) - dV_0(t) \\ \dot{V}_C(t) &= \nu V_0(t) - \sigma V_C(t) - dV_C(t) \\ \dot{S}_2(t) &= -\beta \frac{S_2(t) I(t)}{N(t)} - dS_2(t) + \sigma(R_C(t) + V_C(t)).\end{aligned}$$

The method of lines can be applied to the full model (V1) as well[6]. To this purpose it is necessary to discretize the boosting probability $p(z, \tilde{z})$ (this is expressed by the parameters ξ and θ below). Incorporating the boosting effect, the result is the following system of ODEs.

$$
\begin{aligned}
\dot{S}(t) &= (1-\alpha)b(N(t)) - \phi S(t) - \beta\frac{S(t)I(t)}{N(t)} - dS(t) \\
\dot{I}(t) &= \beta\frac{I(t)}{N(t)}\left(S(t)+S_2(t)\right) - (\gamma + d + d_I)I(t) \\
\dot{R}_F(t) &= \gamma I(t) - \mu R_F(t) - dR_F(t) + \beta\frac{I(t)}{N(t)}\left(R_W(t) + (1-\theta)R_C(t)\right) \\
\dot{R}_W(t) &= \mu R_F(t) - \nu R_W(t) - dR_W(t) + \beta\frac{I(t)}{N(t)}(\theta R_C(t) - R_W(t)) \\
\dot{R}_C(t) &= \nu R_W(t) - \sigma R_C(t) - dR_C(t) - \beta\frac{I(t)}{N(t)}R_C(t) \\
\dot{V}_R(t) &= \beta\frac{I(t)}{N(t)}\left(V_0(t) + (1-\xi)V_C(t)\right) - \mu V_R(t) - dV_R(t) \\
\dot{V}_0(t) &= \phi S(t) + \alpha b(N(t)) + \mu V_R(t) - \nu V_0(t) - dV_0(t) \\
&\quad + \beta\frac{I(t)}{N(t)}\left(\xi V_C(t) - V_0(t)\right) \\
\dot{V}_C(t) &= \nu V_0(t) - \sigma V_C(t) - dV_C(t) - \beta\frac{V_C(t)I(t)}{N(t)} \\
\dot{S}_2(t) &= -\beta\frac{S_2(t)I(t)}{N(t)} - dS_2(t) + \sigma(R_C(t) + V_C(t)).
\end{aligned}
$$

The linear chain of ODEs provides a rough approximation of the PDEs in model (V1). Indeed, with the method of lines we approximate the PDE dynamics considering only changes at the grid points (z_{min}, z_W, z_F), whereas the dynamics remains unchanged in each immunity interval $[z_j, z_{j+1}]$. We consider as representative point of the interval the lowest boundary z_j - for this reason we do not have a differential equation for $R_{z_{max}}(t)$ or $V_{z_{max}}(t)$.

4.2. *Connection to DDE models*

Delay models with constant delay can be recovered from special cases of model (V1). We show here how to obtain the classical SIRS model with delay studied by Taylor and Carr[36], or extensions thereof.

In the following we neglect boosting effects and vaccination. Further we do not distinguish between originally susceptibles and host who have lost immunity, hence w.r.t. model (V1) we identify the classes S and S_2. From our assumptions, the disease-induced immunity lasts for a fix time, $\tau > 0$

years, given by

$$\int_{z_{min}}^{z_{max}} \frac{1}{g(x)}\, dx = \tau.$$

We can express the total immune population at time t as the number of individuals who recovered in the time interval $[t-\tau, t]$,

$$R(t) = \gamma \int_{t-\tau}^{t} I(y) e^{-d(t-y)}\, dy = \gamma \int_{0}^{\tau} I(t-x) e^{-dx}\, dx.$$

Differentiation with respect to t yields

$$\dot{R}(t) = \gamma I(t) - \gamma I(t-\tau) e^{-d\tau} - dR(t). \tag{10}$$

On the other side, we have the definition in terms of distribution of immune individuals,

$$R(t) = \int_{z_{min}}^{z_{max}} r(t,z)\, dz.$$

Differentiate the last relation and compare with (10):

$$g(z_{max}) r(t, z_{max}) = \gamma I(t), \qquad g(z_{min}) r(t, z_{min}) = \gamma I(t-\tau) e^{-d\tau}.$$

This means that individuals with maximal level of immunity are those who recover from infection. If a host who recovers at time t_1 survives up to time $t_1 + \tau$, he exits the R class and enter S. In turn, we find a delay term in the equation for S too, and have a classical SIRS model with constant delay

$$\begin{aligned}
\dot{S}(t) &= b(N(t)) - \beta \frac{S(t)I(t)}{N(t)} - dS(t) + \gamma I(t-\tau) e^{-d\tau} \\
\dot{I}(t) &= \beta \frac{S(t)I(t)}{N(t)} - (\gamma + d + d_I) I(t) \\
\dot{R}(t) &= \gamma I(t) - \gamma I(t-\tau) e^{-d\tau} - dR(t),
\end{aligned}$$

which was studied by Taylor and Carr[36].

Now we can include again vaccination and the class S_2 as in the general model (V1). We assume that vaccine-induced immunity lasts for a time $\tau_v > 0$,

$$\tau_v := \int_{z_{min}}^{z_{vax}} \frac{1}{g(x)}\, dx \; < \; \int_{z_{min}}^{z_{max}} \frac{1}{g(x)}\, dx =: \tau.$$

With similar arguments as for the immune population, we obtain the relations

$$g(z_{vax})v(t, z_{vax}) = \alpha b(N(t)) + \phi S(t),$$
$$g(z_{min})v(t, z_{min}) = (\alpha b(N(t-\tau_v)) + \phi S(t-\tau_v))\, e^{-d\tau_v},$$

and find a system with two constant delays

$$\begin{aligned}
\dot{S}(t) &= (1-\alpha)b(N(t)) - \phi S(t) - \beta\frac{S(t)I(t)}{N(t)} - dS(t)\\
\dot{I}(t) &= \beta\frac{I(t)}{N(t)}(S(t) + S_2(t)) - (\gamma + d + d_I)I(t)\\
\dot{R}(t) &= \gamma I(t) - \gamma I(t-\tau)e^{-d\tau} - dR(t)\\
\dot{V}(t) &= \alpha b(N(t)) + \phi S(t) - (\alpha b(N(t-\tau_v)) + \phi S(t-\tau_v))\, e^{-d\tau_v} - dV(t)\\
\dot{S}_2(t) &= -\beta\frac{S_2(t)I(t)}{N(t)} - dS_2(t) + \gamma I(t-\tau)e^{-d\tau}\\
&\quad + (\alpha b(N(t-\tau_v)) + \phi S(t-\tau_v))\, e^{-d\tau_v}.
\end{aligned}$$

Acknowledgments

Authors were supported by the ERC Starting Grant No 259559. MVB was supported by the European Union and the State of Hungary, co-financed by the European Social Fund in the framework of TÁMOP-4.2.4. A/2-11-1-2012-0001 National Excellence Program. GR was supported by Hungarian Scientific Research Fund OTKA K109782 and TÁMOP-4.2.2.A-11/1/KONV-2012-0073 "Telemedicine focused research activities on the field of Mathematics, Informatics and Medical sciences".

References

1. I. J. Amanna, N. E. Carlson and M. K. Slifka, Duration of humoral immunity to common viral and vaccine antigens, *New Eng. J. Med.* **357**(19), 1903–1915 (2007)
2. R. Antia, V. V. Ganusov and R. Ahmed, The role of models in understanding CD8+ T-cell memory, *Nat. Rev. Immunol.* **5**(2), 101–111 (2005)
3. N. Arinaminpathy, J. S. Lavine and B.T. Grenfell, Self-boosting vaccines and their implications for herd immunity, *Proc. Natl. Acad. Sci. U. S. A. Early Ed.* **109**(49), 154–159 (2012)
4. J. Arino, K. Cooke, P. van den Driessche and J. Velasco-Hernández, An epidemiology model that includes a leaky vaccine with a general waning function, *Discr. Cont. Dyn. Sys.* **B4**(2), 479–495 (2004)

5. J. Arino and P. van den Driessche, Time delays in epidemic models, In: *Delay differential equations and applications, NATO Science Series* **205**, Springer, New York, 539–578 (2006)
6. M. V. Barbarossa and G. Röst, Immuno-epidemiology of a population structured by immune status: a mathematical study of waning immunity and immune system boosting, *preprint* **arXiv:1411.3195** (2014)
7. S. Bhattacharya and F. R Adler, A time since recovery model with varying rates of loss of immunity, *Bull. Math. Biol.* **74**(12), 2810–2819 (2012)
8. K. B. Blyuss and Y. N. Kyrychko, Stability and bifurcations in an epidemic model with varying immunity period, *Bull. Math. Biol.* **72**(2), 490–505 (2010)
9. F. Brauer and C. Castillo-Chávez, Mathematical models in population biology and epidemiology, *Springer*, New York (2001)
10. M. P. Dafilis, F. Frascoli, J. G. Wood and J. M. McCaw, The influence of increasing life expectancy on the dynamics of SIRS systems with immune boosting, *ANZIAM J.* **54**(1-2), 50–63 (2012)
11. K. Glass and B. Grenfell, Antibody dynamics in childhood diseases: waning and boosting of immunity and the impact of vaccination, *J. Theor. Biol.* **221**(1), 121–131 (2003)
12. K. Glass and B. Grenfell, Waning immunity and subclinical measles infections in England, *Vaccine* **22**(29), 4110–4116 (2004)
13. K. Glass, J. Kappey and B. Grenfell, The effect of heterogeneity in measles vaccination on population immunity, *Epidemiol. Infect.* **132**(4), 675–683 (2004)
14. K. Glass, Y. Xia and B. Grenfell, Interpreting time-series analyses for continuous-time biological models: measles as a case study, *J. Theor. Biol.* **223**(1), 19–25 (2003)
15. R. A. Goldsby, T. J. Kindt, B. A. Osborne and J. Kubi, Immunology 5th ed., *W. H. Freeman and Company*, London (2003)
16. J. M. Heffernan and M. J. Keeling, An in-host model of acute infection: Measles as a case study, *Theor. Popul. Biol.* **73**(1), 134–147 (2008)
17. J. M. Heffernan and M. J. Keeling, Implications of vaccination and waning immunity, *Proc. R. Soc. B: Biol. Scie.* **276**(1664), 2071–2080 (2009)
18. H. W. Hethcote, H. W. Stech and P. van den Driessche, Nonlinear oscillations in epidemic models, *SIAM J. Appl. Math.* **40**(1), 1–9 (1981)
19. H. W. Hethcote and P. van den Driessche, Two SIS epidemiologic models with delays, *J. Math. Biol.* **40**(1), 3–26 (2000)
20. W. Katzmann and K. Dietz, Evaluation of age-specific vaccination strategies, *Theo. Popul. Biol.* **25**(2), 125–137 (1984)
21. Y. N. Kyrychko and K. B. Blyuss, Global properties of a delayed sir model with temporary immunity and nonlinear incidence rate, *Nonlinear Anal. Real* **6**(3), 495–507 (2005)
22. J. S. Lavine, A. A. King and O. N. Bjørnstad, Natural immune boosting in pertussis dynamics and the potential for long-term vaccine failure, *Proc. Nat. Acad. Sci.* **108**(17), 7259–7264 (2011)
23. S. Li, N. Rouphael, S. Duraisingham et al., Molecular signatures of antibody responses derived from a systems biology study of five human vaccines. *Nat.*

Immunol. **15**(2), 195–204 (2013)
24. Z. Luo, H. Shi, H. Zhang et al., Plasmid DNA containing multiple CpG motifs triggers a strong immune response to hepatitis B surface antigen when combined with incomplete freund's adjuvant but not aluminum hydroxide, *Mol. Med. Rep.* **6**(6), 1309–1314 (2012)
25. M. Martcheva and S. S. Pilyugin, An epidemic model structured by host immunity, *J. Biol. Sys.* **14**(02), 185–203 (2006)
26. A. McLean and R. Anderson, Measles in developing countries Part II. The predicted impact of mass vaccination, *Epidemiol Inf.* **100**(3), 419–442 (1988a)
27. A. R. McLean and R. M. Anderson, Measles in developing countries Part I. Epidemiological parameters and patterns, *Epidemiol Inf.* **100**(1), 111–133 (1988b)
28. A. R. McLean and S. M. Blower, Imperfect vaccines and herd immunity to HIV, *Proc. R. Soc. Lond.* **B253**, 9–13 (1993)
29. S. M. Moghadas, M. E. Alexander ME and B. M. Sahai, Waning herd immunity: a challenge for eradication of measles, *Rocky Mountain J Math.* **38**(5), 1587-1607 (2008)
30. J. Mossong and C. P. Muller, Modelling measles re-emergence as a result of waning of immunity in vaccinated populations, *Vaccine* **21**(31), 4597–4603 (2003)
31. J. Mossong, D. J. Nokes, W. J. Edmunds et. al., Modeling the impact of subclinical measles transmission in vaccinated populations with waning immunity, *Am. J. Epidemiol.* **150**(11), 1238–1249 (1999)
32. T. C. Reluga, J. Medlock and A. S. Perelson, Backward bifurcations and multiple equilibria in epidemic models with structured immunity, *J. Theor. Biol.* **252**(1), 155–165 (2008)
33. V. Rouderfer, N. G. Becker and H. W. Hethcote, Waning immunity and its effects on vaccination schedules, *Math. Bioscie.* **124**(1), 59–82 (1994)
34. W. E. Schiesser, The numerical method of lines. Academic Press, San Diego (1991)
35. C. A. Siegrist, Vaccine immunology. In: Plotkin SA, Orenstein WA and Offit PA (eds.) *Vaccines*, Elsevier Inc, Philadelphia, 1736 (2008)
36. M. L. Taylor and T. W. Carr, An sir epidemic model with partial temporary immunity modeled with delay, *J. Math. Biol.* **59**(6), 841–880 (2009)
37. D. Wodarz, Killer cell dynamics: mathematical and computational approaches to immunology, *Springer*, New York (2007)
38. Y. Yuan and J. Bélair, Threshold dynamics in an SEIRS model with latency and temporary immunity, *J. Math. Biol.* **69**(4), 875–904 (2013)

WHAT IS THE OPTIMAL LEVEL OF INFORMATION DISSEMINATION DURING AN EPIDEMIC?*

MAREK LASKOWSKI
Lassonde School of Engineering, York University, Toronto, Ontario M3J 1P3, Canada

PREETI DUBEY
Department of Mathematics, Birla Institute of Technology & Science, Pilani 333031, Rajasthan, India

MURRAY E. ALEXANDER
Department of Physics The University of Winnipeg, R3B 2E9, Canada

SHANNON COLLINSON AND JANE M. HEFFERNAN
Modelling Infection and Immunity Laboratory Centre for Disease Modelling, Department of Mathematics & Statistics, York University, Toronto, M3J 1P3, Canada

SEYED M. MOGHADAS
Agent-Based Modelling Laboratory, York University, Toronto, M3J 1P3, Canada

*This work is supported by NSERC (Canada).

In the event of an emerging disease, a number of factors could influence the course of an epidemic spreading through a society of human beings. Besides preventive and control measures implemented by public health, individual characteristics play an important role in disease transmission and control. These characteristics may relate to the health statuses of the individuals, but also to the perception of risk and severity of infection that could influence behavioural responses and social norms. Such responses may change disease dynamics at both the individual and population levels. Here, we propose three modelling structures, namely: deterministic, stochastic, and agent-based, to encapsulate the effect of external factors (such as information dissemination through media and other sources) on the behavioural changes of susceptible individuals during the course of the epidemic. The modelling frameworks proposed here will be used to explore the optimal level of information dissemination to reduce the burden of disease, and avoid the state of "media-fatigue" that may be caused by disseminating a vast amount of information.

1. Introduction

Human responses, both individual and social, to disease outbreaks play a crucial role in the spread of infection through the population[1]. An example of such a response is the imposition of quarantine, requiring a well co-ordinated public health response. Other forms of behavioural responses to disease awareness include social distancing, wearing of protective masks, and even global measures such as closing of national borders to international traffic – most recently, in regard to the Ebola hemorrhagic fever epidemic in West Africa[2]. Behavioural responses to disease epidemics, individually or socially, may be triggered and encouraged by a number of means and sources in the population, including public health messaging, social networking, and media coverage, all of which aim to increase awareness and may therefore influence behavioural responses and social norms. Behavioural responses to the presence of a disease have the potential to reduce the size of the outbreak or even prevent one, as in the case of the 2003 SARS epidemic[3,4]. Rumours (sometimes called "an infection of mind")[5] are an important form of social communication affecting human behaviour, and their spreading can shape public responses, including fear response to disease outbreaks[6].

In the past few years, mathematical models have attempted to capture the contact patterns between individuals, in order to better understand behavioural responses during human disease outbreaks[7]. Models that incorporate 'awareness' essentially analyze the interplay between information spreading of the awareness of a disease, and the epidemic infection itself.

One approach is the use of 'multiplex networks'[8], consisting of a 2-layer network: one where the dynamics of awareness evolves, the other where the epidemic itself spreads; demonstrating that the epidemic threshold is defined by the 'awareness dynamics'. Another approach[7] captures the ephemeral nature of information-passing by tracking a loss of quality: first-hand information about a disease leads to more decisive action than subsequent, 'word of mouth' information, and correspondingly more effective reduction of susceptibility. It was found that, in a well-mixed population, while the size of outbreak is reduced, the epidemic threshold is unaffected. It may be the case that, though the entire population is aware of the risk of an epidemic, only a portion tries to reduce the risk of infection or seek early treatment[9,10]. It was also shown that, provided information dissemination occurs rapidly enough, the disease can be eradicated. A configuration network model[11] with a *static* connectivity describes that the local and global awareness can be determined by the number of contacts an individual has, suggesting that individual behaviours greatly influence epidemic dynamics. Studies of the effects of information dissemination or mass media reports on epidemics include the 2003 SARS epidemic [4,12,13], the human influenza and the 2009 H1N1 pandemic [14,15,16], the spatial spread of infectious diseases as applied to air transportation [17], and even donation epidemics [18]. In a recent study of the effects of media fatigue – the development of resistance to information [19,20,21] – it was found that variations in the rates of information provision and media fatigue, can greatly affect the outcome of an epidemic [14,15].

In most relevant modelling efforts, the focus has been on the effect of behavioural responses on the epidemic size and pattern. Despite its importance for behavioural changes, little attention has been paid to the optimal level of information dissemination required during an outbreak to reduce infection spread [14,15]. In the present study, we develop models of epidemic spread in a population to address the effect of information dissemination on epidemic size. We consider three modelling structures: (i) deterministic, in which the susceptible sub-population of individuals responds in different ways to global information of the risk of infection. In this model, some practise protective behavioural responses, while others develop a resistance to this information ("media fatigue"); (ii) stochastic, where events associated with behavioural responses randomly occur with specified probabilities; and (iii) agent-based, representing heterogeneity and evolution of contact networks of the population with detailed individual responses at both the local and global levels. By means of a discrete-time simulation, incorporating a Markov-chain compartmental disease state, the agent-based model takes into account detailed spatial interactions between 'agents', whereas the deterministic model assumes uniform mixing of the individuals and tracks only the 'mean-field' behaviour of the population. Comparison of these models enables us to ascertain the importance of the diversity of interaction types allowed by the agent-based structure, relative to the simplified uniform mixing assumption, and therefore the utility and accuracy of the deterministic models. The outcome of principal interest is the final epidemic size derived from these models, and how this is affected by the fraction of susceptible individuals who develop resistance to information and behavioural changes[22] caused by media fatigue[23,24,14,15]. Details of the models and simulation results are described in the following sections. We close the paper with a discussion of our findings and their implications for public health.

2. Modelling Frameworks

We applied a susceptible-infected-recovered compartmentalization of the population to develop the modelling frameworks in this study. Considering a short-term epidemic scenario, we ignored the effect of demographic factors (e.g., birth, death) on the population dynamics. We also assumed that individuals who are recovered from infection will be immune against re-infection during the same epidemic episode. The effect of information dissemination is assumed to influence the behaviour of susceptible individuals in order to reduce disease transmission. Media fatigue is built into the model through the development of resistance to the use information in susceptible individuals.

2.1. *Deterministic Structure*

In the deterministic structure, we divided the population into four compartments: (i) susceptible individuals who may use information to implement protective behavioural responses (S); (ii) susceptible individuals who are currently practising protective behavioural responses (S_p); (iii) susceptible individuals who have developed resistance to the use of information (S_r); and (iv) infected individuals (I). Using a mass action disease incidence, we express the epidemic dynamics with the following system of differential equations:

$$S' = -\beta SI - f(\kappa, I)S + (1-\xi)\delta S_p \quad (1)$$
$$S'_p = (1-p)f(\kappa, I)S - \beta_p S_p I - \delta S_p \quad (2)$$
$$S'_r = pf(\kappa, I)S + \xi\delta S_p - \beta S_r I \quad (3)$$
$$I' = (\beta S + \beta_p S_p + \beta S_r)I - \gamma I \quad (4)$$

where β is the baseline disease transmission rate of unprotected susceptible individuals; β_p is the reduced transmission rate for susceptible individuals practicing protective behavioural responses; and γ is the rate of recovery from infection. We assumed that individuals in S_p class leave this class at a rate δ by either developing resistance to the use of information (moving to S_r) or returning to the regular behaviour (moving to S). The parameter ξ represents the fraction of individuals in S_p class who develop resistance to information, and therefore will no longer be influenced by information on protective behavioural responses, while being fully susceptible to the disease. The effect of information dissemination on behavioural changes of the susceptible population is included in $f(\kappa, I)$, defined as a functional response with the following specific properties:

$$f(\kappa, 0) = 0; \quad f(\kappa, I) \geq 0; \quad \frac{\partial f(\kappa, I)}{\partial I} \geq 0; \quad \lim_{I \to \infty} f(\kappa, I) = \kappa, \quad (5)$$

where κ is the maximum rate of information dissemination influencing the behaviour of susceptible individuals. This functional form suggest that as the number of infected individuals increases, the rate at which susceptible individuals are exposed to information on the use of protective behavioural changes increases, and saturates at its maximum value (κ). Some specific forms of this functional response are given by $\kappa I(t)/(a + I(t))$ (Holling-type II), and $\kappa(1 - \exp(-aI(t)))$ (Ivlev-type), where a is the saturation constant [25,26,27]. The parameter p represents the fraction of susceptible individuals who develop resistance to information and move directly to S_r.

For the system at equilibrium, there is a disease-free state given by $E_0 = (S^*, 0, S_r^*, 0)$, where S^* and S_r^* are the initial population sizes of unprotected susceptible individuals who are sensitive and resistant to information on behavioural changes, respectively. Using the next generation technique[28], one can easily obtain the condition $R_0 < 1$ for stability of the E_0, where

$$R_0 = \frac{\beta(S^* + S_r^*)}{\gamma}.$$

Table 1. Movements of individuals between compartments according to the transition rates. The parameter r is a random variate drawn from the uniform distribution on the unit interval $(0, 1)$. Other parameters are the same as those defined for the deterministic structure.

Event	Transitions	Transition rates
infection of an S-individual	$S \to S - 1$ $I \to I + 1$	βSI
infection of an S_p-individual	$S_p \to S_p - 1$ $I \to I + 1$	$\beta_p S_p I$
infection of an S_r-individual	$S_r \to S_r - 1$ $I \to I + 1$	$\beta S_r I$
movements of an S-individual due to behavioural changes	$S \to S - 1$ either $S_p \to S_p + 1$ or $S_r \to S_r + 1$	$f(\kappa, I)S$ if $r \geq p$ if $r < p$
movements of an S_p-individual due to behavioural changes	$S_p \to S_p - 1$ either $S \to S + 1$ or $S_r \to S_r + 1$	δS_p if $r \geq \xi$ if $r < \xi$
recovery of an I-individual	$I \to I - 1$	γI

2.2. *Stochastic Structure*

Applying the same compartmentalization, we developed the stochastic structure for the model (1)-(4). We assumed that events in this model are defined by the rates of movement between different compartments in the deterministic structure. In this framework, new infections are binomially distributed with the number of trials equal to the number of susceptibles at time t and the probability of success (i.e., infection occurs). The state of the system is defined by the integer number of individuals in each compartment and changes discretely whenever an event occurs. These changes are summarized in Table 1. Transition rates in Table 1 can be converted to probabilities of the corresponding event happening by considering

$$P(\text{Event } i) = \frac{a_i}{\sum_i a_i},$$

where a_i is the transition rate of the event i. In this formulation, the time to the next event (τ) is exponentially distributed with the parameter equal to the sum of the rates for all possible events. The probability density function is given by

$$f(\tau) = \Big(\sum_i a_i\Big) \exp\Big(-\tau \sum_i a_i\Big).$$

Using inverse transform sampling[29], we can estimate the time to the next event: for a given random variate r drawn from the uniform distribution on the unit interval $(0, 1)$, τ can be estimated as $-\ln(r)/\sum_i a_i$. The state of the system is updated according to the transitions given in Table 1 for the event occurring during the transition time τ.

2.3. *Agent-Based Structure*

To include heterogeneity in the system dynamics, we also developed an agent-based model of disease spread in the population based on the deterministic framework, and included age structure. There are two basic entities in this model[30,31,33]: (i) agents that represent individuals living in the simulated population; and (ii) the environment that represents the place in which agents interact. The interaction of agents is determined by modular behavioural components, each containing state information relative to that behaviour. For example, each agent's disease component maintains the state of disease specific to that agent, and movements are driven by schedules, leading to indirect relationships between agents by virtue of being co-located at a particular position and at a particular time step in the environment. Disease may spread from infected to uninfected agents that are located at the same position on a particular time step according to the disease state component.

The agent-based approach taken here is a discrete-time simulation model[31,32], based on the movement of individuals (i.e., agents) that are distributed in a scaled lattice representation of the environment. In implementation of the agent-based model, a Markov-chain compartmental disease state is maintained for each agent[32]. The underlying structure of the epidemic dynamics includes the epidemiological statuses of individuals as susceptible and sensitive to changing their behavioural response (S); susceptible and practicing protective behavioural responses (S_p); susceptible and resistant to information on behavioural changes (S_r); and infected individuals (I) spreading the disease. Movements of individuals between these classes occur as a result of a change in their behaviour or epidemiological status over time. Assuming a relatively simple agent schedule for individuals' mobility, the model captures work, school, and household contacts with random mixing for between-agent interactions at these places. Disease transmission occurs as a result of contact between susceptible and infectious agents. We assumed a standard Markov-chain for disease progression in the model compartments. In our model, each time-step in the simulation is associated with an independent Bernoulli trial for disease transmission.

To parameterize the model, one may use estimated ranges from the published literature, or sample from known ranges and distributions. For transmission between each susceptible-infected pair of agents[32], we considered the probability:

$$P_{\text{transmission}} = 1 - (1 - \beta_{\text{base}})^t, \tag{6}$$

where β_{base} is the baseline transmission probability per unit time, and t is the amount of time spent in each contact. For individuals in the S_p class, the probability of transmission may be reduced due to the effect of protective behavioural responses. In a deterministic structure, this effect was included as a reduction factor $0 \leq \alpha < 1$, where $\beta_p = \alpha\beta_{base}$. In the agent-based model, we include this effect as a reduction in the number of potentially infectious contacts between susceptible and infected individuals, and therefore consider α as the probability that the contact will take place.

In the deterministic model, individuals in the S_p class will change their behaviour at the rate δ. This rate corresponds to an average time duration of $1/\delta$ that a susceptible individual practises protective behavioural responses. In the agent-based model, we may assume a time distribution with the mean of $1/\delta$, and sample the duration of time staying in the S_p class independently for each individual. For a given probability ξ, and each individual in the S_p class on each simulation time-step, a random number r is drawn from a uniform distribution on the unit interval $(0, 1)$ to determine whether the individual is moving to S_r (if $r < \xi$) or remain in S_p (if $r \geq \xi$). If transition to S_r does not occur, the individual will return to S class after the sampled duration has elapsed. A similar process was implemented for the movements of susceptible individuals to S_p or S_r due to the effect of information dissemination.

For the effect of information on behavioural changes, we considered κ to be the time-unit probability (i.e., hourly probability in our simulations) of a susceptible individual moving to S_p class to practise protective behavioural response, when there is only one infected individual is introduced into the population. This probability is multiplied by the number of infected individuals as the functional response in the agent-based model at each simulation time-step during the course of epidemic.

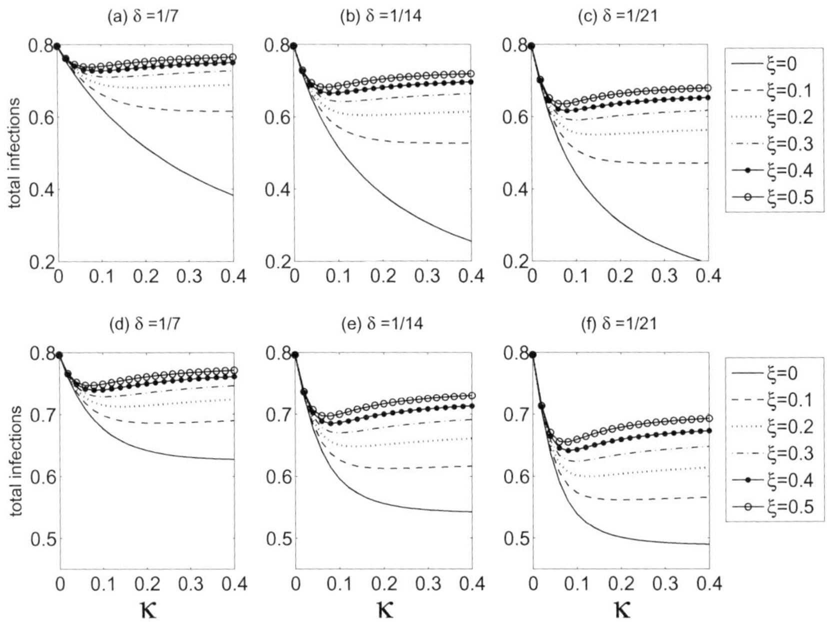

Figure 1. Epidemic final size in the deterministic model as a function of κ for different values of ξ. Parameter values are: $R_0 = 2$, $\gamma = 1/3.38$, and (a-c): $p = 0$; (d-f): $p = 0.1$. The parameter β was calculated based on the expression of R_0 with the initial conditions: $S_0 = 10^4 - 1$, $S_p = 0$, $S_r = 0$, and $I_0 = 1$.

3. Simulation Experiments

To compare the outcomes of the three modelling structures developed here, we performed simulations for a set of parameter values, while considering a range of values for parameters describing information dissemination and probability of developing resistance to information. In all simulations conducted here, the reduction of transmissibility due to protective behavioural responses was assumed to be 50% and $R_0 = 2$. The mean infectious period was $1/\gamma = 3.38$ days in the deterministic and stochastic models, but was sampled independently for each infected individual from a log-normal distribution with the mean of 3.38 days[32].

In the case of deterministic structure, the simulations show that in absence of behavioural resistance ($\xi = 0$, $p = 0$), increasing the rate of information dissemination reduces the epidemic size (i.e., total number of infections). This reduction is clearly more pronounced as the average amount

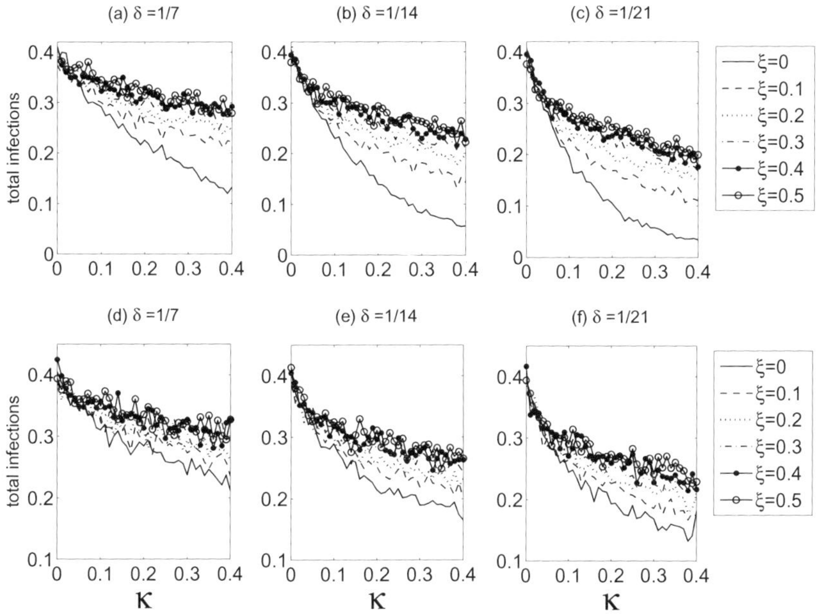

Figure 2. Epidemic final size in the stochastic model as a function of κ for different values of ξ. Parameter values are: $R_0 = 2$, $\gamma = 1/3.38$, and (a-c): $p = 0$; (d-f): $p = 0.1$. The parameter β was calculated based on the expression of R_0 with the initial conditions: $S_0 = 10^4 - 1$, $S_p = 0$, $S_r = 0$, and $I_0 = 1$.

of time $(1/\delta)$ that susceptible individuals practise protective behavioural changes increases (Figures 1a-1c, solid curves). While these outcomes are expected, the epidemic size pattern changes in the presence of resistance to information. Given a fixed value of ξ, the epidemic size decreases for low values of κ. Depending on the value of ξ, two outcomes are possible: (i) the epidemic size reaches a plateau at which increasing κ will induce no further reduction in epidemic size; (ii) there is a threshold of κ at which the epidemic size exhibits a local minimum, and increasing κ beyond the threshold leads to larger epidemic sizes (Figures 1a-1f). These observations remain valid whether resistance can occur before $(p > 0)$ or only after $(p = 0)$ practicing protective behavioural changes.

For the stochastic structure, we observed different outcomes for the epidemic size (Figure 2a-2f). The most conspicuous is the level of epidemic size which is considerably lower than that predicted by the deterministic structure in all scenarios, regardless of the presence of resistance to information. Similar to the deterministic case, the sharpest decline in epidemic size occurs for low values of κ; however, we do not find a specific threshold of κ in the simulated range, at which the epidemic size exhibits a local minimum. Nor do we observe a clear pattern of increase in the epidemic size when κ increases and resistance to information is present. This is notwithstanding the fact that increasing κ is associated with a decelerating rate of decrease in the epidemic size.

In the agent-based simulations (Figure 3), we observed a significant difference in the epidemic final size between the scenario in which resistance to information is absent ($\xi = 0$) and scenarios in which resistance can occur ($\xi > 0$). The sharpest decline for the final size reduction occurs for small values of κ in the simulated range. The rate of reduction changes drastically as κ increases. When $\xi > 0$, the variations in the epidemic final size provide no significant difference between simulated values of ξ. However, as κ increases, there is a decreasing trend for the final size in all scenarios, and similar to stochastic simulations, we do not observe any specific threshold of κ at which a unique local minimum of the final size can occur. Overall, simulations of the stochastic and agent-based models show qualitatively similar trends for the epidemic final size, which is in contrast to those observed in deterministic structure for sufficiently high value of ξ.

4. Concluding Remarks

In this study, we proposed three different types of model to implement a strategy for behavioural responses during an epidemic: a compartmental, deterministic one which assumes uniform mixing of the individuals in the population; a stochastic version of the compartmental model; and an agent-based model which tracks the detailed behaviour of individuals in a population. The epidemic final size was simulated when the parameters associated with behavioural changes varied in their respective ranges within each model structure. In all cases examined, the deterministic model predicted significantly larger epidemic sizes than either the stochastic or agent-based model. However, in all models, the absence of resistance to protective behavioural changes resulted in a significantly smaller epidemic size – most notably for the agent-based model. We have not explored the underlying reasons for the difference between the patterns of the final sizes derived from the deterministic and stochastic or agent-based models. However, it may be attributed to a measure of covariance between the subpopulations [14,15,37]. Such covariances are inherently included in the stochatic or agent-

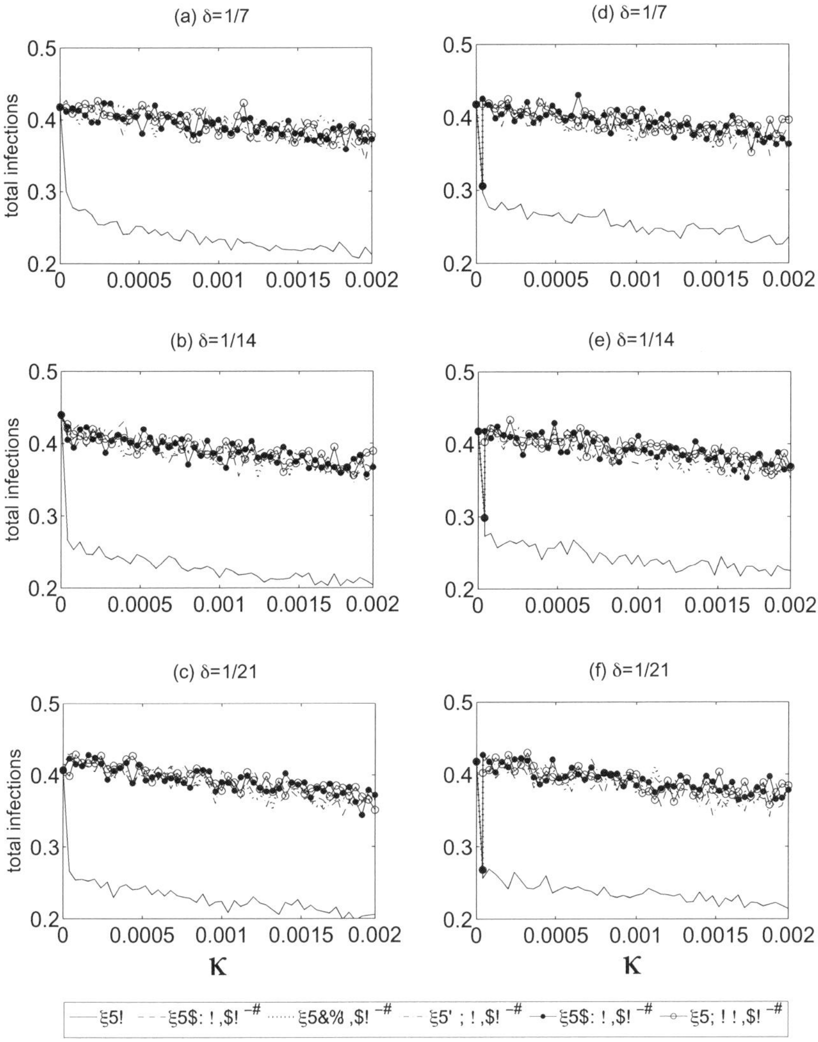

Figure 3. Epidemic final size in the agent-based model as a function of κ for different probabilities ξ of developing resistance to information. Parameter values are: $R_0 = 2$, and infectious period γ was sampled from a log-normal distribution with the mean of 3.38. Probability of developing resistance to information for a susceptible individuals, is (a-c): $p = 0$; and (d-f): $p = 0.1\kappa$. The parameter β was found by calibrating the agent-based model to achieve $R_0 = 2$ over the average of 1000 independent realizations, with the initial conditions: $S_0 = 10^4 - 1$, $S_p = 0$, $S_r = 0$, and $I_0 = 1$. With these parameters, the daily probability of each S-class individual moving to S_p-class in the presence of only one infected case ranges from $0 - 4.69\%$.

based model; however, in the deterministic framework they are ignored. This can be seen in a derivation of equations (1)-(4) from first prinicples, assuming independent populations. These considerations warrant further study.

The differences between the agent based and comparmetnal models (deterministic and stochastic) suggest that *local interactions* and *fluctuations* play an important role in determining the epidemic outcome, since these characteristics are well-captured by the agent-based model, whereas the compartmental model assumes uniform mixing. If this is so, then implementing the behavioural algorithm on a *network* model would be of particular interest, as it would capture variations in contact structure (represented by the degree distribution of the nodes) through the population. In addition, both local and global behavioural responses could be included[6]. For example, local interactions may depend on the number of infected nodes adjacent to a given node[34]. Epidemic thresholds depend on local interactions[35], and implementing behavioural responses in heterogeneous environments complement those provided by globally-available information, as was considered in this paper.

Based on the findings of this study, the question "What is the optimal level of information dissemination during an epidemic?" may be addressed in two different contexts. Compared to the scenario in which information on epidemic status is absent, any amount of information dissemination is useful and will potentially result in a lower epidemic size, even in the deterministic model. This is due to the fact that developing resistance to behavioural changes will not change the level of susceptibility of the population even if all susceptible individuals develop resistance to information; but it could reduce disease spread based on the reduction of transmissibility achieved in a fraction of the susceptible population practising protective behavioural responses before developing resistance. However, comparison of the epidemic sizes for two different levels of information dissemination depends on several parameters, as well as the population structure as shown in our simulations. While stochastic and agent-based models studied here do not agree with the prediction of the deterministic model with regards to the increase of the epidemic size for large amounts of information dissemination, all models suggest that the rate of change in epidemic size could be very minimal for high levels of information dissemination. We observed that the sharpest decline of the epidemic final size is associated with small values of parameter (κ) encouraging protective behavioural responses. The results suggest that the optimal level of information dissemination may not necessarily correspond to a local minimum for the epidemic final size, but rather to the relative reduction in the epidemic final size.

Our focus in this paper has been on quantifying the effect of information dissemination on behavioural responses, which ultimately changes the epidemic dynamics. Our models are subject to limitations and assumptions. For example, we have not included any other intervention strategies for disease control; however, a combination of public health measures (e.g., treatment of infectious cases, or vaccination of susceptible individuals[36]) is likely to be implemented (when available) during an epidemic. Furthermore, we considered behavioural responses only for the susceptible population; however, infectious individuals could also change their behaviour – for example, by practising self-isolation to reduce disease transmission in the population [14,15]. Finally, we used a function $f(\kappa, I)$ to modify the uptake of protective behavioural practices and development of information resistance. The form of this functional response, however, can affect the final size of an epidemic [27]. Equations (1)-(4) can be extended to incorporate mass media and information dissmenation data explicitly [14,15,27]. Extension of models proposed here to include possible interventions and other forms of behavioural responses is the course for future work.

References

1. E.P. Fenichel, C. Castillo-Chavez, *et al. Proc. Nat. Acad. Sci.* **108**(15), 6306-6311 (2011).
2. Africa, U.N. Step Up Effort to Fight Ebola Epidemic. *Wall Street J.* August 7, (2014).
3. Consensus document on the epidemiology of SARS. *World Health Organization.* Geneva, Switzerland, (2003).
4. J. Cui, X.Tao, and H. Zhu. *Rocky Mount. J. Math.* **38**(5), 1323-1334,(2008).
5. M. Nekovee, Y. Moreno, G. Bianconi and M. Marsili, *Physica A.* **374**, 457-470 (2007).
6. S. Funk, M. Salathé and V. A. A. Jansen, *J. R. Soc. Interface.* **7**(50), 1247-1256 (2010).
7. S. Funk, E. Gilad, C. Watkins, and V. A. A. Jansen, *Proc. Nat. Acad. Sci.* **106**(16), 6872-6877 (2009).
8. C. Granell, S. Gómez, and A. Arenas, *Phys. Rev. Letts.* **111**, 128701 (2011).
9. I.Z. Kiss, J. Cassell, M. Recker and P.L. Simon, *Math. Biosci.* **225**, 1-10 (2010).
10. V. Hatzopoulos, M. Taylor, P.L. Simon and I.Z. Kiss, *Math. Biosci.* **231**, 197-209 (2011).
11. Y. Shang, *J. Biol. Phys.* **39**, 489-500 (2013).
12. J. Cui, Y. Sun and H. Zhu, *J Dyn. Diff. Equ.* **20**, 31-53 (2008).
13. Y. Li, C. Ma, and J. Cui. *Rocky Mount. J Math.* **38**, 1437/1455 (2008).

14. S. Collinson. *Modelling the effects of mass media on influenza transmission*, York University Press (2013).
15. S. Collinson and J.M. Heffernan. Submitted.
16. J.M. Tcheunche, N. Dube, *et al. BMC Public Health.* **11**, S5 (2011).
17. C. Sun, W. Yang, *et al. Math. Biosci.* **230**, 87-95 (2011).
18. F. Schweitzer and R. Mach. *PLoS One.* **3**: e1458 (2008).
19. M. Brodie, E. Hamel, L.A. Brady, J. Kates and D.E. Altman, *Supplement to Columbia Journalism Review.* March/April 2004.
20. A. Vartti, P. Jallinoja and A. R. Aro, *Intern. J. Health.* **6**(1), (2007).
21. B. M. Spadacini, *Global AIDSLink.* **79**, 12-13 (2003).
22. K.A. Fanti, *et al. Agg. Beh.* **35**(2), 179-187 (2009).
23. K. N. Kinnick, D. M. Krugman and G. T. Cameron, *Journalism & Mass Commun. Quart.* **73**(3), 687-707 (1996).
24. S. Galea, *New York: Springer.* (2007).
25. S. M. Moghadas, M. E. Alexander and B. D. Corbett, *Phys D.* **188**, 134-151 (2004).
26. M. Hesaaraki and S. M. Moghadas, *Ecol. Modell.* **142**(12), 1-9 (2001).
27. S. Collinson and J.M. Heffernan. *BMC Public Health.* **14**, 376 (2014).
28. P. van den Driessche and J. Watmough, *Math. Biosci.* **180**, 29-48 (2002).
29. R.Y. Rubinstein and B. Melamed, *New York: Wiley.* (1998).
30. C. W. Reynolds, Individual-based modeling. **1999**, (1999).
31. L. C. Mostao-Guidolin, *et al. In: Biomedical Engineering, Trends, Researches and Technologies, In-Tech.* **Chapter 33**, 641-664 (2011).
32. M. Laskowski, L.C. Mostaco-Guidolin, A.L. Greer, J. Wu and S.M. Moghadas, *Sci. Rep.* **1**(105), 1-7 (2011).
33. M. Laskowski and S.M. Moghadas, *BIOMAT Proc.*, 318-339 (2013).
34. F. Bagnoli, L. Pietro, and S. Luca, *Phys. Rev. E.* **76**, 061904 (2007).
35. R. Olinky and L. Stone, *Phys. Rev. E.* **70**, 030902 (2004).
36. R. Kalan, *et al. BMJ Open.* **4**, 8: e005130 (2014).
37. J.M. Heffernan and L.M. Wahl. *J Theoret. Biol.* **243**, 191-204 (2006).

RICH DYNAMICS OF HEPATITIS C VIRAL INFECTION WITH LOGISTIC PROLIFERATION*

SANDIP BANERJEE

Department of Mathematics,
Indian Institute of Technology Roorkee (IITR),
Roorkee 247667, Uttaranchal, India
E-mail: sandofma@iitr.ac.in

A mathematical modeling of Hepatitis C Virus (HCV) dynamics has been presented in this paper. The proposed model, which involves four coupled ordinary differential equations, describes the interaction of target cells (hepatocytes), infected cells, infectious virions and non-infectious virions. In the model, density dependent proliferation terms of both uninfected and infected hepatocytes have been used. The model takes into consideration the addition of ribavirin to interferon therapy and explains the dynamics regarding biphasic and triphasic decline of viral load in the model. A critical drug efficiency parameter has been defined and it is shown that for efficiencies above this critical value, HCV is eradicated whereas for efficiencies lower this critical value, a new steady state for infectious virions is reached, which is lower than the previous steady state.

1. Introduction

Hepatitis C virus (HCV) is the cause of the infectious viral disease Hepatitis C, which infects an individual through blood. Worldwide, individuals between 200 to 300 millions are currently infected with HCV [1]. In earlier days, people suffering from Hepatitis C are often mistaken to be suffering from Hepatitis non-A or non-B or even alcoholic liver disease. But with time, as the dynamics of Hepatitis C becomes clear, it is understood that the disease is a major threat for hepatocellular carcinoma.

The combination therapy with two drugs, namely, pegalyted interferon and ribavirin is the current treatment for HCV infection [2]. However, only 50% of the patient treated with combination therapy show positive

*This work is supported by the Initiation Grant a (Grant no. iitr/sric/100518) from the Indian Institute of Technology, Roorkee, India.

response[2,3] and for non-responders in combination therapy, no effective alternative exits [3]. Hence, there is need to identify markers of disease progression and a refinement in the therapeutic protocols, so that, there is an enhancement of long-term response rates.

Mathematical models have been used extensively to understand the dynamics of hepatitis C, which have provided insights into the pathogenesis of HCV in vivo [4,5,6,7,8] and succeeded in explaining the confounding patterns of the viral load changes in HCV infected patients undergoing therapy. Neumann et al. [4], used the basic model of viral dynamics to HCV, which assumes a simplified view of HCV infection by taking into account of IFN-α monotherapy. The model consists of three populations, namely, uninfected hepatocytes (the target cell), productively infected hepatocytes and free HCV virions and describes the response to interferon therapy. Though the basic model by Neumann [4] provides insights into the HCV dynamics in vivo but it fails to predict the long term response rates observed during combination therapy as it ignores the influence of ribavirin. Dixit et al. [5] made an advancement in the basic model by taking into account the role of ribavirin action along with interferon. The dynamics of model suggests that ribavirin has very little role to play during the first phase decline induced by interferon, as ribavirin does not alter the viral production. However, ribavirin enhances the second phase slope when interferon effectiveness is small, which is in close agreement with experiments [9].

One of the drawback of the model by Dixit et al. [5] is that non-responders and patients with triphasic decay patterns are not studied and is restricted to biphasic responses only. A modified model, which include the proliferation of uninfected and infected cells driven by liver homeostatic mechanisms was studied by Dahari et al. [6]. The model predicts the triphasic decline and succeeds in explaining the origins of non-response. Chakrabarty and Joshi [7] presented a model where they used deterministic control theory to obtain an optimal treatment strategy using interferon and ribavirin. The optimal treatment which they obtained succeeded in reducing the levels of viral load and at the same time keeping the side effects of the drug to a minimal. Banerjee et al. [8] studied the dynamics of hepatitis C virus by attributing different measures for combined drug efficacies, interferon and ribavirin.

Here, the model under consideration follows after the existing models that appear in [4,5,6] and is a modification of the model in the article by Chakrabarty and Joshi [7]. The model is extended to account for the proliferation of hepatocytes, as was used to model HCV RNA kinetics in primary infection in chimpanzees [10]. It has been shown that the modified model can explain both biphasic and triphasic viral decays, the response seen in *flat partial responders* in which virus decay is characterized by a flat second phase, as well as the rebound of virus to its pretreatment or baseline level on the cessation of therapy. It has also been shown that there exists a critical drug efficacy such that for efficacies below the critical value a flat partial response will occur, whereas for efficacies above the critical value ultimate HCV elimination should occur.

2. Mathematical Model

The following system of coupled ordinary differential equations represent the model as:

$$\frac{dT}{dt} = s + rT\left(1 - \frac{T+I}{T_{max}}\right) - d_1 T - (1 - c\eta_1)\alpha V_I T \tag{1}$$

$$\frac{dI}{dt} = (1 - c\eta_1)\alpha V_I T + rI\left(1 - \frac{T+I}{T_{max}}\right) - d_2 I \tag{2}$$

$$\frac{dV_I}{dt} = \left(1 - \frac{\eta_r + \eta_1}{2}\right)\beta I - d_3 V_I \tag{3}$$

$$\frac{dV_{NI}}{dt} = \left(\frac{\eta_r + \eta_1}{2}\right)\beta I - d_3 V_{NI} \tag{4}$$

In this model, target cells (T) are produced from a source at a constant rate s. The density dependent proliferation term is introduced for both target cells (T) and infected hepatocytes (I). The uninfected (T) and infected (I) hepatocytes can proliferate with maximum proliferation rate r, under a blind homeostasis process, in which there is no distinction between infected and uninfected cells in the density dependent term. Because HCV infection is non-cytopathic and viral production rates in vivo appear to be low, there is no reason to suppose that infected cells cannot proliferate.The total number of hepatocytes, $T + I$, can attain the maximum value T_{max}, at which point cell proliferation stops. Here η_1 denotes the effectiveness of interferon in blocking the release of new virions, $c\eta_1$ is fraction of effectiveness ($0 < c < 1$) and $(1 - c\eta_1)$ is the ineffectiveness of interferon. $(1 - c\eta_1)\alpha V_I T$ is conversion term which happens when target cell gets infected by the infectious virus. d_2 is the natural death of infected cells. In the third equation, the first term, namely, $(1 - \frac{\eta_r+\eta_1}{2})$, is ineffectiveness of combined therapy by interferon and ribavirin, which leads to the growth of the infectious virions; η_r is the effectiveness of ribavirin, β is infection rate constant and d_3 is the natural death of the infectious and non-infectious virions. In the fourth equation, $(\frac{\eta_r+\eta_1}{2})$, is the combined effect of ribavirin and interferon, which restricts the growth of infectious virions and results in the growth of non-infectious virions. Note that $0 \leq \eta_1 \leq 1, 0 \leq \eta_r \leq 1$ and $0 \leq \frac{\eta_1+\eta_r}{2} \leq 1$. System (2.1) has to be analyzed with the following initial conditions: $T(0) > 0, I(0) > 0, V_I(0) > 0, V_{NI}(0) > 0$. All system parameters are positive.

3. Qualitative Analysis of the model

3.1. *Positivity and boundedness*

Theorem 3.1. *The solutions of the system (1-4) are positive for all $t > 0$.*

Proof. Let $(T(t), I(t), V_I(t), V_{NI}(t))$ be a solution of system (2.1). Let for some $t_1 > 0, T(t_1) = 0$ and if t_1 be the first such time, then $\frac{dT(t_1)}{dt} \leq 0$. But at $t = t_1$, $\frac{dT(t_1)}{dt} = s > 0$ (from (1)), which leads to a contradiction. Therefore, $T(t) > 0$, for all $t > 0$.
It is again observed that $I(t) > 0$ for $t \geq 0$ as

$$\frac{dI}{dt} \geq -d_2 I \Rightarrow I(t) \geq I(0)e^{-d_2 t} \tag{5}$$

Similarly, equations (3) and (4) of the system, gives

$$V_I = \left(1 - \frac{\eta_r + \eta_1}{2}\right) \beta e^{-d_3 t} \int_0^t I(s) e^{-d_3 s} ds > 0 \tag{6}$$

$$\text{and } V_{NI} = \left(\frac{\eta_r + \eta_1}{2}\right) \beta e^{-d_3 t} \int_0^t I(s) e^{-d_3 s} ds > 0, \tag{7}$$

for all $t \geq 0$. Thus, it is concluded that $(T(t), I(t), V_I(t), V_{NI}(t)) > 0$ for all $t \geq 0$. □

Theorem 3.2. *The solutions of the system (1-4) are bounded.*

Proof. Let $(T(t), I(t), V_I(t), V_{NI}(t))$ be any solution of system (1-4).

$$\begin{aligned}
\frac{dT}{dt}+\frac{dI}{dt} &= s+rT-\frac{rT^2}{T_{max}}-\frac{rTI}{T_{max}}-d_1T+rI-\frac{rTI}{T_{max}}-\frac{rI^2}{T_{max}}-d_2I \\
&\le s+\frac{rT_{max}}{2}-\left(\frac{\sqrt{r}T}{\sqrt{T_{max}}}-\frac{\sqrt{rT_{max}}}{2}\right)^2-d_1T-d_2I \\
&-\left(\frac{\sqrt{r}I}{\sqrt{T_{max}}}-\frac{\sqrt{rT_{max}}}{2}\right)^2 \\
&\le \left(s+\frac{rT_{max}}{2}\right)-A(T+I) \qquad (8)
\end{aligned}$$

where $A=\min(d_1,d_2)$. It follows from (8) that

$$\lim_{t\to+\infty}\sup(T+I)(t)\le\frac{\left(s+\frac{rT_{max}}{2}\right)}{A}=M^*.$$

Therefore, there exist positive constants $M>M^*$ and $T_1>0$ such that if $t\ge T_1, T(t)+I(t)<M$. Also, we can say that $I(t)<M$. □

From (3), we get

$$\frac{dV_I}{dt}=\left(1-\frac{\eta_1+\eta_r}{2}\right)\beta I-d_3V_I\le\left(1-\frac{\eta_1+\eta_r}{2}\right)\beta M-d_3V_I.$$

$$\lim_{t\to+\infty}\sup V_I(t)\le\frac{\left(1-\frac{\eta_1+\eta_r}{2}\right)\beta}{d_3}=M_I^*.$$

Therefore, there exist positive constants $M_I>M_I^*$ and $T_2\ge T_1$ such that if $t\ge T_2, V_I(t)<M_I$.

Similarly, from (4) equation of (2.1)

$$\lim_{t\to+\infty}\sup V_{NI}(t)\le\frac{\left(\frac{\eta_1+\eta_r}{2}\right)\beta}{d_3}=M_{NI}^*.$$

Therefore, there exist positive constants $M_{NI}>M_{NI}^*$ and $T_3\ge T_2$ such that if $t\ge T_3, V_{NI}(t)<M_{NI}$. This gives the complete proof.

3.2. *Equilibria*

System (1-4) has two steady states; an uninfected steady state, $E_1(\overline{T},0,0,0)$, where total number of uninfected hepatocytes,

$$\overline{T}=\frac{T_{max}}{2r}\left[(r-d_1)+\sqrt{(r-d_1)^2+\frac{4rs}{T_{max}}}\right]\quad(r>d_1)$$

and an infected steady state, $E_2(T^*, I^*, V_I^*, V_{NI}^*)$, where

$$T^* = \frac{1}{2E}\left(\frac{r(r-d_1)}{T_{max}} - (r-d_2)D + \sqrt{\frac{r(r-d_1)}{T_{max}} - (r-d_2)D^2 + \frac{4rsE}{T_{max}}}\right),$$

$$I^* = \frac{s + (r - d_1 - \frac{rT^*}{T_{max}})T^*}{DT^*}, \quad V_I^* = \frac{\left(1 - \frac{\eta_r+\eta_1}{2}\right)\beta I^*}{d_3}, \quad V_{NI}^* = \frac{\left(\frac{\eta_r+\eta_1}{2}\right)\beta I^*}{d_3}$$

and

$$D = \frac{r}{T_{max}} + \frac{(1-c\eta_1)\left(1 - \frac{\eta_r+\eta_1}{2}\right)\alpha\beta}{d_3}, \quad E = \left(\frac{(1-c\eta_1)\left(1 - \frac{\eta_r+\eta_1}{2}\right)\alpha\beta}{d_3}\right)^2.$$

3.3. *Stability Analysis*

3.3.1. *Local Stability of uninfected steady state* E_1

The variational matrix for uninfected steady state E_1 is
$$\begin{pmatrix} -\frac{s}{\overline{T}} - \frac{r\overline{T}}{T_{max}} & -\frac{r\overline{T}}{T_{max}} & -(1-c\eta_1)\alpha\overline{T} & 0 \\ 0 & r - d_2 - \frac{r\overline{T}}{T_{max}} & (1-c\eta_1)\alpha\overline{T} & 0 \\ 0 & (1-\frac{\eta_r+\eta_1}{2})\beta & -d_3 & 0 \\ 0 & (\frac{\eta_r+\eta_1}{2})\beta & 0 & -d_3 \end{pmatrix}$$
whose characteristic equation is

$$\left(\frac{s}{\overline{T}} + \frac{r\overline{T}}{T_{max}} + \lambda\right)(d_3+\lambda)(\lambda^2 - C_0\lambda + C_1) = 0 \tag{9}$$

where

$$C_0 = r - d_2 - d_3 - \frac{r\overline{T}}{T_{max}}, \quad C_1 = (d_2 - r + \frac{r\overline{T}}{T_{max}})d_3 - (1-\eta)\alpha\beta\overline{T}$$

Clearly from (9), two roots of characteristic equation are negative. The other two roots will be negative if

$$r < \frac{(d_2+d_3)T_{max}}{\overline{T} - T_{max}} \text{ and } \alpha\beta < \frac{\left(d_2 - r + \frac{r\overline{T}}{T_{max}}\right)d_3}{(1-\eta)\overline{T}}$$

3.3.2. ***Local Stability of endemic equilibrium point*** E_2

The variational matrix for infected steady state E_2 is

$$\begin{pmatrix} -\frac{s}{T^*} - \frac{rT^*}{T_{max}} & -\frac{rT^*}{T_{max}} & -(1-c\eta_1)\alpha T^* & 0 \\ (1-c\eta_1)\alpha V_I^* - \frac{rI^*}{T_{max}} & -\frac{(1-c\eta_1)\alpha V_I^* T^*}{I^*} - \frac{rI^*}{T_{max}} & (1-c\eta_1)\alpha T^* & 0 \\ 0 & (1-\frac{\eta_r+\eta_1}{2})\beta & -d_3 & 0 \\ 0 & (\frac{\eta_r+\eta_1}{2})\beta & 0 & -d_3 \end{pmatrix}$$

The characteristic equation is given by

$$(\lambda + d_3)(\lambda^3 + A_1\lambda^2 + A_2\lambda + A_3) = 0 \tag{10}$$

where

$$\begin{aligned} A_1 &= d_3 + \frac{s}{T^*} + \frac{rI^*}{T_{max}} + \frac{rT^*}{T_{max}} + \frac{(1-c\eta_1)\alpha T^* V_I^*}{I^*} \\ A_2 &= \frac{sd_3}{T^*} + \frac{rd_3 I^*}{T_{max}} + \frac{rsI^*}{T_{max}T^*} + \frac{sd_3 T^*}{T_{max}} \\ &+ (1-c\eta_1)\alpha\left(\frac{sV_I^*}{I^*} + \frac{d_3 T^* V_I^*}{I^*} + \frac{rT^* V_I^*}{T_{max}}\right) \\ &+ (1-c\eta_1)\alpha\frac{r(T^*)^2 V_I^*}{T_{max}I^*} - (1-\eta)\alpha\beta T^* \\ A_3 &= \frac{rsd_3 I^*}{T_{max}T^*} - (1-\eta)(1-c\eta_1)\alpha^2\beta T^* V_I^* \\ &+ (1-c\eta_1)\alpha d_3 V_I^*\left(\frac{s}{I^*} + \frac{rT^*}{T_{max}} + \frac{r(T^*)^2}{I^* T_{max}}\right) \\ &- (1-\eta)\alpha\beta\left(s + \frac{rI^*T^*}{T_{max}} + \frac{r(T^*)^2}{T_{max}}\right) \end{aligned}$$

Clearly, one root is negative and by the Routh-Hurwitz criterion, the other roots will be negative if $A_1 > 0, A_2 > 0, A_3 > 0$ and $A_1A_2 - A_3 > 0$. Thus, the endemic equilibrium point $E_2(T^*, I^*, V_I^*, V_{NI}^*)$ is locally asymptotically stable, provided the above mentioned conditions are satisfied.

4. Critical Drug Efficacy

In system (1-4), the effectiveness of interferon and ribavirin are given by the terms $(1 - c\eta_1)$ and $\left(1 - \frac{\eta_r+\eta_1}{2}\right)$. Here, these terms are combined into a single term as $(1 - c\eta_1)\left(1 - \frac{\eta_r+\eta_1}{2}\right) = 1 - \eta$, where η represents the overall drug efficiency. Clearly, before treatment $\eta_1 = \eta_r = 0$, that is, drug efficiency is zero and $0 < \eta_1 < 1, 0 < \eta_r < 1$ during antiviral therapy. The stability criteria for uninfected steady state ($C_1 > 0$) is

$$(d_2 - r + \frac{r\overline{T}}{T_{max}})d_3 - (1-\eta)\alpha\beta\overline{T} > 0,$$

$$\Rightarrow \ (1-\eta) < \frac{(d_2 - r + \frac{T_{max}}{2}\left[(r-d_1) + \sqrt{(r-d_1)^2 + \frac{4rs}{T_{max}}}\ \right])d_3}{\alpha\beta\frac{T_{max}}{2r}\left[(r-d_1) + \sqrt{(r-d_1)^2 + \frac{4rs}{T_{max}}}\ \right]}$$

This clearly indicates that there exists a point that acts as a point of separation between the region of stability for uninfected steady state and the region of stability for infected steady state. This point can be termed as a transcritical bifurcation point, which is given by

$$1-\eta = \frac{(d_2 - r + \frac{T_{max}}{2}\left[(r-d_1) + \sqrt{(r-d_1)^2 + \frac{4rs}{T_{max}}}\ \right])d_3}{\alpha\beta\frac{T_{max}}{2r}\left[(r-d_1) + \sqrt{(r-d_1)^2 + \frac{4rs}{T_{max}}}\ \right]}$$

Accordingly, the critical drug efficacy is defined as

$$\eta_\varepsilon = 1 - \frac{(d_2 - r + \frac{T_{max}}{2}\left[(r-d_1) + \sqrt{(r-d_1)^2 + \frac{4rs}{T_{max}}}\ \right])d_3}{\alpha\beta\frac{T_{max}}{2r}\left[(r-d_1) + \sqrt{(r-d_1)^2 + \frac{4rs}{T_{max}}}\ \right]}$$

During antiviral therapy, if $\eta > \eta_\varepsilon$, the drug therapy is successful and the viral load can be eradicated. On the other hand, if $\eta < \eta_\varepsilon$, the viral load and the infected cells converge to a new steady state with lower values.

5. Numerical Simulation

System (1-4) has been simulated for viral kinetics during successful antiviral treatment with the parameter values $s = 1.0$ cell ml^{-1} day^{-1}, r = 1.99 day^{-1}, $T_{max} = 3.6 \times 10^7$ cells ml^{-1}, $d_1 = 0.01$ / day, $\alpha = 2.25 \times 10^{-7}$ ml day^{-1} virions^{-1}, $d_2 = 1.0$ /day, $\beta = 2.9$ virions/cell/day, $d_3 = 6.0/day$ [5,6]. Numerical value of critical efficacy is 0.745. Figure 1 shows of hepatitis C viral dynamics. Figure 1A shows the viral decline due to interferon monotherapy, which is biphasic in nature. The first phase decline is quite sharp and occurs for (24-48) hours after the administration of the drug interferon and the second phase starts after 48 hours and viral decline is slower than the first phase, which was also observed clinically [11].

Figure 1B again shows the biphasic viral decay, with lower interferon efficacy and varying ribavirin efficacy. During the combined drug therapy

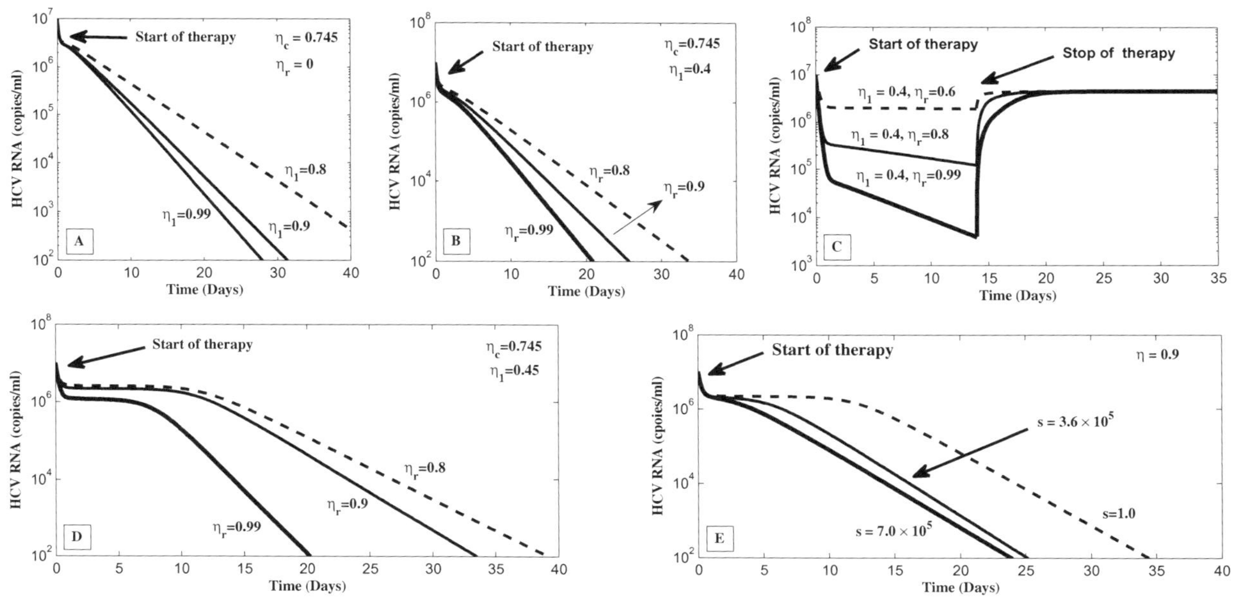

Figure 1: The figure shows different viral kinetics of hepatitics C virus.

(as shown in the figure), the viral load in the blood falls below the detection limit during therapy (100 copies/ml) in 21, 25, 32 days depending on the combined drug efficacy. The virus remains undetected even after the therapy is stopped, that is, the patient has attained sustained virological responses, which means that they are generally cured of the HCV infection. This too has been observed clinically [12].

Figure 1C shows the viral kinetics after the therapy cessation. A virus resurgence post therapy cessation has been simulated for the system (1-4), with different drug efficacy $(\eta_1, \eta_r) = ((0.4, 0.6); (0.4, 0.8); (0.4, 0.9))$ from time 0 to 14 days and then set η_1 and η_r to zero for the rest of the simulation. After 14 days, the virus resurges to pretreatment levels within (7-7.5) days of post therapy cessation. Thus, the model given by system (1-4), predicts resurgences to pretreatment levels after cessation of therapy.

For certain parameter values, the viral decay is triphasic (see figure 1D). The first phase shows a rapid viral decline, followed by a flat or shoulder phase, in which the viral load remains constant and a third phase of faster viral decay. Here, the flat phase, where the constant infected cell level is maintained, is the focus of our discussion. During this period, the loss of the infected cell level is compensated by the infected cell proliferation. As they are in quasi-steady state with the infected cells that produce virus, the viral levels remain constant. Also, during the flat phase, the target cells also proliferates as well as generated (by the constant term s) and when they reach the level of infected cells, they succeeded in decreasing the proliferation rate of the infected cells, due to density dependent form of the proliferation term. Since, there is a decrease in the proliferation rate of the infected cells, their level falls, resulting in the end of the flat or shoulder phase. Triphasic decline has been observed in some patients [13]. Figure 1E shows that higher influx rates of new hepatocytes (s), shrinks the flat phase and even eliminates it giving rise to biphasic viral decline.

6. Conclusion

Hepatitis C virus is affecting millions of people with high levels of morbidity and mortality (NIH, 2002). The combined therapy of interferon-α and ribavirin has been able to eradicate the virus in a substantial number of patients. In this model, density dependent proliferation terms of both uninfected and infected hepatocytes have been used. The model succeeded in explaining the virus resurges to pretreatment levels within a week of post therapy cessation, the biphasic decline of viral load, the triphasic decline in viral decay, including the flat or shoulder phase. The model presents a framework for rational therapy optimization and make prediction of the long term outcomes of the therapy. We conclude by saying that the inclusion of the proliferation terms in hepatocytes make the model more realistic and can predict better viral kinetics in chronic HCV patients during and after antiviral therapy. We sincerely hope that systematic application of the model of this type will lead to a better understanding of the dynamics of hepatitis C virus and the success and failure of its treatment.

References

1. World Health Organization, *World Health Org. Weekly Epidemiol. Rec.* **75**, 18 (2000).
2. National Institutes of Health, *NIH Consens. Statements* **19**, 1 (2002).
3. S.L. Tan, A. Pause, Y. Shi and N Sonenberg, *Nat. Rev. Drug Discov.* **1**, 867 (2002).
4. A.U. Neumann, N.P. Lam, H. Dahari, D.R. Gretch and T.E. Wiley, *Sci.* **282**, 103 (1998).
5. N.M. Dixit, J.E. Layden-Almer, T.J. Layden and A.S. Perelson, *Nat.* **432** 922 (2004).
6. H. Dahari, A. Lo, R.M. Ribeiro and A.S. Perelson, *J. Theor. Biol.* **247** 371 (2007).
7. S.P. Chakrabarty and H.R. Joshi, *J. Biol. Syst.* **17 (1)** 97 (2009).
8. Sandip Banerjee, Ram Keval and Sunita Gakkhar, *Math. Biosci.* **245**, 235 (2013).
9. Stefan Zeuzem and Eva Herrmann, *Ann. hepatol.* **1(2)** 56 (2002).
10. H. Dahari, M. Major, X. Zhang, K. Mihalik, C.M. Rice, A.S. Perelson, S.M. Feinstone and A.U. Neumann, *Gastroenterol.* **128 (4)** 1056 (2005).
11. N.P. Lam, A.U. Neumann, D.R. Gretch, T.E. Wiley, A.S. Perelson and A.J. Layden, *Hepatol.* **26** (1997).
12. J.J. Feld and J.H. Hoofnagle, *Nat.* **436** (2005).
13. E. Herrmann, J.H. Lee, G. Marinos, M. Modi and S. Zeuzem, *Hepatol.* **37(6)** 351 (2003).

MARKOV CHAINS AS A TOOL MEASURING EFFECTIVENESS OF A PSYCHOTHERAPY PROCESS

PIOTR SLIWKA
Faculty of Mathematics and Natural Sciences, S. Cardinal Wyszynski University, 1/3 Woycickiego,01-938, Warsaw; Poland
E-mail: p.sliwka@uksw.edu.pl

WITOLD SIMON
Department of Neurotic Disorders and Psychotherapy, Institute of Psychiatry and Neurology
Sobieskiego 9, 02-957 Warsaw; Poland
E-mail: witold.simon@wp.pl

To evaluate the effectiveness of a therapeutic process usually measurement or statistical tests based on parametric distributions are used. The disadvantage of this approach is its verification based on assumptions of the parametric distribution, condition which in fact is rarely met. We propose an alternative and not previously used method to measure the effectiveness of the therapeutic process based on a homogeneous or inhomogeneous Markov chains depending on the nature of the data. The Outcome Questionnaire (OQ-45.2) was used for empirical verification of the proposed methods. Its results were obtained during the course of psychotherapy of Polish and American patients. A control group of Polish students was used as a benchmark. Preliminary results indicate the usefulness of the proposed method, which does not depend directly on the assumptions about the parameters of probability distributions.

1. Introduction

Psychometric measures of effectiveness of a therapeutic process usually relate to the static situation. This article proposes an approach based on Markov chain that inherently take into account the dynamics of the process and therefore allow a dynamic assessment of effectiveness of the therapeutic process session-by-session. The proposed method was applied to study the effectiveness of empirical data derived due to the Outcome Questionnaire (OQ-45.2).

The OQ-45.2 is a 45-item self-report measure developed specifically for the purpose of tracking and assessing patient outcomes in a therapeutic setting. Reliability of the OQ is acceptable for all subscales and the total score. Lambert[1] reported internal consistency reliability for the OQ-45 of .93 and a 3-week testretest reliability value of .84 for the OQ-45 total score. Concurrent validity of the OQ-45 total score has been examined by correlating it with numerous measures, e.g., the Symptom Checklist-90 (SCL-90)[2], or Beck Depression Inventory (BDI)[3], and yield significant results at the .01 level with a range of rs[1] from .50 to .85.

2. Methods

The article structure gauge the effectiveness of therapy evaluation (and the rate of change) is based on a Markov chain, with the assumption that assessment of a patient at time t depends directly on evaluation of the patient at time $t-1$ and does not depend directly on observation at the moments $t-2$, $t-3$, ... Finite (countable) number of assessments as well as random nature of data point out to Markov chain[4], which allow to describe the patient trajectory from the first stage (i.e. start of therapy) to the last stage (i.e. the end of therapy).

From a mathematical point of view patients can be identified with a sequence of random variables $X_0, X_1, \ldots, X_n$, assuming values in the finite set of space state $S = \{0, 1, \ldots, N-1\}$. If $X_n = j$ depends directly only upon the immediate past $X_{n-1} = i$ and not upon the remote past $X_0 = m, \ldots, X_{n-2} = k$, then X_n satisfies the Markov property

$$\begin{aligned} P(X_n = j | X_{n-1} = i, X_{n-2} = k, \ldots, X_0 = m) &= P(X_n = j | X_{n-1} = i) \\ &= p_{ij} \quad i, j, k, \ldots, m \in S (1) \end{aligned}$$

If X_n satisfies (1) for all $n \geq 1$ and the transition probabilities p_{ij} are time-homogeneous, than the sequence of random variables is called a finite homogenous Markov chain, while in case when p_{ij} are not time-homogeneous is called non-homogenous.

Having the precise numbers of the individuals (number of patients "moving" between states) for which $X_{t-1} = i$ and $X_t = j$ we can apply the microdate[5]. The microdate allows us, based on the maximum likelihood estimator (ML), to estimate the parameters $\hat{p}_{ij}$ of the Markov chain by estimator in homogenous and non-homogeneous case respectively[5]:

$$\hat{p}_{ij} = \frac{\sum_{t \in T} v_{ij}(t)}{\sum_{t \in T} \sum_{j \in S} v_{ij}(t)} \tag{2}$$

$$\hat{p}_{ij} = \frac{v_{ij}(t)}{\sum\limits_{t \in T} \sum\limits_{j \in S} v_{ij}(t)} \tag{3}$$

with assumptions:
$\sum\limits_{i=1}^{s} p_{ij} = 1$,
$\hat{P} = [\hat{p}_{ij}]_{i,j \in S}$,
$1 \geq p_{ij} \geq 0$,
$t \in T$ – the space of the time,
$i, j, s \in S$ – the space of the states (in the case of OQ answers: 0, 1,..., 4 identifying with states: 1, 2, 3, 4, 5),
$v_{ij}(t)$ – the number of individuals for which $X_{t-1} = i$ and $X_t = j$ at time $(t-1, t]$.

Lack of homogeneity of Markov chain prevents from defining the form of a stationary distribution. Therefore, and taking into account a sufficient number of observations of patients moving between stages of the therapeutic process, compartmental methods of ergodic distribution of Markov set-chain were used.

In Markov set-chain the intervals for each transitions probability p_{ij} are created by an interval defined by the minimum and maximum possibility of achieving probability of replacement (i.e., in our case: the minimum and maximum number of transitions between states: 1, 2, 3, 4, 5). The proposed method[6] allows to approximate the interval probability transitions. The transitions probability intervals are expressed in two nonnegative matrices **P** and **Q** (**P** $\leq$ **Q**, element-wise). If **P** and **Q** (constructed by rows) satisfy

$$p_{ij} = \min_{A \in [P,Q]} a_{ij} \quad and \quad q_{ij} = \max_{A \in [P,Q]} a_{ij} \quad \forall i, j$$

then [**P**,**Q**] is called tight[7] (full description of the method: see refs. 7, 8).

The process of bounds construction was the following. First we constructed each element of the matrix **P** (the lower bound) as a minimum of all reported numbers of substitutions: 1 to 1, 1 to 2,..., 1 to 5, 2 to 1,...,4 to 5, 5 to 5 etc., and each element of matrix **Q** (the upper bound) as a maximum of reported substitutions written above. Based on matrices: **P** and **Q** we applied Markov set-chain algorithm to calculate the bounds of stationary distributions taking into account all the OQ items (TOTAL), original American subscales as designated by Lambert and colleagues[9] (SD, IR, MS), and Polish normalized subscales established by Simon, Sliwka and colleagues[10] of four groups of respondents (non-patients, day-patients, inpatients, outpatients).

Responses: "never", "seldom", ..., "always" was replaced with numbers: 0, 1, ...,4 and equated with the states of Markov chain. From the point of view of therapeutic process it means the moving of patients into the direction of total recovery (in this case state 0). Also, the achievement of the envisaged chain stationary (ergodic) distribution (the distribution after a sufficiently long duration of the process) allows the therapist to see what percentage of patients reaches states: 0, 1, ..., 4. In other words, the clinician is able to track these patients who regardless of a treatment duration do not reach "0", i.e. do not recover or even deteriorate.

3. Results

Assuming that each patient is characterized by identical prerequisites prior to therapy and achieves more or less the same effects after each session one could apply homogeneous Markov chain in order to establish transition probability matrix of patients between different states of constant recovery on session-by-sessions basis. However, such situation does not occur in any therapeutic practice. The above thesis is confirmed by the classical χ^2 goodness of fit test with respect to distributions of transition probability matrix $\mathbf{P_{i\_j}}$ between sessions i and j $(i, j = 1, .., 8$, e.g.: $\mathbf{P_{1\_2}}$ - transition probability matrix between the first and second session) and matrix $\mathbf{P_M}$ based on the average number of patients in each state with respect to all patients (TOTAL). The obtained value of the test shows that the use of a homogeneous Markov chain is not validated (the p-value for the TOTAL terms of two sample groups: Polish day-patients as DA8 and American experimental patients as EXP8 in Table 1).

Table 1. p-value of the χ^2 test: DA8 (8 session) and EXP8 (8 session).

Therapy groups	$\mathbf{P_{1\_2}}$ ver $\mathbf{P_M}$	$\mathbf{P_{2\_3}}$ ver $\mathbf{P_M}$	$\mathbf{P_{3\_4}}$ ver $\mathbf{P_M}$	$\mathbf{P_{4\_5}}$ ver $\mathbf{P_M}$	$\mathbf{P_{5\_6}}$ ver $\mathbf{P_M}$	$\mathbf{P_{6\_7}}$ ver $\mathbf{P_M}$	$\mathbf{P_{7\_8}}$ ver $\mathbf{P_M}$
DA8	4.99E-41	8.91E-14	0.0031	0.0241	1.38E-08	8.60E-12	4.97E-21
EXP8	1.59E-15	4.72E-10	0.0018	0.0061	<0.1E-16	<0.1E-16	<0.1E-16

Low values of the p-value (smaller then 0.05) suggests a rejection of H_0: appropriate probability distributions are not consistent in each case. Lowering the level of significance for the 0.001 would accept the χ^2 statistics relative $\mathbf{P_{3\_4}}$ matrix (between 3 and 4 sessions), and $\mathbf{P_{4\_5}}$ (4 and 5 session) in relation to the $\mathbf{P_M}$, but also small level of significance does not ensure homogeneity in the rest of sessions. Acceptance testing two consecutive

transition probability matrix ($\mathbf{P_{i\_i+1}}$ versus $\mathbf{P_{i+1\_i+2}}$) results in slightly higher values for the first and last matrices $\mathbf{P}$, but none of these achieves p-value > 0.001. The rest of therapeutic groups with regard to four (polish: non-patients, day-patients, inpatients and U.S.: eating disorder inpatients, experimental patients, treatment as usual patients) or five sessions (polish: day-patients, inpatients and U.S.: eating disorder inpatients, experimental patients, treatment as usual patients) achieved similar results. These results indicate a heterogeneous rate of movement of the patient during subsequent therapy sessions. Probably this is conditioned by uncontrolled variable of therapy type. Therefore a more appropriate tool to evaluate effectiveness of the therapy would be a non-homogeneous Markov chain. Unfortunately, direct use of non-homogeneous Markov chain to obtain a final stationary distribution is not possible for at least two reasons:
a) a sufficient number of respondents, the usage of the ML estimator $p_{ij}(t)$, and large number of observations still do not allow to meet a satisfactory accuracy of the stationary distribution (in particular ergodic);
b) a sufficient number of completed questionnaires (e.g., from the 1-th session to 15-th sessions for day-patients) does not allow to differentiate patients who did not meet the requirements for the minimum number of cases necessary to the correct interpretation of the results of the χ^2 goodness-of-fit test using the ML estimator (a minimum of 100 persons).

In such a situation the only remaining option designate the upper and lower limits (upper and lower bounds) coordinate vector ergodic distribution based on Markov-set chains (MSC). The MSC application does not allow for the appointment of an explicit form of the matrix $\mathbf{E}$ or ergodic distribution, although this accuracy dependent on the distance between the upper and lower limits of each coordinate of the ergodic distribution: the smaller the distance, the better the precision of the estimate. On the basis of these constraints conclusions regarding the proportion of patients with respect to each of the states for all patients (TOTAL) were obtained. Moreover, based on obtained bounds the measure to evaluate the interval rate of convergence matrix $\mathbf{P}$ to ergodic matrix $\mathbf{E}$ were proposed.

Bounds are based on the n = 4, 5 or 8 full sessions (each patient participating in n subsequent therapy sessions filled n questionnaire) for the following groups of patients: Polish non-patients ($n = 448$ for 4 sessions - abbreviation: NON), Polish day-patients ($n = 131$ for 5 sessions, $n = 107$ for 8 sessions, abbreviation: DA5 and DA8 respectively), Polish in-patients ($n = 160$ for 5 sessions, $n = 43$ for 8 sessions, abbreviation: IN5 and IN8 respectively), Polish day- and inpatients together ($n = 150$ for 8 sessions, abbreviation: DAIN8), U.S. American eating disorder inpatients ($n = 114$ for 5 sessions, abbreviation: AED5), U.S. experimental patients ($n = 176$ for 5 sessions , $n = 173$ for 8 sessions, abbreviation: EXP5 and EXP8 respectively) and U.S. treatment as usual - patients ($n = 97$ for 5 sessions, abbreviation: TAU5). To compare with Polish non-patients group (as a benchmark) additional 4-session group for: DA4 ($n = 201$), IN4 ($n = 233$), AED4 ($n = 127$), EXP4 ($n = 193$), TAU4 ($n = 150$) were determined.

The bounds of ergodic distributions for analyzed subscale TOTAL computed by Markov set-chains presented in Table 2 (due to lack of space for 4-session and 8-session bounds only presented).

Table 2. 4- and 8-sessions bounds of ergodic distributions for Polish and US patients group.

Patients group	**Bounds**	$\mathbf{TOT_0}$	$\mathbf{TOT_1}$	$\mathbf{TOT_2}$	$\mathbf{TOT_3}$	$\mathbf{TOT_4}$
NON	LO	0.32	0.33	0.148	0.044	0.007
	HI	0.45	0.42	0.21	0.071	0.017
DA4	LO	0.21	0.21	0.24	0.15	0.03
	HI	0.32	0.26	0.31	0.22	0.059
IN4	LO	0.19	0.18	0.2	0.1	0.038
	HI	0.41	0.25	0.34	0.22	0.11
AED4	LO	0.13	0.17	0.27	0.16	0.064
	HI	0.25	0.24	0.35	0.24	0.15
EXP4	LO	0.19	0.21	0.27	0.14	0.041
	HI	0.27	0.28	0.34	0.19	0.07
TAU4	LO	0.12	0.18	0.25	0.11	0.083
	HI	0.29	0.3	0.37	0.22	0.12
EXP8	LO	0.084	0.14	0.18	0.073	0.012
	HI	0.41	0.34	0.39	0.32	0.18
DA8	LO	0.15	0.15	0.2	0.12	0.019
	HI	0.37	0.31	0.34	0.3	0.11
IN8	LO	0.13	0.15	0.15	0.08	0.03
	HI	0.46	0.31	0.39	0.28	0.17
DAIN8	LO	0.18	0.19	0.21	0.12	0.028
	HI	0.36	0.29	0.34	0.23	0.09

Results presented in Table 2 indicate the minimum and maximum limits of coordinate of the stationary distribution vector with respect to the considered groups: non-patients (4 sessions) and patients (no. with a group of patients indicates the number of sessions used to determine the LO and HI bounds , e.g. DA4 day-patients and 4 sessions). Of course, for LO: $\sum_{i=0,\ldots,4} TOT_i < 1$ and for HI: $\sum_{i=0,\ldots,4} TOT_i > 1$ for any group of patients and non-patients. Interpretation e.g. in the case of patients DA4 for the full questionnaire (TOTAL), who after the therapeutic process reach state 0 (the most "desirable" from the point of view of the therapist) lower limit (LO) is 0.21 and the upper limit (HI) of the same coordinate is 0.32. This means that after completion of the therapy, the percentage of patients in state 0 (declared in the questionnaire the value "0") will not be lower than 0.21 (about every fifth patient) and will not be higher than 0.32 (about every third patient). The most desirable situation would be to get the very narrow limits between LO and HI and a maximum large percentage of patients in the states with the lowest numbers (0, 1 , ...). The analysis of the control group (non-patients) shows that in the states 0 and 1 is from 65 % to 87 % non-patients and it is the largest (as expected) part of the survey participants compared to the same number with respect to an identical four sessions (the lowest relative span of the class 0: (0.45-0.32)/0.32 $\approx$ 41 %).

Moreover other components of the same stationary vector show small percentage share in the whole vector (in particular TOT_3 and TOT_4). Also in this case the important role played a relatively small span between the lower (LO) and the upper (HI) and definitely the lowest reduction in classes 2, 3 and 4 relative to the bounds obtained on the basis of 4 session for all groups of patients. Analyzing values presented in Table 2 it is easy to notice that in other cases of the groups of patients both spans are larger (with some exceptions) and in most cases the dominant states are 0, 1, 2 and 3 (with a slight predominance of state 2 for both the lower and the upper limit), which may suggest that the treatment process is not yet completed. The situation is similar for the groups of patients who completed the questionnaire for 5 subsequent sessions. The situation has changed a little in the case of state 0 among groups of patients who completed the questionnaires for 8 consecutive sessions (e.g. inpatients group for state 0: 0.19-0.41 in the case of IN4, 0.17-0.35 in the case of IN5, 0.13-0.46 in the case of IN8; inpatients group for state 1: 0.18-0.25 in the case of IN4, 0.18-0.27 in the case of IN5, 0.15-0.31 in the case of IN8, etc.).

Noteworthy, the increased number of observations narrows bounds (IN8, DA8 and combined subgroup DAIN8), which seems to be consistent with the property of a designated maximum likelihood estimator methods formula in (1) of ref. 5. In this context, for groups of 8 sessions the large spread between LO and HI for each coordinate in the group EXP8 in relation to such combined DAIN8 groups is surprising. This number is roughly comparable to and more importantly composed of two heterogeneous groups of patients. Increased span might suggest greater (relative to DAIN8) differentiation transition probability matrix between the states, and consequently the greater diversity in the therapeutic group ("deteriorating" patient selection criteria for a group) or their expectations, needs, etc. relation to the therapeutic process. Similarly on the level 5 comparing AED5 with DA5 and IN5 with EXP5 (groups similar in terms of the cardinality) can be seen that more patients will be in the states 0 and 1 and less in the states 3 and 4 in the case of DA5 to AED5 (in particular upper bounds for the DA5 in the worst states is equal 0.06, means 1 per 16-17 patients remains in state 4). In the case EXP5 to IN5 situation is not so clear. It may be due to the fact that patients in the IN5 in the vast majority finished therapy after 5 sessions ($n5 = 160$ to $n8 = 43$) in contrast to EXP8 ($n8 = 176$ to $n8 = 173$).

TAU5 is the smallest group. Potential span around a coordinate of stationary vector (the largest in the majority of cases in relation to AED5 and DA5) seem to be consistent with the expectations for such a number of sessions. The only exception is the lower limit for condition 2 (slightly higher) and upper limits of coordinates in the states 3 and 4 (extremely lower relative to AED5).

Not knowing the true value of the ergodic distribution (or stationary) Markov chains can not unambiguously determine the expected time of returning to the state i measured by means of steps between the output of the chain at the state i and the other way round to the first time to state i. It is a similar case with determination of the average transition times and the average of the return of the chain. Therefore, for the respective groups of patients, the first-pass time set maximum and minimum bounds on the basis of the upper (HI) and lower (LO) are met assuming the relevant ergodic distribution is between these limitations. The results are presented in Table 3.

As expected, the non-patients group is characterized by the longest times first return to classes "worst" (from the therapeutic point of view: grade 4 and 3), while the shortest sometimes return for classes "best" (class 0) starting with them HI_4 $\sim$ 142 periods and LO_4 $\sim$ 59 periods and periods HI_0 $\sim$ 3.13 and $\sim$ 2.22 LO_0 periods with respect to groups of patients (4 sessions). Moreover, the control group is characterized by a generally lowest difference between the time of return in the case of HI LO terms in relation to all groups of patients for classes "best" (e.g., 3.13 for LO_0 and 2.22 for HI_0, which is (3.13-2.22)/2.22 $\approx$ 41%; 3.03 for LO_1 and 2.38 for HI_1, i.e. (3.03-2.38)/2.38 $\approx$ 27%). Noteworthy, treatment groups and IN4 sometimes have lower DA4 than desired classes 0, 1, and 2 to both the HI and LO with respect to AED4 and TAU4, EXP4 (except TAU4 for

Table 3. The first-pass time: 4- and 8-sessions bounds of ergodic distributions for Polish and U.S. patients group.

Patients group	Bounds	$\mathbf{TOT_0}$	$\mathbf{TOT_1}$	$\mathbf{TOT_2}$	$\mathbf{TOT_3}$	$\mathbf{TOT_4}$
NON	LO	3.13	3.03	6.76	22.73	142.86
	HI	2.22	2.38	4.76	14.08	58.82
DA4	LO	4.76	4.76	4.17	6.67	33.33
	HI	3.13	3.85	3.23	4.55	16.95
IN4	LO	5.26	5.56	5.00	10.00	26.32
	HI	2.44	4.00	2.94	4.55	9.09
AED4	LO	7.69	5.88	3.70	6.25	15.63
	HI	4.00	4.17	2.86	4.17	6.67
EXP4	LO	5.26	4.76	3.70	7.14	24.39
	HI	3.70	3.57	2.94	5.26	14.29
TAU4	LO	8.33	5.56	4.00	9.09	12.05
	HI	3.45	3.33	2.70	4.55	8.33
EXP8	LO	16.39	8.33	5.88	14.71	90.91
	HI	2.44	2.86	2.27	2.86	5.26
DA8	LO	5.26	5.26	4.55	8.33	43.48
	HI	2.86	3.33	3.03	4.35	12.50
IN8	LO	7.69	6.67	6.67	12.50	33.33
	HI	2.17	3.23	2.56	3.57	5.88
DAIN8	LO	5.56	5.26	4.76	8.33	35.71
	HI	2.78	3.45	2.94	4.35	11.11

class 1 and 2, which may be more weight patients concentrated in these "middle" classes). Similar conclusions can be drawn in relation to patient groups, at 8 session group usually IN8, DA8 and INDA8 are characterized by lower classes, sometimes return "better" is comparable with the same in the case of EXP8.

The next crucial measure verifying efficiency of therapeutic process is the expected return times m_{ij} from the state i to the state j of the measured number of cycles of therapy (weekly). In Table 4 expected time to move from stage i to stage j is presented by assumptions that the lack of homogeneity of Markov chain (each session can be characterized by a distinct matrix $\mathbf{P}$ - state transitions, the product of which designate $\mathbf{P_{ERG}}$): average ($\mathbf{P_{ERG-MEAN}}$) of the lower and upper limit of the stationary distribution, the lower limit (m_{ij-LO}) and the upper limit (m_{ij-HI}).

Thus, the value of $m_{21} = 3.01$ (second row for TOT_1, first column for TOT_0) matrix m_{ij} for $\mathbf{P_{MICRO}}$ in the non-patients group describes the average number of weeks that must overcome the host unit at a time value of 1 ("good") and wanting to reach a value of 0 ("very good") for an average time of arrival ("mean") $m_{21\_LO} = 3.41$ for m_{ij_LO} means that the same control unit need no more than 3.41 a week and $m_{21\_HI} = 2.62$: at least 2.62 a week. As expected, in the control group (and similarly in the groups of patients) transition from "worse" to "better" requires fewer periods than the transition from the "better" to "worse" (both in the case $\mathbf{P_{MICRO}}$ and $\mathbf{P_{ERG}}$ for $i > j$: $m_{ij} < m_{ji}$ or simply: the value of the respective upper-triangular matrix **P** are much higher than the corresponding values of the lower-triangular matrix **P**, for example, $m_{21} = 3.01 < 3.8 = m_{12}$). An analogical conclusion can be drawn in relation to the rest of the matrix of expected transition times, regardless of the type of group and the duration of treatment. In addition, it can be seen that the time to return to class a "better" class "inferior" ($i > j$, i.e. the lower triangular matrix) takes a bit shorter in the case of groups DA4 and IN4, than in the case of groups AED4, EXP4 and TAU4 (e.g. $\mathbf{P_{MICRO-MEAN}}$: m_{21_IN4}=5.2, m_{21_DA4}=5.9, m_{21_AED4}=8.4, m_{21_EXP4}=6.4, m_{21_TAU4}=8.2; similarly for $\mathbf{P_{ERG}}$). A similar conclusion can be drawn considering similar in terms of the number of sessions of Polish and U.S. patients, and the greater the number of sessions, the disparity between the time of return in the case of groups of patients with the Polish and the U.S. is increasing (e.g., in the case of 8 sessions, $\mathbf{P_{MEAN}}$: m_{21_IN8}=7.3, m_{21_DA8}=6.6, m_{21_EXP8}=12; $\mathbf{P_{ERG}}$: m_{21_IN8}=5, m_{21_DA8}=4.1, m_{21_EXP8}=8.4). Due to lack of space only for the group of non-patients (4 sessions) exemplary matrices presented in Table 4 for the non-patients and Table 5 for the day-patients.

Table 4. Expected time to move from stage i to stage j non-patients (NON).

$\mathbf{P_{ERG-MEAN}}$	i \ j	$\mathbf{TOT_0}$	$\mathbf{TOT_1}$	$\mathbf{TOT_2}$	$\mathbf{TOT_3}$	$\mathbf{TOT_4}$
		m_{i1}	m_{i2}	m_{i3}	m_{i4}	m_{i5}
	m_{1j-LO}	0	4.4	6.9	15	70
$\mathbf{TOT_0}$	m_{1j}	0	3.8	6.7	17	72
	m_{1j-HI}	0	3.2	6.6	18	73
	m_{2j-LO}	3.41	0	5.06	11.7	56.2
$\mathbf{TOT_1}$	m_{2j}	3.01	0	5.01	13.2	58.7
	m_{2j-HI}	2.62	0	4.95	14.7	61.2
	m_{3j-LO}	3.86	2.93	0	10.7	53.7
$\mathbf{TOT_2}$	m_{3j}	3.54	2.77	0	12.9	59.7
	m_{3j-HI}	3.22	2.6	0	15.1	65.7
	m_{4j-LO}	4.21	3.23	4.2	0	53.7
$\mathbf{TOT_3}$	m_{4j}	3.84	3.03	4.39	0	59.5
	m_{4j-HI}	3.46	2.82	4.58	0	65.4
	m_{5j-LO}	3.87	3.41	4.46	9.83	0
$\mathbf{TOT_4}$	m_{5j}	3.5	3.15	4.57	11.6	0
	m_{5j-HI}	3.13	2.88	4.69	13.3	0

Table 5. Expected time to move from stage i to stage j day-patients (DA4).

$\mathbf{P}_{\mathbf{ERG-MEAN}}$	i \ j	$\mathbf{TOT_0}$	$\mathbf{TOT_1}$	$\mathbf{TOT_2}$	$\mathbf{TOT_3}$	$\mathbf{TOT_4}$
		m_{i1}	m_{i2}	m_{i3}	m_{i4}	m_{i5}
	m_{1j-LO}	0	4.4	6.9	15	70
$\mathbf{TOT_0}$	m_{1j}	0	3.8	6.7	17	72
	m_{1j-HI}	0	3.2	6.6	18	73
	m_{2j-LO}	3.41	0	5.06	11.7	56.2
$\mathbf{TOT_1}$	m_{2j}	3.01	0	5.01	13.2	58.7
	m_{2j-HI}	2.62	0	4.95	14.7	61.2
	m_{3j-LO}	3.86	2.93	0	10.7	53.7
$\mathbf{TOT_2}$	m_{3j}	3.54	2.77	0	12.9	59.7
	m_{3j-HI}	3.22	2.6	0	15.1	65.7
	m_{4j-LO}	4.21	3.23	4.2	0	53.7
$\mathbf{TOT_3}$	m_{4j}	3.84	3.03	4.39	0	59.5
	m_{4j-HI}	3.46	2.82	4.58	0	65.4
	m_{5j-LO}	3.87	3.41	4.46	9.83	0
$\mathbf{TOT_4}$	m_{5j}	3.5	3.15	4.57	11.6	0
	m_{5j-HI}	3.13	2.88	4.69	13.3	0

Comparing Table 4 with Table 5 is easy to notice that the values of the expected time to move for the control group are typically lower for stages 0, 1, and 2 and higher for stages 3 and 4 than for the group of patients. Similar conclusions can be drawn in relation to other groups of patients. Moreover, Polish patients are generally characterized by lower values than U.S. patients.

An important element in the analysis of phenomena modeled by Markov chain is the rate of convergence of the matrix $\mathbf{P}$ to the ergodic matrix $\mathbf{E}$. For this purpose usually tested eigenvalues of $\mathbf{P}$. It is well known[11] that an important role is played by $|\lambda_2|$ - the Second Largest Eigenvalues Module (SLEM). SLEM is the basis of most of the measures of the rate of convergence[11,12]: the lower value $|\lambda_2|$ implies a shorter time to reach the ergodic matrix E. Table 6 shows $|\lambda_2|$ for products of the matrix $\mathbf{P_{i,i+1}}$ ($\Pi_{i=1,2}\mathbf{P_{i,i+1}}, \ldots, \Pi_{i=1,\ldots,7}\mathbf{P_{i,i+1}}$) for all 4-sessions groups.

Table 6. Second Largest Eigenvalues Module 4 sessions groups.

Group\\$\lvert\lambda_2\rvert$	$\mathbf{P_{1\_2}}$	$\mathbf{P_{1\_2}P_{2\_3}}$	$\mathbf{P_{1\_2}P_{2\_3}P_{3\_4}}$
NON4	0.67	0.47	0.35
IN4	0.72	0.54	0.41
DA4	0.76	0.58	0.45
AED4	0.78	0.61	0.49
EXP4	0.74	0.57	0.44
TAU4	0.68	0.52	0.42

Table 6 results confirm the expected thesis that, a group of non-patients earlier than group of patients reaches ergodic state on each level of therapeutic process.

To measure the number of steps n necessary to achieve the ergodic distribution by homogeneous Markov chain at a given accuracy usually the coefficients of ergodicity or measures of the rate of convergence are used[11,12]. Unfortunately, these measures do not work in the case of a MSC. Future studies may explore this issue.

4. Conclusions

The article examines the usefulness of Markov chains and Markov set-chains in assessing the efficacy of the therapeutic process. The proposed approach takes into account the dynamic changes occurring in the course of the psychotherapy and therefore allows to constantly adjust the therapy process. Depending on the nature of the date an approach based on homogeneous and non-homogeneous Markov chain was proposed. In the first case, the efficiency of the process can be controlled by the transition probability matrix $\mathbf{P}$ and related measures of the type: ergodic distribution, SLEM, first-pass time, expected time to move from the stage i to j, whereas in the second case on the basis of Markov set-chains and related intervals estimates. As expected, the results for the non-patients group were significantly different when compared to the results obtained for both Polish and American groups of patients. It seems that proposed approach can be useful for psychotherapy practitioners.

References

1. M. J. Lambert, J. J. Morton, C. Harmon, S. Hamilton, R.C. Reid, K. Shimokowa, C. Christopherson and G.M. Burlingame, *Administration and Scoring Manual for the Outcome Questionnaire -45.* Orem, UT: American Professional Credentialing Services (2004).
2. L. R. Derogatis, *The SCL-90 manual: Scoring, administration and procedures for the SCL-90.* Baltimore: Johns-Hopkins University School of Medicine, Clinical Psychometrics Unit (1977).
3. A. T. Beck, R. A. Steer and Garbin, M. G. *Clinical Psychology Review, 8,* 77100 (1988).
4. J. G. Kemeny, and J. L. Snell, *Finite Markov Chains*, New York (1960).
5. T. C. Lee, G. G. Judge, and A. Zellner, *Estimating the Parameters of the Markov Probability Model from Aggregate Time Series*, North-Holland, Amsterdam (1970).
6. P. Sliwka, M. Dudkiewicz, *Biomat 2008*, 235-251 (2009).

7. D. J. Hartfiel, Markov set-chains, Springer, New York (1998).
8. D. J. Hartfiel, *J. Satist. Comput. Simulation.* **38**, 947 (1994).
9. M. J. Lambert, G. M. Burlingame, V. Umphress, N. B. Hansen, D. A. Vermeersch, G. C. Clouse and C. Christopherson, *Clinical Psychology and Psychotherapy, 3*, 249258, (1996).
10. W. Simon, P. Sliwka, J. A. Sobanski, K. Klasa, P. Sala, W. Zak, G. Busath, and M. J. Lambert In press.
11. E. Seneta, *Non-negative Matrices and Markov Chains*, Springer (1981).
12. P. Sliwka, *Statystyka aktuarialna - stan i perspektywy rozwoju w Polsce*, 403-423, Wrocaw (2006).

IN SILICO MANIPULATION OF SINGLE DNA MOLECULES

R. CORTINI, M. BARBI

Laboratoire de Physique Théorique de la Matière Condensée,
Université Pierre et Marie Curie, CNRS UMR 7600
4, Place Jussieu, 75252 Paris Cedex 05, France
E-mail: cortini@lptmc.jussieu.fr barbi@lptmc.jussieu.fr

P. CARRIVAIN

Institute of Human Genetics, CNRS UPR 1142,
Montpellier 34000, France
E-mail: pascal.carrivain@igh.cnrs.fr

We recently developed a new powerful method to reproduce single-molecule manipulation experiments *in silico*[1]. As a proof of principle, we demonstrate that flexible polymers like DNA can be simulated at nanometer resolution using physics engines thanks to an original implementation of Langevin Dynamics in the open source library called Open Dynamics Engine. We moreover implement a global thermostat which accelerates the simulation sampling by two orders of magnitude. We retrieve force-extension as well as rotation-extension diagrams of reference experimental studies. We are currently implementing this methodology to simulate circular DNA and the protein-DNA assemblies that organize DNA in cell nuclei, the nucleosomes.

1. Introduction

Holonomic constraints in statistical mechanics have been a long standing subject of debate, both theoretically and numerically (for a review see Frenkel and Smit textbook[2]). They originate from the need to deal with frozen degrees of freedom in coarse grained modeling, as in articulated mechanical systems. Constraint algorithms as SHAKE[3] and its derivatives have been designed for molecular dynamics, subsequently extended to Langevin dynamics[3,4] and since then have been widely used in simulations of macromolecules, polymers and proteins. These algorithms are all based on the introduction of additional forces into the system to maintain the constraints, and are typically based on an iterative approach: the equations of motion of the unconstrained system are first solved, then additional forces are calculated and applied to the system iteratively until the constraint equations are verified to a given tolerance. In 1996 Baraff first proposed the use of non-iterative linear-time simulation methods to articulated rigid bodies in the context of computer graphics, thus avoiding the numerous time-consuming correction steps of iterative methods[5]. In the wake of Baraff approach, "velocity based" algorithms have been developed. Instead of using additional forces to enforce the holonomic constraints, these algorithms compute additional forces that allow to adjust the velocities of the rigid bodies so that the errors on the constraints are corrected in one time step. This approach presents important advantages, as the possibility of increasing the time step size and preventing indeterminacy problems. An exhaustive review can be found in Ref.[6]. The introduction of velocity–based algorithms has impulsed the development of new powerful tools, called physics engines, that have been developed by the mechanics and robotics community to reproduce the dynamic behavior of articulated systems of rigid bodies. Physics engines are acquiring an increasing importance, notably in the fields of computer graphics and video games, where they are used to simulate rigid body motion under realistic conditions and in real-time. Open Dynamics Engine (ODE) is one of the most popular rigid-body dynamics implementations for robotics applications[7]. As other physics engines, ODE simulates in an optimal way the kinematics of articulated systems of any shape by using permanent joints that impose holonomic constraints, instead of bond potentials (as typically used in bead spring models). The same method is used to manage collisions: when an overlapping between bodies is detected, a temporary joint is locally created that reproduces the action of the contact forces, again without the need of explicit permanent 2-body interaction potentials.

These extremely efficient simulators have not, up to now, been used in statistical mechanics. In our previous work[1], we shown how to implement a Langevin dynamics in a physics engine velocity based algorithm. By extending the Bussi-Parinello "global thermostat" algorithm[8] we further obtained a considerable acceleration of the simulation sampling. A clear demonstration of the efficiency our method can be given by the simulation of the behavior of a single DNA molecule in the same conditions as in magnetic tweezers experiments. These "in silico" single molecule manipulations have a potentially important role for the interpretation and prediction of experimental results. The interpretation of experiments often requires, indeed, an important modeling effort, particularly in the more complex cases where DNA is associated with proteins, as for instance in chromatin assemblies[9,10]. The technique presented in this paper should improve significantly the modeling effort, especially in the most complex conditions (forthcoming publication).

2. DNA Single Molecule Experiments Using Magnetic Tweezers

2.1. *DNA topology and elasticity*

In the nucleus of eukaryotic cells, DNA compaction is achieved through a highly complex hierarchical architecture, which moreover should fulfill functional properties in order to regulate DNA accessibility and gene transcription, in a tissue-specific and time dependent manner.

This architecture has to deal with the mechanical and topological properties of DNA, which are at play during all important biological processes. During transcription, in particular, DNA has to be screwed inside the transcription factory and DNA (with its associated proteins) should accommodate the torsional and mechanical constraints upstream and downstream of the transcription site.

A precise definition of the DNA topology is therefore necessary to describe its behavior in such conditions[11]. The mathematical definition of the three relevant topological quantities, i.e. the twist Tw, the writhe Wr and the linking number Lk has been precisely introduced in the context of DNA conformations by Fuller[12,13]. To introduce them simply, imagine to rotate the extremity of a partially stretched ribbon around its local tangent. The ribbon may modify its geometrical conformation by accumulating the imposed rotation in two different forms: its twist and the writhe. The ribbon twist is its excess torsion around its own axis, as observed for instance if the ribbon is stretched sufficiently strongly to keep it straight. If the stretching force is reduced, the ribbon may also assume a curved shape. The ribbon writhe, which only depends on its geometrical path in space, measures the curling of the resulting path.

For a filament whose ends are oriented in fixed directions, if one of them is rotated around its tangent, then the sum of the twist and the writhe variations will always correspond to the number of turns applied. Fuller showed that those quantities satisfy indeed the famous Călugăreanu-White-Fuller theorem $\Delta Lk = \Delta Tw + \Delta Wr$, where Lk is the filament *linking number*. For a closed ribbon, the linking number is therefore a topological invariant.

Associated with these topological deformations of DNA, there are evidently some energy costs, which depend on the DNA elasticity with respect to different strains: bending elasticity for writhe, torsional elasticity for twist. The DNA stiffness with respect to bending or twisting is commonly described by assigning the corresponding bending and twist persistence lengths, L_P^B and L_P^T, defined as the length over which the chain loses its orientational or torsional correlation. The energy costs can then be simply estimated, in the linear limit, as quadratic functions of the local bending and twist angles, with proportionality parameters depending on L_P^B and L_P^T, respectively.

2.2. *Magnetic tweezers*

Because DNA response to mechanical constrains is a relevant features in several *in vivo* processes, it has been the object of experimental investigations by the biophysics community. Since single-molecule manipulation techniques have become available[14,15], a large amount of experimental data have been accumulated on DNA[15,16,17,18]. In particular, magnetic tweezers allow for the application of a controlled torsion on a single DNA molecule (or DNA-protein assembly) at constant stretching force. In this experiments, a super-paramagnetic bead is tethered to one end of a DNA molecule, whose second end is tethered to a glass coverslip. Magnets allow to exert a controlled pulling force on the bead and to rotate it, hence to apply controlled mechanical constraints to the DNA molecule (Figure 1). Importantly, in these experiments the only physical variable that can be measured directly is the DNA extension, *i.e.* the distance between its two ends. The interpretation of experimental results requires to model the DNA behavior in order to evaluate its end-to-end distance dependence on the external constrains. If theoretical approaches may be successful in some case[19,20,21], simulations are often crucial tests of the proposed model validity.

2.3. *Force extension curves*

The simplest investigation that this experimental tool make accessible is the measure of the elastic behavior of DNA when subject to a pulling force[14]. At low forces, the DNA molecule is still considerably disordered in space due to thermal fluctuations. Such fluctuations tends to shorten its end-to-end distance against the applied force, resulting in an elastic behavior of purely entropic origin. This behavior is observed up to forces of the order of 0.1 pN. Above this value, DNA starts to be aligned, and its intrinsic elasticity comes into play.

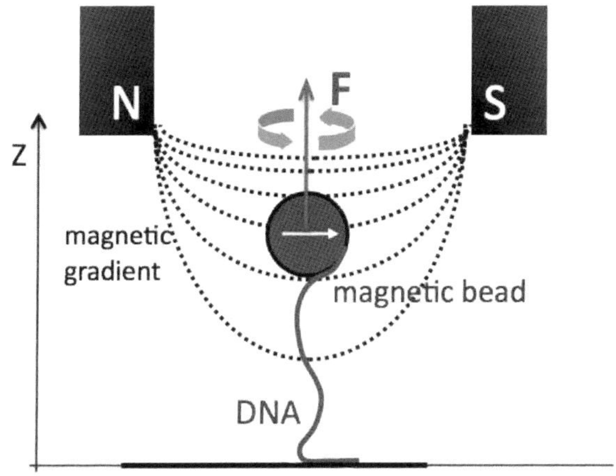

Figure 1. Schematic view of the magnetic tweezers set up, allowing to manupulate a single DNA molecule.

The behavior observed at low forces can be approximated by modeling DNA as an ideal chain (or freely-jointed chain, FJC), i.e. by assuming that it is equivalent to a chain of rigid rods of a fixed length with completely independent orientations and no interactions between monomers. However, this model fails at larger forces (and extensions). A much better fit of the experimental data is obtained by a Worm-Like Chain model, that considers DNA as a continuous, elastic rod, and takes into account the bending stretching energy involved in distorting the chain[22].

Above approximately 10 pN, the stretching modulus of DNA becomes important, and data deviates from the WLC behavior[a]. In this regime, the intrinsic molecular elasticity comes into play and a linear stretching behavior becomes more pertinent[24,25].

For force larger than 65 pN, finally, the DNA molecule suddenly overstretches to about 1.7 times its contour length[26], a transition that involves important rearrangements of its internal structure and leads to the so called S-DNA.

[a]For a critical discussion of the WLC model and a comparison with a three-dimensional continuum mechanics approach, see Ref.[23].

2.4. *Rotation-extension curves*

In the characteristic rotation-extension "hat curves" the DNA extension decreases linearly once reached a critical point where the resulting applied torque crosses a critical buckling value, Γ_c. This linear decrease corresponds to the formation of plectonemes, i.e. loops of helices twisted together (see an example in Figure 3). Any additional turn beyond the buckling transition is absorbed in the growing plectoneme without changes in the torque[21]. Since standard magnetic tweezers cannot measure torques, the buckling torque can be only indirectly deduced by integrating the change of the molecule extension with respect to the applied force[27,28]. However, new setups have recently been proposed to measure and impose the torque[29,30,31,32]. These innovative experiments confirm that the torque stays constant during the plectoneme formation, and allow to investigate the dependence of Γ_c on the applied stretching force. However, these techniques allow only to *impose* the torque via a retroaction loop, and it is therefore not possible to measure the torque fluctuations when the magnetic bead is held fixed (at a given torsion).

The linear decrease of the end-to-end distance is observed, for positive torques, up to pulling forces of the order of 3 pN. For negative torsions, however, the unpairing of DNA base pairs (DNA denaturation) is induced approximately above 0.3 pN, and a coexistence of paired and open base pairs is observed. In this regime, the additional torsion is adsorbed in the increase of the unpaired (and untwisted) DNA segment, and this results in a constant length behavior[33]. Again, for very high forces (and for positive torsions), a strong deformation of the molecule can be induced, leading to a new molecular arrangement that has been named P-DNA[34].

3. Simulation Tools

3.1. *Physics engines and rigid bodies modeling*

Open Dynamics Engine (ODE) is a simulation tool where the state of a system consisting of N rigid bodies is described by the positions $\mathbf{r}_i$ of their centers of mass, a quaternion representation of their orientations q_i, and their linear and angular velocities $\vec{v}_i$ and ω_i respectively, collected in the column vector $\vec{\mathcal{V}} = \{\vec{v}_1, \omega_1, \ldots, \vec{v}_N, \omega_N\}^T$. The vector $\vec{\mathcal{L}} = \mathcal{M}\vec{\mathcal{V}}$ collects all linear and angular momenta, where $\mathcal{M} = \{m_1, \mathcal{I}_1 \ldots m_N, \mathcal{I}_N\}$ is a $6N \times 6N$ block diagonal matrix whose elements are the mass matrices $_i$ and inertia matrices $\mathcal{I}_i$ of the N bodies (with the 3×3 identity matrix). The Newtonian dynamics equation then reads $\dot{\vec{\mathcal{L}}} = \vec{\mathcal{F}}$ where the generalized force $\vec{\mathcal{F}}$ is a vector collecting forces and torques applied to the system.

These may be external or internal forces and torques. In particular, in articulated systems as polymer models, the rigid bodies that compose the system are connected by mechanical joints, defined as relationship that is enforced between two bodies so that they can have only certain positions and orientations relative to each other. Different types of joints can be implemented in ODE leading to different kinds of articulations. In our model, we will use ball-in-socket joints that, mathematically, impose some holonomic constraint between both connected bodies, that reads $\vec{\delta} = 0$ where $\vec{\delta}$ is the distance between both joint bearings. Mechanical joints thus exert reaction forces and torques on the joint bearings that are collected into a generalized constraint force $\vec{\mathcal{F}}_c$. Since bodies are rigid, the principle of virtual work $\vec{\mathcal{F}}_c^T \vec{\mathcal{V}} = 0$ reads $\vec{\mathcal{F}}_c = \mathcal{J}^T \vec{\lambda}$ where $\vec{\lambda}$ is a vector of Lagrange multipliers that precisely accounts for the reaction forces and torques coming from the joint bearings[6].

The Newtonian dynamics equation therefore reads $\dot{\vec{\mathcal{L}}} = \vec{\mathcal{F}}_e + \vec{\mathcal{F}}_c$ where $\vec{\mathcal{F}}_e$ and $\vec{\mathcal{F}}_c$ are the external and internal generalized force respectively and, using the previous relation, becomes an equation for $\vec{\lambda}$ that reads $\dot{\vec{\mathcal{L}}} = \vec{\mathcal{F}}_e + \mathcal{J}^T \vec{\lambda}$. Solving this equation for $\vec{\lambda}$ should satisfy the holonomic constraints $\vec{\delta} = 0$ at every timestep t. However the discretization used in the numerical calculation results in errors on $\vec{\delta}$. In order to keep both $\vec{\delta}$ and $\mathcal{J}\mathcal{V}$ close to zero at every timestep, ODE introduces an error reduction parameter[7] that has to be adjusted to some optimal value between 0 (no correction at all) and 1 (complete correction of $\vec{\delta}$ in one timestep). Recommended values are between 0.1 and 0.8.

3.2. *Detection of collisions*

When dealing with rigid bodies, collisions needs to be correctly detect and the bodies trajectories to be corrected accordingly. The way how collision detection is implemented is one of the most interesting aspects of ODE. ODE uses indeed a powerful software called *libccd*[35] to detect collisions between two convex shapes, which is based on a simple and smart principle: whenever an overlapping is detected between two rigid bodies, a temporary joint, called a “contact joint”, is created between them, for each contact point. The contact joint also accounts for extra information about the contact, as the friction present at the contact surface, its softness, etc. The kinematic constraint imposed by the contact joint allow to determine the

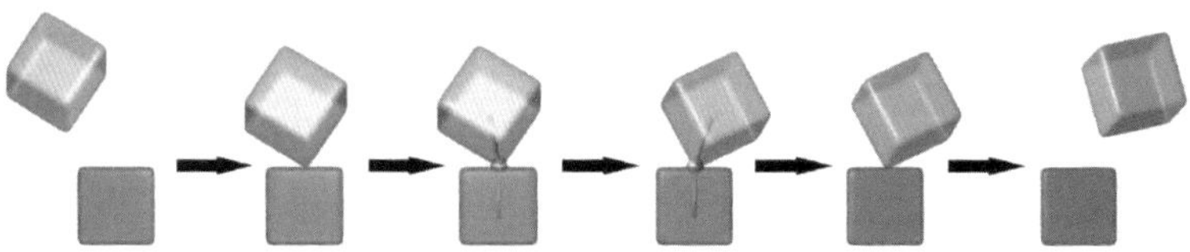

Figure 2. Principle of collision handling in a rigid body dynamics simulation.

associated constraint force. The motion solver gives then the new positions and orientations of the articulated bodies at time $t + \Delta t$, so that the bodies effectively behave as in presence of a colliding contact between two rigid surfaces with given rigidity and roughness. Contacts joints are then destroyed before moving to the next step. The sequence of events is summarized in Figure 2.

3.3. *Thermal bath implementation*

Having been developed to model macroscopic mechanical systems, ODE does not provide any tool for the simulation of microscopic systems in equilibrium with a thermal bath or in any case animated by thermal fluctuations. In a previous work[1], we implemented Langevin-Euler equation in the ODE algorithm. A "standard" way to do this is to add to mechanical forces an additional, thermal contribution $\vec{\mathcal{G}} = -\boldsymbol{\Sigma}\vec{\mathcal{L}} + \boldsymbol{\Xi}\dot{\vec{\mathcal{W}}}$ containing a frictional term $-\boldsymbol{\Sigma}\vec{\mathcal{L}}$, with $\boldsymbol{\Sigma}$ the matrix of the coupling frequencies to the thermostat and a random force vector $\boldsymbol{\Xi}\dot{\vec{\mathcal{W}}}$ obtained by applying the matrix $\boldsymbol{\Xi}$ of white noise amplitudes to a generalized vector $\dot{\vec{\mathcal{W}}}$ of normalized and independent Wiener processes.

In such local thermostats, therefore, a correction force including both a frictional term and a stochastic term is exerted on each particle to drive the system to the canonical distribution at a prescribed temperature. Global schemes of Langevin dynamics are designed to minimize the perturbation introduced by the thermostat on the Hamiltonian trajectory (so called "disturbance" as defined originally in Ref.[36]), hence on the dynamical properties, such as autocorrelation functions, and related quantities, such as diffusion coefficients.

Two main global algorithms have been designed so far: (a) Stochastic Velocity Rescaling methods, most notably the "global thermostat" introduced by Bussi and Parrinello[8], and (b) the Nosé-Hoover thermostat[37]. In both these global approaches the stochastic term of the correction force acting on each particle is proportional to the momentum of that particle. The scheme designed by Bussi and Parinello preserves the dynamical properties and allows for a faster sampling of the phase-space[8]. The Bussi Parinello algorithm is, however, designed for the translational degrees of freedom only. In order to apply it to a articulated rigid body system, we extend it to the rotational degrees of freedom and adapt it to the ODE algorithm. To this aim, we replace the traditional Langevin correction force (local thermostat) $\vec{\mathcal{G}}$ by a corresponding global version $\tilde{\vec{\mathcal{G}}}$, which reads

$$\tilde{\vec{\mathcal{G}}} = \Sigma^\star \left[\frac{\langle E \rangle}{E} \left(1 - \frac{1}{2\beta \langle E \rangle} \right) - 1 \right] \vec{\mathcal{L}} + \sqrt{\frac{\Sigma^\star}{\beta E}} \dot{\mathcal{W}} \vec{\mathcal{L}}. \tag{1}$$

Equation (1) shows that $\tilde{\vec{\mathcal{G}}}$ is proportional to $\vec{\mathcal{L}}$, so that the stochastic generalized forces associated with the thermostat are in the same direction as $\vec{\mathcal{L}}$. The global version of the Langevin dynamics minimizes the disturbance induced by the thermostat on the Hamiltonian trajectory (equal to $\tilde{\vec{\mathcal{G}}}^T \mathcal{M}^{-1} \tilde{\vec{\mathcal{G}}}$[1,8]), nevertheless retaining the same thermalization speed as usual Langevin dynamics.

Interestingly, when used in the framework of ODE, the global thermostat presents an additional advantage. The global Langevin contribution $\tilde{\vec{\mathcal{G}}}$ is indeed decoupled from the constraint forces, in the sense that it cancels out in the equations for $\vec{\mathcal{F}}_c$. Hence, $\tilde{\vec{\mathcal{G}}}$ not only minimizes the disturbance of the Hamiltonian trajectory, but also does not disturb the generalized constraint force $\vec{\mathcal{F}}_c$. Both effects cooperate and lead to an important acceleration of the simulation sampling, that is, in the case of our model, approximately 100 times faster than with the local thermostat[1].

4. DNA Chain Modeling

We model the DNA molecule as a linear chain of rigid cylinders of length $l = 3.34$ nm, corresponding to 10 base pairs each. The total DNA contour length L corresponds to a number $N = L/l$ of cylinders. It is well known that DNA-DNA electrostatic repulsion can be accounted for by an effective DNA diameter[20,38,39]. We chose therefore the effective radius r of the cylinders as to reproduce its mechanical and electrostatic properties in given salt conditions, a radius of $r = 2$ nm (used in the following here) corresponding to what observed at concentration of the order of 100 mM monovalent salt.

We connect the cylinders to each other by ball-in-socket joints, i.e. freely rotating points linking the two bodies and needing three holonomic constraints. The local chain bending is then defined by angle θ_i with $\cos\theta_i = \boldsymbol{t}_i^T \boldsymbol{t}_{i+1}$ where $\boldsymbol{t}_i$ is the tangent vector to cylinder i. The corresponding bend energy cost is then of the form $E_b = g_b\theta^2/2$, where the bending rigidity constant g_b is related to the polymer persistence length L_P^B through the equilibrium condition stating that the thermal average of the local chain bending should equal its ensemble average obtained from the square end-to-end distance[40].

The twist energy reads $E_t = g_t\phi^2/2$, with ϕ the local polymer twist. The twist rigidity constant g_t is defined similarly to its bending analogue. Torsional and bending ridigities have been implemented in our ODE DNA model by defining appropriate bending and twist resisting torques in order to mimic the polymer rigidity. These definitions should be coherent with the previous expressions for the bending and twist energy and, since the energy only depends on the orientation of two successive segments, consistent definitions of resisting torques are obtained in Ref.[1].

5. Results

We will describe here simulation results aiming to to reproduce the behavior of a DNA molecule under magnetic tweezers simulations, either in force-extension and in extension-rotation conditions. The same kind of simulations have been presented in Ref.[1], where a detailed discussion concerning the determination of the buckling torque leading to the transition to plectonemes.

We simulate a DNA of contour length $L = 1\ \mu$m anchored, at one end, to a surface, and at the other end, to a bead, as in magnetic tweezers experiments [41]. A snapshot of the simulation is shown in Fig. 3. In the conformation represented in the figure, a large number of turns has been applied to the magnetic bead, this leading to the formation of two plectonemes.

5.1. *Force-extension curves*

With this numerical set up we were able to simulate both force-extension and extension-rotation experiments, and to compare them to experimental results. In Fig. 4, the averaged extension $\langle\epsilon\rangle = z/L$ of the DNA molecule is plotted as a function of the force F applied to the DNA molecule. The simulation well reproduces the WLC behavior, shown for comparison, and confirms the validity of our approach to the modeling of the DNA bending persistence.

Figure 3. Snapshot of a magnetic tweezers simulation with two plectonemes formed following the application of a positive torsion to the bead.

5.2. *Extension-rotation curves*

We then simulate the extension-rotation behavior of our DNA model by imposing the bead rotation and comparing to experimental data[28], where the relative DNA extension ϵ is given for a given stretching force as a function of the relative overtwist $\sigma = \Delta \mathrm{Lk}/\mathrm{Lk}_0$ (Fig. 5)[b].

We notice that an evident limitation of the model developed here is that it cannot reproduce large modifications of the double helix structure, as base pair opening or S-DNA and P-DNA transitions, observed in magnetic or optical tweezers experiments at high force and torsion. However, we can conveniently describe all situations involving DNA supercoiling, that are of main importance for many protein-DNA complexes relevant *in vivo.*

[b] Lk_0 represents the intrinsic linking number of the relaxed DNA double helix.

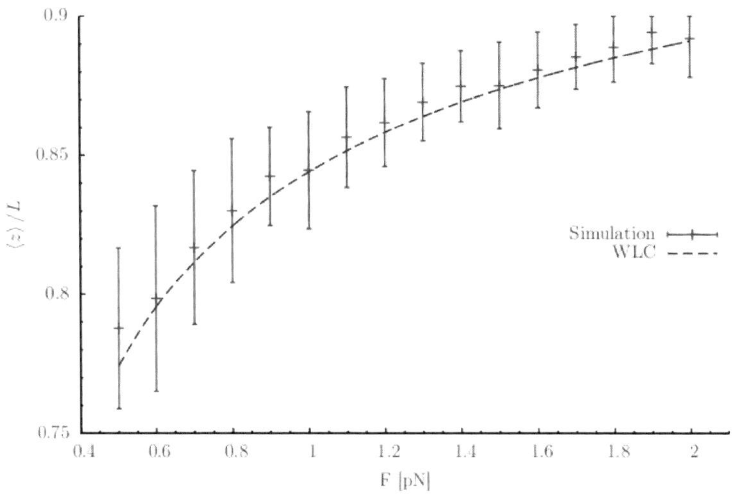

Figure 4. Force-extension behavior of the model with bending persistence length $L_P^B =$ ..., total length $L =$ .... Error bars are... Dashed line: Worm-Like-Chain formula with bending persistence length $L_P^B =$ ....

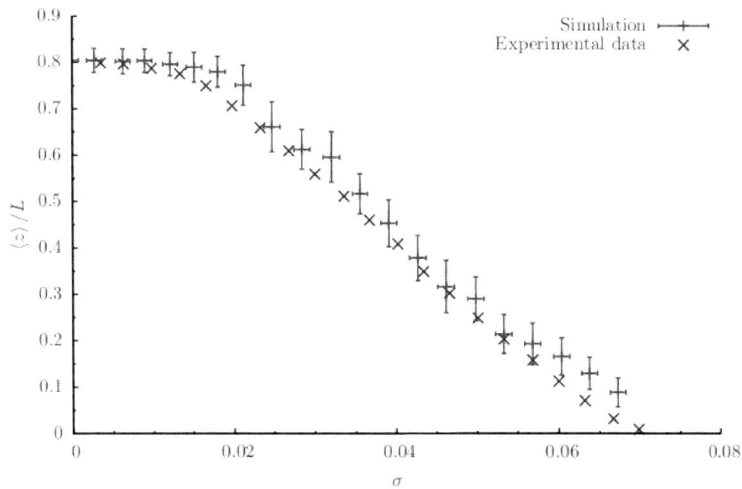

Figure 5. Relative mean average extension as a function of the overtwist σ for a stretching force f of ... pN. We superimpose to the simulation points the experimental results from Mosconi (2009) (crosses). Error bars ...

6. Perspectives: Toward a Numerical Modeling of the Architectural Protein-DNA Complexes

6.1. *The nucleosome: a metastable structure*

The basic units of the hierarchical architecture that allows the functional and compact arrangement of DNA in the cell nucleus is the nucleosome. A nucleosome is a spool of 8 proteins (histones) on which approximately 150 DNA base pairs (bp) are wrapped in about 1.7 turns, thus forming a left-handed super-helix. Nucleosomes are more or less regularly spaced on the genome every 200 bp, forming a bead on a string array, that is further compacted and folded in a still largely unknown way.

In 2006, Bancaud et al. obtained the first magnetic tweezers manipulation of nucleosome arrays, i.e. DNA stretches on which up to 30 regularly-spaced nucleosomes were reconstituted[9,10,42]. Compared to DNA, nucleosome arrays are shorter and the center of their rotation-extension curves are shifted to negative values. This shift is the expected consequence of the introduction of approximately one *negative* superhelical turn per nucleosome with respect to the corresponding naked DNA. The observed shortening, of about 50 nm per nucleosome, results from DNA wrapping around the histone core.

A very unexpected result of these experiments is the observation of a hysteretic behavior when the nucleosome array is submitted to extensive positive supercoiling[10]. This hysteresis has been interpreted as a consequence of the trapping of positive turns in individual nucleosomes through their transition to an altered form, called reversome (for reverse-nucleosome). The altered form should be metastable in order to explain the presence of a hysteresis, and has to be stabilized by the application of a large positive torque. Moreover, the reversome must contribute to the fiber linking number with a positive value, in order to explain the trapping of positive turns. The structure of this altered nucleosome is *a priori* not known, but it should contribute to the array total linking number (rotation) for approximately an opposite contribution (i.e. approximately one *positive* superhelical turn) with respect to standard nucleosomes. This leads to the suggestion that the reversome could be a right-handed, chiral symmetric of the nucleosome, obtained through a chiral inversion of the nucleosome and implying the breaking of internal protein-protein bonds and a dramatic rearrangement of the proteins in the core particle[10].

Assuming the existence of these two nucleosome states and of an energy barrier between them, we have been able to to built a minimal statistical mechanical model able to reproduce the observed behavior[42,43]. The main mechanism underlying the appearance of an hysteresis is that the free energy profile of the two-state system is deformed by the applied positive torque, leading to the stabilization of the reversome state, which is initially metastable. An appropriate choice of the model parameters leads to a convenient fit of the experimental data[43].

6.2. *Toward a numerical modeling of the nucleosome assembly*

If the hypothesis of a metastable reversome succeeded in reproducing the observed behavior under strong positive torsion, proving the plausibility of the chiral transition at the molecular level remains challenging. In order to address this important question, it may be useful to built a rigid-body model of the nucleosome in the framework of our ODE simulation tool, accounting, one one side, for its geometrical arrangement, and, on the other side, for its flexibility and the existence of weak bonds that can be broken under the action of an external mechanical constraint. To this aim, we have first performed a normal mode analysis of the inner part of the nucleosome, called the tetrasome ($\sim$ 50 basepairs wrapped on the inner four proteins) to identify the three main modes corresponding to internal deformations linked to the most collective movements of this structure[10]. A study of the internal structure of the assembly, together with different experimental results on the nucleosome mechanical response to various stresses, can be then used to propose a nucleosome model as an assembly of articulated rigid bodies with some internal breakable bonds[44,45]. Preliminary investigations allowed us to reproduce a first series of experimental observations and to show that a chiral inversion is indeed possible[10,45], and we are now able to start simulating nucleosome arrays in the same condition as in experiments (forthcoming paper).

6.2.1. *Concluding remarks*

The method developed here gives an unparalleled opportunity to study much more complex biological systems, as protein-DNA complexes, as modeling of magnetic tweezers response of nucleosome assemblies[10], by associating our *in silico* DNA model with a rigid body representation of the histone octamer. This task is challenging since a correct representation of such large scale molecular assemblies requires solving several problems. The first one is to model a polymer, i.e. an articulated chain, with the correct twisting and bending elasticities. The effective diameter of DNA (depending on electrostatic conditions) should be taken into account and, when proteins are present, one needs moreover to model their shape and excluded volume. Hence, since steric hindrance is taken into account, it is also necessary to deal with collisions, notably in order to reproduce DNA supercoiled structures and steric effects in DNA-protein assemblies. Finally, the model should include statistical mechanics features to account for thermal motion. As we have shown here, all these diverse tasks can be accounted for by our approach, combining articulated rigid body dynamics, special elastic restoring forces, an efficient collision detection, and an efficient thermal bath simulation.

References

1. P. Carrivain, A. Cournac, C. Lavelle, A. Lesne, J. Mozziconacci, F. Paillusson, L. Signon, J.-M. Victor and M. Barbi, *Soft Matter* **8**, 9285 (2012).
2. F. D. and S. B., *Understanding Molecular Simulation*, 2 edn. (San Diego: Academic Press, 2002).
3. J.-P. Ryckaert, G. Ciccotti and H. J. Berendsen, *Journal of Computational Physics* **23**, 327 (1977).
4. W. van Gunsteren, H. J. Berendsen and J. Rullmann, *Molecular Physics* **44**, 69 (1981).
5. D. Baraff, Linear-time dynamics using Lagrange multipliers, in *Proceedings of the 23rd annual conference on Computer graphics and interactive techniques*, SIGGRAPH '96 (ACM, New York, NY, USA, 1996).
6. K. Erleben, Stable, Robust, and Versatile Multibody Dynamics Animation, PhD thesis, The Department of Computer Science, University of Copenhagen, Denmark. 2005.
7. R. Smith, Open dynamic engine (2008).
8. G. Bussi and M. Parrinello, *Computer Physics Communications* **179**, 26 (2008) Special issue based on the Conference on Computational Physics 2007 - CCP 2007.
9. A. Bancaud, N. Conde e Silva, M. Barbi, G. Wagner, J.-F. Allemand, J. Mozziconacci, C. Lavelle, V. Croquette, J.-M. Victor, A. and Prunell, J.-L. Viovy, *Nat Struct Mol Bio* **13**, 444 (2006).
10. A. Bancaud, G. Wagner, N. Conde e Silva, C. Lavelle, H. Wong, J. Mozziconacci, M. Barbi, A. Sivolob, E. Le Cam, L. Mouawad, J.-L. Viovy, J.-M. Victor, *Molecular cell* **27**, 135 (2007).
11. J. Vinograd, J. Lebowitz, R. Radloff, R. Watson, P. Laipis, *Proc. Natl. Acad. Sci. USA* **53**, 1104 (1965).
12. F. B. Fuller, *Proc. Natl. Acad. Sci. USA* **68**, 815 (1971).
13. F. B. Fuller, *Proceedings of The National Academy of Sciences* **75**, 3557 (1978).
14. S. Smith, L. Finzi and C. Bustamante, *Science* **258**, 1122 (1992).

15. T. R. Strick, J.-F. Allemand, D. Bensimon, A. Bensimon, V. Croquette, *Science* **271**, 1835 (1996).
16. Bryant Zev, Stone Michael D., Gore Jeff, Smith Steven B., Cozzarelli Nicholas R. and Bustamante Carlos, *Nature* **424**, 338 (jul 2003), 10.1038/nature01810.
17. Strick Terence R., Croquette Vincent and Bensimon David, *Nature* **404**, 901 (apr 2000), 10.1038/35009144.
18. Koster Daniel A., Croquette Vincent, Dekker Cees, Shuman Stewart and Dekker Nynke H., *Nature* **434**, 671 (mar 2005), 10.1038/nature03395.
19. J. D. Moroz and P. Nelson, *Proc. Natl. Acad. Sci. USA* **94**, p.14418 (1997).
20. S. Neukirch, *Phys. Rev. Lett.* **93**, p.198107 (Nov 2004).
21. J. F. Marko, *Phys. Rev. E* **76**, p.021926 (Aug 2007).
22. O. Kratky and G. Porod, *Rec. Trav. Chim. Pays-Bas* **68**, 1106 (1949).
23. R. W. Ogden, G. Saccomandi and I. Sgura, *Proceedings of the Royal Society A: Mathematical, Physical and Engineering Science* **462**, 749 (2006).
24. C. Bustamante, S. B. Smith, J. Liphardt and D. Smith, *Current Opinion in Structural Biology* **10**, 279 (2000).
25. T. Odijk, *Macromolecules* **28**, 7016 (1995).
26. S. B. Smith, Y. Cui and C. Bustamante, *Science* **271**, 795 (1996).
27. H. Zhang and J. F. Marko, *Phys. Rev. E* **77**, p.031916 (Mar 2008).
28. F. Mosconi, J. F. Allemand, D. Bensimon and V. Croquette, *Phys. Rev. Lett.* **102**, p.078301 (Feb 2009).
29. A. Celedon, I. M. Nodelman, B. Wildt, R. Dewan, P. Searson, D. Wirtz, G. D. Bowman and S. X. Sun, *Nano Letters* **9**, 1720 (2009).
30. S. Forth, C. Deufel, M. Y. Sheinin, B. Daniels, J. P. Sethna and M. D. Wang, *Phys. Rev. Lett.* **100**, p.148301 (2008).
31. Deufel Christopher, Forth Scott, Simmons Chad R, Dejgosha Siavash and Wang Michelle D, *Nat Meth* **4**, 223 (mar 2007), 10.1038/nmeth1013.
32. M. Wang, H. Yin, R. Landick, J. Gelles and S. Block, *Biophysical Journal* **72**, 1335 (1997).
33. T.R. Strick, J.-F. Allemand, D. Bensimon and V. Croquette, *Biophysical Journal* **74**, 2016 (1998).
34. J. F. Allemand, D. Bensimon, R. Lavery and V. Croquette, *Proceedings of the National Academy of Sciences* **95**, 14152 (1998).
35. D. Fiser, Library for collision detection between complex shapes.
36. H. J. C. Berendsen, J. P. M. Postma, W. F. van Gunsteren, A. DiNola and J. R. Haak, *The Journal of Chemical Physics* **81**, 3684 (1984).
37. A. Samoletov, M. Chaplain and C. Dettmann, *Journal of Statistical Physics* **128**, 1321 (2007).
38. N. Clauvelin, B. Audoly and S. Neukirch, *Biophysical Journal* **96**, 3716 (2009).
39. D. Argudo and P. K. Purohit, *Acta Biomaterialia* **8**, 2133 (2012).
40. M. D. Frank-Kamenetskii, A. V. Lukashin and V. V. Anshelevich, *J. Biomol. Struct. Dynam.* **2**, 1005 (1985).
41. A. Vologodskii, *Macromolecules* **27**, 5623 (sep 1994).
42. P. Recouvreux, C. Lavelle, M. Barbi, N. Conde e Silva, E. Le Cam, J-M. Victor and J-L Viovy, *Biophysical J.* **100**, 2726 (2011).

43. M. Barbi, J. Mozziconacci, H. Wong and J.-M. Victor, *Journal of Mathematical Biology* **68**, 145 (2014).
44. H. Wong, Modèle moléculaire multi-échelles de l'architecture de la chromatine: nouveaux outils, nouvelles perspectives, PhD thesis, Laboratoire de Physique Théorique de la Matière Condensée (CNRS), Université Pierre et Marie Curie 2008.
45. P. Carrivain, Des moteurs de jeux á la physique des chromosomes, PhD thesis, Laboratoire de Physique Théorique de la Matière Condensée (CNRS), Université Pierre et Marie Curie 2012. (in french).

VERY HIGH SYNCHRONY IN EVOLUTION OF ORGANELLES AND HOST GENOMES

A. I. CHERNYSHOVA AND Yu. A. PUTINTSEVA
Siberian federal university, Institute of fundamental biology and biotechnology
660051 Russia, Krasnoyarsk,
Svobodny prosp., 79
E-mail: anna2121651@gmail.com; kinomanka85@yandex.ru

M. G. SADOVSKY
Institute of computational modelling of SB RAS
660036 Russia, Krasnoyarsk, Akakdemgorodok
E-mail: msad@icm.krasn.ru

A relations between the triplet frequency composition of mitochondria and chloroplasts genomes, and the phylogeny of their bearers was studied. To do that, the clusters in 63-dimensional space were identified through K-means, and the clade composition of those clusters has been investigated. Very high regularity in genomes distribution among the clusters was found, with strong correlation to taxonomy. The strong synchrony in evolution manifests through this correlation: the proximity in frequency space was determined over the organelle genomes, while the proximity in taxonomy was determined morphologically.

1. Introduction

Study of statistical properties of nucleotide sequences brings a lot to a researcher. Statistical patterns observed in a sequence correlate to a function encoded by a sequence, or a taxonomy of a bearer of that latter. Here we study those correlations between the structure, and the taxonomy.

A variety of patterns to be found in a nucleotide sequence falls beyond any imagination. A consistent and comprehensive study of frequency dictionaries answers the questions concerning the statistical and information properties of DNA sequences. A frequency dictionary, whatever one understands for it, is rather multidimensional entity. A frequency dictionary of a nucleotide sequence is supposed to be the simplest structure of that type. Further, we concentrate on frequency dictionaries [8,9,10] of the thickness $q = 3$; in other words, the triplet composition only will be taken into consideration. Here we study the *structure – taxonomy* relations for mitochondrion vs. host genomes, and chloroplast vs. host genomes.

More strictly, consider a continuous symbol sequence from four-letter alphabet $\aleph = \{\mathsf{A}, \mathsf{C}, \mathsf{G}, \mathsf{T}\}$ of the length N corresponding to some genetic entity (genome, chromosome, etc.); the length N is the total number of symbols in it. We stipulate that no other symbols or gaps in the sequence take place. Any coherent string $\omega = \nu_1\nu_2\ldots\nu_q$ of the length q makes a word. A set of all the words occurred within a sequence yields the support of that latter. Counting the numbers of copies n_ω of the words, one gets a finite dictionary; changing the numbers for the frequency

$$f_\omega = \frac{n_\omega}{N}$$

one gets the frequency dictionary W_q of the thickness q. This is the main object of our study.

Thus, any genetic entity[a] is represented by a point in (formally) 64-dimensional space. Obviously, two genetic entities with identical frequency dictionaries $W_3^{(1)}$ and $W_3^{(2)}$ are mapped into the same point in the space. Reciprocally, the absolute congruency of two frequency dictionaries $W_3^{(1)}$ and $W_3^{(2)}$ does not guarantee a complete coincidence of the original sequences. Nonetheless, such two sequences are indistinguishable from the point of view of their triplet composition.

Definitely, few entities may have very proximal frequencies of all the triplets, but few others may have not, thus making a distribution of the points in 64-dimensional space inhomogeneous. So, the key question here is what is the pattern of this distribution of genomes in that space? Do they yield some discrete clusters, and if yes is there a correlation to a phylogeny of the genome bearers and clusters? Some results preliminary answering this question could be found in [8,5,6].

To address the questions, we have implemented an unsupervised classification of both mitochondrion and chloroplast genomes, in (metric) space of frequencies of triplets. There were implemented a series of clusterizations, for two, three, four, ..., eight clusters, for both types of genomes. Then, the taxa composition of the classes has been studied; moreover, the relation between the clusters was specially studied, when we changed a clusterization in K clusters for that one in $K - 1$ clusters. Besides, a considerable correlation in taxa composition was found, for the observed clusterizations.

[a]That is mitochondrion or chloroplast genome, in our case.

Briefly speaking, these correlations prove the high synchrony in the evolution of two (physically) independent genetic systems: somatic one, and the organelle one.

2. Material and Methods

2.1. *Genetic sequences*

All sequences were downloaded from EMBL–bank. In this paper, the release of March, 2011 has been used, for mitochondrion genomes; the release of October, 2013 has been used to retrieve the chloroplast genomes. Some entries in both releases contain the "junk" symbols (those that fall beyond the original alphabet $\aleph$). All such symbols have been omitted, and the subsequences appeared due to the elimination of the junk symbols, have been concatenated into a bounded entity.

Originally, the release used for mitochondria contained $\sim 3.4 \times 10^3$ entries. The final database used in our study enlists 1132 entries. Similarly, the full list of chloroplast genomes at the release used for them exceeded five hundred entries; the study on chloroplast clusterization has been carried out with 251 genomes. This discrimination comes from the (not obvious) constraint: we had to eliminate from the study the entries which represent rather highly ranked clade solely, as the single species in the clade. A highly order clade presented with a single genome (that is a single species) yields a "signal" strong enough to deteriorate a general pattern, but weak one to produce distinguishable details in the distribution pattern. Thus, we enlist into the final databases the entries representing an order (and higher clades) with five species or more, for mitochondria. Similar cut-off number for chloroplast genomes was equal to 3 species.

The stricture of the final mitochondrion database is following: it contains the genomes of animals, only, with 988 entries of *Chordata* and 144 entries of *Arthropoda* phyla. *Arthropoda* phylum consists of 83 entries of *Endopterygota*, 30 entries of *Paraneoptera* and 30 entries of *Orthopteroidea*. Since the total number of chloroplast genomes under consideration is significantly less, in comparison to the mitochondria list, the clade composition of that former seems to be less apparent; meanwhile, the chloroplast database includes 157 temperate broadleaf species against 94 conifer ones.

2.2. *Clusterization methods*

Unsupervised classification by K-means has been used to develop classes (see details and a lot of examples in [2,7,11]). To cluster, we had to reduce the data space dimension to 63: this reduction comes from the fact that the sum of all frequencies is equal to 1. Formally speaking, any triplet could be excluded from the data set; practically, we excluded the triplets (specific, for mitochondria, and for chloroplasts) that yield the least standard deviation, over the set of genomes under consideration. This choice is evident: such triplets make least contribution into the distinguishing the entities, in the space of frequencies. Thus, the triplet GCG was eliminated, for mitochondria database (standard deviation $\sigma_{\mathsf{GCG}} = 0.001399$), and the triplet GAC was eliminated, for chloroplast database (standard deviation $\sigma_{\mathsf{GAC}} = 0.000540$).

A K-means implementation may be based on a number of distances; here we used Euclidean distance. Also, no class separability has been checked. All the results were obtained with *ViDaExpert* software by A. Zinovyev[b].

2.3. *"Downward" vs. "upward" classification*

There are two versions for classification implementation developed due to K-means; we call them "downward" and "upward" ones, respectively. They both are based on a standard K-means technique, and the difference is in the mutual interaction between the clusterizations developed for different number of clusters.

2.3.1. *"Downward" classification*

This kind of classification is designed to follow the classical morphology based classification. It starts from the clusterization of the entire set of genomes (frequency dictionaries) into the minimal number $\mathfrak{M}_c$ of clusters with the given stability of the clusterization. That latter is understood as the given number of volatile genomes, i. e. genomes that may change their cluster attribution with any new clustering realization. Then each of the clusters is to be separated into the similar (i. e. minimal stable subclusters) set of subclusters, etc. The procedure is to be trunked at the given "depth" of the cluster separation, usually determined by the volatility of a significant part of genomes.

Thus, a "downward" classification yields the structure that is a tree, so making it close to a standard morphological classification.

[b]http://bioinfo-out.curie.fr/projects/vidaexpert/

2.3.2. *"Upward" classification*

On the contrary, the upward classification consists in the separation of the entire set of genomes, sequentially, into the series of clusters

$$\mathfrak{C}_2, \mathfrak{C}_3, \mathfrak{C}_4, \ldots, \mathfrak{C}_{K-1}, \mathfrak{C}_K .$$

Here we assume that the clusterization $\mathfrak{C}_2$ is stable. Again, the series is to be trunked at the given number K; we put $K = 8$ in this study with chloroplasts.

The key question here is the mutual relation between the members of a cluster $\mathfrak{C}(l)_j$ from $\{\mathfrak{C}(i)_j\}$ $(1 \leqslant i \leqslant j)$ clusterization with the clusters from $\{\mathfrak{C}(m)_{j-1}\}$ $(1 \leqslant m \leqslant j-1)$ clusterization. Here the index l enlists the clusters at the $\{\mathfrak{C}(i)_j\}$ clusterization. There could be (roughly) three options:

- A cluster $\mathfrak{C}(n)_j$ is entirely embedded into the cluster $\mathfrak{C}(l)_{j-1}$, with some l and j;
- The greater part of the members of a cluster $\mathfrak{C}(n)_j$ is embedded into the cluster $\mathfrak{C}(l)_{j-1}$, but the minor part is embedded into the other cluster $\mathfrak{C}(m)_{j-1}$;
- A cluster $\mathfrak{C}(n)_j$ is almost randomly spread between the set of clusters $\mathfrak{C}(l)_{j-1}$, $l = 1, 2, \ldots, l^*$.

Thus, an upward classification yields a pattern that is a graph with cycles; at the worst case, the graph is fully connected, and no essential structuredness is then observed. If the graph has rather small number of cycles, then it reveals the relations between the clusters (determined through the proximity in frequency space), and the taxonomy (determined over the nuclear genome).

3. Results

We studied the relation between the structure defined in the terms of a triplet composition of organelle genomes, and the taxonomy determined according to a morphology, for two types of organelles: chloroplasts and mitochondria. Subsection 3.1 shows the results of this study for chloroplasts; similar results for mitochondria are presented in Subsection 3.2.

3.1. *Chloroplasts*

Figure 1 shows the graph of embedments for the clusters, where the number of these latter changed from 8 to 3. We developed the clusterizations for

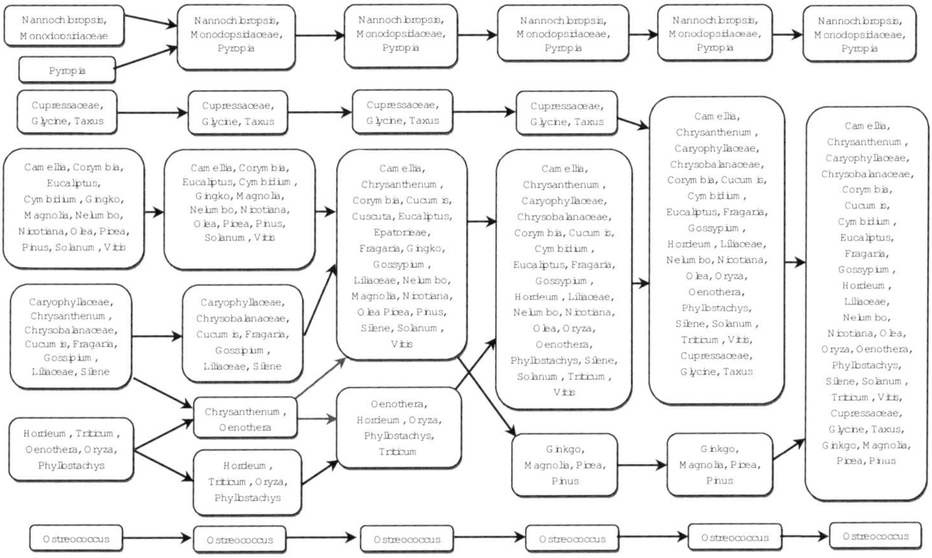

Figure 1. Pattern of the "upward" embedment of various clades of plants in the upward classification, developed over chloroplast genomes.

three, four, ..., eight clusters, and studied the composition of each cluster, at the each "depth". The key question was whether the species tend to keep together, when the number of clusters in a clusterization is decreased.

Figure 1 answers distinctly and apparently the question: the clades in the boxes correspond to genera, while the species (not shown in the figure) always make a solid group, when changing the number of clusters. Thus, boxes having two upright arrows showing the transfer of entities from $\mathfrak{C}_l$ clusterization to $\mathfrak{C}_{l-1}$ one contan two groups of species belonging the same genus (or family).

3.2. *Mitochondria*

Also, we have studied the clusterization of mitochondria, while here the three-cluster and two-cluster patterns have been explored. It should be stressed, that the clusterizations for both patterns was extremely stable; it exceeded significantly the stability of similar clusterizations observed for chloroplasts.

Table 1 shows the results of this classification implementation. Turtles occupy the same class, with no exclusions. Also, this class is occupied by fossils (*Archosauria* and *Lepidosauria*). Three clades (*Neoptera* division, mammalia and fossils) are distributed between two classes, only, and the

Table 1. Clade distribution among the clusters, for the three-clusters clusterization.

Taxon	N	I	II	III
Actinopterygii	510	464	46	0
Amphibia	65	40	17	8
Archosauria and *Lepidosauria*	177	1	176	0
Mammalia	212	0	1	211
Neoptera	143	0	4	139
Testudines	25	0	25	0

distribution is extremely biased: there are two single genomes of mammalia and fossils, respectively, belonging to the opposite class. Here turtles and fossils occupy the came class, that is not a matter of surprise. Less clear is an amalgamation of two rather distinct clades (these are *Mammalia* and *Neoptera*) into the single class. A rate of escaped genomes here is less than 0.5 % and 1.5 %, respectively.

976 genomes from *Chordata* phylum form another class, with 12 genomes escaping the opposite class. Again, the escaping genomes set is absolutely stable and includes the following species (accession numbers are in parentheses): *Ranodon sibiricus* (AJ419960), *Aneides flavipunctatus* (AY728214), *Ensatina eschscholtzii* (AY728216), *Rhyacotriton variegatus* (AY728219), *Desmognathus fuscus* (AY728227), *Hydromantes brunus* (AY728234), *Geotrypetes seraphini* (AY954505), *Pachyhynobius shangchengensis* (DQ333812), *Onychodactylus fischeri* (DQ333820), *Dermophis mexicanus* (GQ244467), *Dicamptodon aterrimus* (GQ368657), *Hemiechinus auritus* (AB099481).

Classification implementation in three classes with K-means also was very good and stable. Again, 10^3 runs of the classification development have been carried out. Only three patterns of the classification have been observed differing in the abundances of the classes. Namely, the abundance distributions were the following:

(i) $975 \div 147 \div 10$ entries, 18 cases,

(ii) $510 \div 147 \div 475$ entries, 854 cases,

(iii) $511 \div 146 \div 475$ entries, 136 cases.

Again, there were no volatile genomes (i. e. those permanently changing the class attribution).

Amphibia is the only clade occupying all three classes rather remarkably. Moreover, there are 9 genomes of *Caudata* order, and 13 genomes of *Anura* order, and they are separated between two classes of the statistical classification: all *Anura* genomes fall into the second class, while the third

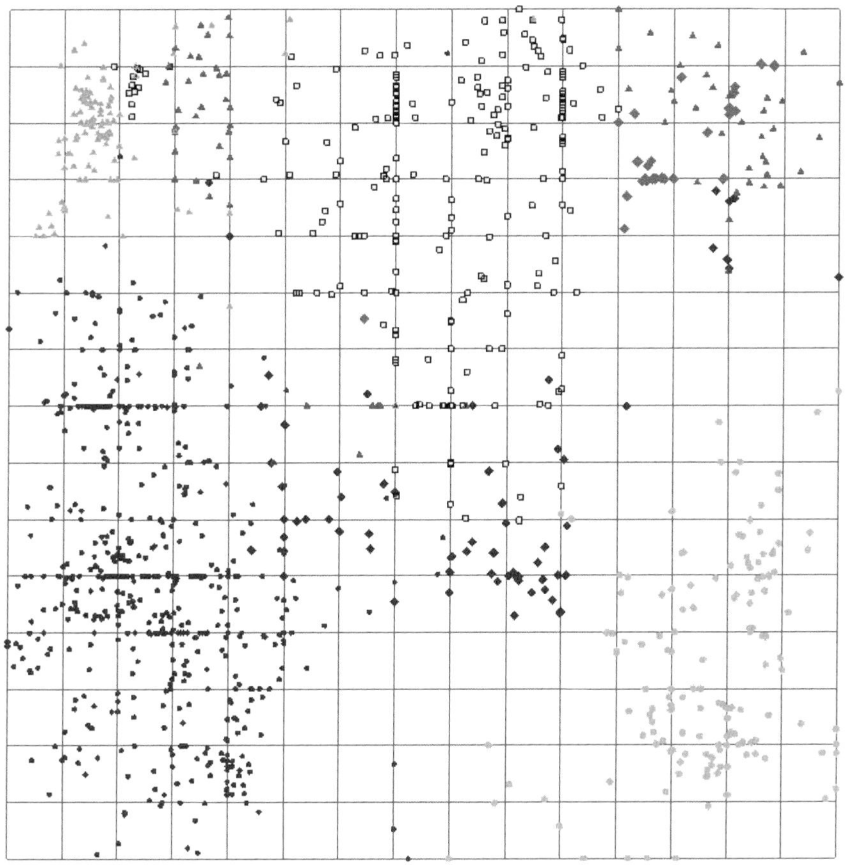

Figure 2. Distribution of genomes in rigid map. Following are the labels of the clades: • – *Actinopterigii*; ⧫ – *Amphibia*; ▲ – *Archosaura*; △ – *Lepidozaura*; □ – *Mammalian*; ¤ – *Neoptera*; ◊ – *Testudines*.

class includes 7 genomes of *Caudata* order, in comparison to two genomes of that latter order found in the second class.

Fig. 2 illustrates the distribution of clades in rigid map (see [2,3] for the details on the rigid map techniques). Labels for clades are in the figure legend.

4. Discussion

A study presented in this paper is done within the scope of population genomics methodology. The most intriguing result of the study is the very high correlation between the statistically identified clusters of genomes, and their taxonomy reference. The key point is that we used organelle genomes to derive the clusterization, while the taxonomy was determined traditionally, through morphology, which is ultimately defined by a nuclear genomes. There is no immediate interaction between the nuclear and organelle genomes. The study has been carried out for both main organelles: chloroplasts and mitochondria.

All mitochondria have the same function; same is true for chloroplasts. Thus, the impact of a function divergence was eliminated in our studies. Probably, a database structure is crucial in this kind of studies. We have used an unsupervised classification technique to develop a distribution of genomes into few groups. The results of such classification are usually quite sensitive to an original database composition [7]. Luckily, the genetic banks are rapidly enriched with newly deciphered genomes of organelles, so the stable and comprehensive results showing the reliable relation between structure and taxonomy could be obtained pretty soon. Moreover, a growth of genetic database may provide a comprehensive implementation of a "downward" classification.

The approach presented above looks very fruitful and powerful. One can expand the approach for the following problems to be solved:

- To study the clusterizations as described above for the database consisting of the genomes of mitochondria, and chloroplasts, of the same species. This study would unambiguously address the question on the relation between structure and function: since the organelle genomes under consideration would belong the same organisms, one may expect an elimination of the taxonomy impact, on the results. Meanwhile, this point should be carefully checked, since the results might be sensitive to the list of species involved into the study;
- To study the clusterization of the frequency dictionaries corresponding to the individual genes (or genes combinations) retrieved from the raw genomes of organelles. The clusterization of such genetic entities would address the question on the mutual interaction in a triadic pattern *structure – function – taxonomy*.

References

1. Bugaenko N. N., Gorban A. N., Sadovsky M. G. *Open Systems & Information Dyn.* **5**, 265 (1998).
2. Gorban A. N., Zinovyev A. Yu. *Int.J. of Neural Systems*, **20**, 219 (2010).
3. Gorban A. N., Kegl B., Wünsch D. C., Zinovyev A. Yu. (eds.) 2007. Principal Manifolds for Data Visualisation and Dimension Reduction, // *Lecture Notes in Computational Science and Engineering*, **58**, Springer, Berlin – Heidelberg – New York. 332 p.
4. Gorban A. N., Popova T. G., Zinovyev A. Yu. *In Silico Biology*, **3**, 471 (2003).
5. Gorban A. N., Popova T. G., Sadovsky M. G., Wünsch D. C. 2001. Information content of the frequency dictionaries, reconstruction, transformation and classification of dictionaries and genetic texts. *Intelligent Engineering Systems through Artificial Neural Networks*, 11 — *Smart Engineering System Design*, N.-Y.: ASME Press, (2001). pp.657 – 663.
6. Gorban A. N., Popova T. G., Sadovsky M. G. *Open Systems & Information Dyn.* **7**, 1 (2000).
7. Fukunaga K., 1990. Introduction to statistical pattern recognition. 2nd edition. Academic Press: London. 591 p.
8. Sadovsky M. G., Shchepanovsky A. S., Putintzeva Yu. A. *Theory in Biosciences* **127**, 69 (2008).
9. Sadovsky M. G. *J.of Biol.Physics* **29**, 23 (2003).
10. Sadovsky M. G. *Bulletin of Math.Biology.* **68**, 156 (2006).
11. Y Shi, Gorban A. N., Yang T. Y. *J. Phys.: Conf. Ser.* **490** 012082, (2014).

A METHOD FOR CLUSTERING HEMAGGLUTININ INFLUENZA PROTEIN SEQUENCES*

X. LI, H. JANKOWSKI, X. WANG
Department of Mathematics and Statistics
York University, 4700 Keele St.
Toronto, Ontario M3J 1P3, Canada
E-mail: lixuan@mathstat.yorku.ca

J.M. HEFFERNAN
Modelling Infection and Immunity Lab
Centre for Disease Modelling
Department of Mathematics and Statistics
York University, 4700 Keele St.
Toronto, Ontario M3J 1P3, Canada
E-mail: jmheffer@yorku.ca

Each year influenza causes widespread disease globally. To combat the virus, vaccination programs are in place, but with the high mutation rate of influenza this vaccine needs to be updated every year. There is a high level of variability in the Hemagglutinin (HA) glycoprotein component of seasonal influenza strains. A better understanding of HA evolution over influenza seasons is needed to better advise vaccine strain development. We have developed a new method for clustering influenza viral sequences. Herewithin, we apply the method to the highly variable influenza A H3N2 HA viral glycoprotein. Our data comprises 1960 viral protein sequences active from 1998 to 2012, and our methodology aggregates these sequences into 23 clusters. Based on these clusters, we provide an investigation of past vaccines and the dominant cluster in each influenza season. We also investigate evolutionary pressures of closely matched circulating and vaccine strains HA glycoproteins. We end with a discussion of future work.

1. Introduction

Influenza viruses are negative stranded and segmented RNA viruses which cause serious and recurrent respiratory disease globally. The influenza season (or 'Flu Season'), which occurs each year in the winter months of each

*This work is supported by NSERC and an Early Researchers Award Ontario.

hemisphere, is associated with significant human mortality and morbidity worldwide[1]. Influenza A and B strains cause seasonal influenza, however, in recent years, influenza A has caused much of the burden[1].

Influenza A viruses are divided into subtypes based the hemagglutinin (HA) and neurominidase (NA) proteins that lie on the surface of the virus. The HA protein has been identified to be the major antigenic component of the influenza virus[2]. Seasonal flu epidemics are normally associated with small mutations in the HA protein (antigenic drift), which allows the virus to evade host immune systems and increases the lifetime susceptibility of the host[2].

Vaccination is the best way to prevent or lessen the severity of seasonal flu. Each year, the seasonal flu vaccine is updated based on a study of the previous year's circulating strains[1]. However, because of the high mutation rate of the seasonal influenza strains it is difficult to determine what the appropriate vaccine strains should be[3]. A study of the HA's evolutionary course can better inform vaccine recommendations.

Phylogenetic analysis of virus strains can help determine evolutionary patterns. Recent advances in molecular biology and computational tools have enabled phylogenetic analysis of small components, to the entire influenza genome[4,5,6,7,8]. However, few pay attention to the influenza swarms or clades regarded as the main target driven by evolutionary forces[9,10]. Additionally, in those studies that do consider the determination of clades, the statistical methods can be improved. For example, in Plotkin et al.[9], some components of the methodology include user-based decisions, so the method is not fully automatic. Also, in Luksa and Lassig[10] the method is computationally complex, and could be improved.

We present a new formal cluster-based technique that can be used to study the evolution of influenza. The method is fully automatic with computational complexity O(N). Herewithin, we employ our methodology to study the evolution of the HA component of the influenza A H3N2 virus, (a major cause of seasonal influenza), and the relationships between this circulating virus and the recommended vaccine strains. We present our data acquisition and methodology in Section 2. In Section 3, we show that our new method can be used to uncover interesting trends in HA evolution, including a relationship between the vaccine and dominant circulating strains. Our results are discussed in Section 4, and this section ends with a critique of our methodology and directions for future work.

2. Data Description and Methodology

2.1. *Data acquisition*

The Influenza Research Database (IRD) is an online repository of influenza sequences obtained through voluntary contributions[11]. It is publicly available online at `www.fludb.org`. The sequences employed in this study were obtained from this database using the criteria listed in Table 1.

Table 1. IRD criteria: All other settings kept default or blank.

Option	Criteria
"Data to return":	protein
"Virus type":	A
"Sub type":	H3N2
"Select segments":	HA
"Complete sequences":	Complete Segments Only
"Date range":	1998 to 2012
"Host":	Human
"Geographic grouping":	All
Advanced options	
"Month Range":	Sep 1998 to July 2012
"Remove Duplicate Sequences":	Yes

This search yields a collection of 1947 sequences of the H3 type HA gene isolated between September 1998 and July 2012 from locations around the globe. Each sequence is made up of 1698 nucleotides plus a stop codon.

To understand the relationship between the observed influenza A strains and the vaccines used, we also include vaccine strains in the data set. The vaccine sequence information was obtained from the World Health Organization[12], and is listed in Table 2. It includes all vaccine strains used from September 1998 to July 2012. Three vaccine strains were already included in the dataset, namely "A/Brisbane/10/2007", "A/Perth/16/2009" and "A/Texas/50/2012". All others were input manually. Note that this resulted in a total of 1960 sequences in the database.

The resulting 1960 sequences were then translated into the corresponding amino acids using Perl or MEGA software[13]. This was done separately for the virus and vaccine strains. The transformation resulted in 566 amino acids in each of the 1960 sequences. We then performed multiple alignment for all of the 1960 protein sequences using MUSCLE software[14]. All sequences were easily aligned with only a few gaps present. Finally, we converted the character records of the amino acids into numerical values

Table 2. Vaccine sequences in the dataset.

Stain Name	Number of sequences	Accession Number
A/Moscow/10/99	2	AY531035, DQ487341
A/Fujian/411/2002	2	CY088483, CY112933
A/California/7/2004	1	CY114373
A/Wisconsin/67/2005	4	CY033646, CY163936 CY114381, EU103823
A/Brisbane/10/2007	3	CY035022, CY039087 EU199366
A/Perth/16/2009	1	GQ293081
A/Victoria/361/2011	1	KC306165
A/Texas/50/2012	2	KC892248, KC892952

using Perl. This results in 1960 observations with 566 categorical variables, each containing 21 categorical states (20 for each kind of amino acid and one to represent a gap).

Each of these 1960 sequences is related to a calendar year, country and city of isolation, inferred from the sequence name (see, for example, Table 2). For the 1947 sequences obtained from the IRD, we can also obtain the date of isolation, which allowed us to partition the data into influenza seasons (October 1st through September 30th).

Files containing both the pre- and post-processed data are provided as supplementary material, and are also available online at `www.math.yorku.ca/~hkj/Research`.

2.2. *Clustering the sequences*

To identify clusters of viral sequences, we use the following methodology.

Distances between any two sequences are calculated using Hamming distance[15]. For two sequences, $A = \{a_1, \ldots, a_{566}\}$ and $B = \{b_1, \ldots, b_{566}\}$ the Hamming distance is defined as

$$d_s(A, B) = \sum_{i=1}^{566} I(a_i \neq b_i), \qquad (1)$$

where $I(E)$ is the indicator function equal to one if E is true, and is zero otherwise. Note that although $a_i, b_i \in \{1, \ldots, 21\}$, they are categorical in nature, and the Hamming distance preserves this property. Heuristically, the Hamming distance between any two HA sequences is the number of locations with different amino acid expressions.

The dimension of our data set is 21^{566}, which is not computationally manageable. We therefore perform some dimension reduction before proceeding with our clustering approach. The main idea behind our dimension reduction is to select a smaller number of sites (from the 566 amino acid sites) to use in our analysis. This is a reasonable approach, as is it well known that parts of HA sequences are well-conserved[16]. To identify the most variable (equivalently, least conserved) sites, we use the notion of entropy.

Suppose that X denotes a categorical random variable with distribution given by $P(X = k) = p_k$, $k = 1, 2, \ldots, 21$. The entropy of X is then defined as

$$H(X) = -\sum_{k=1}^{21} p_k \log p_k \,. \tag{2}$$

In our context, X represents the amino acid state with $k = 1, \ldots, 21$ possible categories, and p_k denotes the observed relative frequency at the site (with frequencies obtained from the 1960 observations). Using (2), we compute the entropy for each of the 566 sites. Note that $H(X)$ is always positive, and larger entropy indicates greater variability at a site. In other words, sites with larger entropy contain more variability in their amino acid states. In our approach, the 566 entropies (one for each site) are sorted increasingly. Sites with entropy equal to zero were removed, as there is no amino acid variability in these sites and hence no useful information for clustering. Sites where 1959 observations were equal, were also removed. Finally, we used a Gaussian mixture model to cluster the remaining entropies[17]. The algorithm results in 5 classes of entropies. We identify the class with the largest entropies as the sites with the greatest variability, and use these sites to cluster the sequences. This class contains 62 sites, which allows us to reduce the dimension to 21^{62}. That is, our data is now made up of 1960 sequences, where each sequence is of length 62.

To cluster the dimension reduced data set, we use the Hamming distance vector (HD vector) algorithm[20]. To understand the approach, consider a general set-up where p nominal categorical attributes are of interest and the jth attribute is categorized by m_j levels. The categorical sample space, Ω, is defined as the collection of all possible p-dimensional vectors of states. For us, $m_j = 21$ for each j, and $j = 1, \ldots, 62$. Therefore, each sequence can be seen as a vector of length 62 ($p = 62$), and each element of the vector is a value taken from one of 21 ($m_j = 21$ for all j) possible categories.

Any given dataset, which in our case can be represented as $\{A_1, \ldots, A_{1960}\}$, gives a distribution of distances on the sample space Ω from a fixed reference position in Ω. We denote this fixed reference position as $S = \{s_1, \ldots, s_p\}$. For a general dataset, we use n to denote the sample size (here, $n = 1960$). Recall the definition of Hamming distance given in (1), and note that it will take values in $0, 1, 2, \ldots, p$. The algorithm relies on the HD vector, which is defined as a $(p+1)$-element vector $U(S) = \{U_0(S), U_1(S), \ldots, U_p(S),\}$ where

$$U_q(S) = \sum_{j=1}^{n} 1(d_s(A_j, S) = q), \quad q = 0, \ldots, p.$$

Thus, $U_q(S)$ counts the number of all observations with Hamming distance to the given reference position S equal to exactly q.

Using the HD vector as a measure of distance, the algorithm proceeds iteratively. First, it uses Pearson's chi-squared statistic to test whether there exist clustering patterns. If all the data points are not uniformly distributed in the sample space, we can use the modified chi-squared statistic to calculate the cluster centre and cluster radius. This proposed HD vector algorithm detects one cluster at a time, and this cluster will be deleted from the existing dataset before the next search. Note that the algorithm is thus fully automatic, selecting both the clusters and the number of clusters.

The results of applying the HD vector algorithm to our data set of sequences is discussed in the following section. This discussion uses the notion of mean Hamming distance between two clusters, which we now define. Consider two clusters, $\mathcal{C}_1$ and $\mathcal{C}_2$. Each cluster is made of up of a number of sequences, say, $\mathcal{C}_1 = \{A_1, \ldots, A_{\kappa_1}\}$ and $\mathcal{C}_2 = \{B_1, \ldots, B_{\kappa_2}\}$. The mean Hamming distance is then

$$d_c(\mathcal{C}_1, \mathcal{C}_2) = \sum_{i,j} \frac{d_s(A_i, B_j)}{\kappa_1 \kappa_2},$$

if $\mathcal{C}_1$ and $\mathcal{C}_2$ are two different clusters. If $\mathcal{C}_1 = \mathcal{C}_2$, then we use instead

$$d_c(\mathcal{C}_1, \mathcal{C}_1) = \sum_{i<j} \frac{d_s(A_i, A_j)}{\kappa_1(\kappa_1 - 1)/2} \quad = \quad \sum_{i \neq j} \frac{d_s(A_i, A_j)}{\kappa_1(\kappa_1 - 1)}$$

This modification is due to the fact that when comparing the same cluster, all distances "along the diagonal" will always be equal to zero.

3. Results

3.1. *Sites of Variability*

As mentioned previously, to reduce the dimension of the system, we identified sites of high variability across the HA sequence. 62 sites were found to

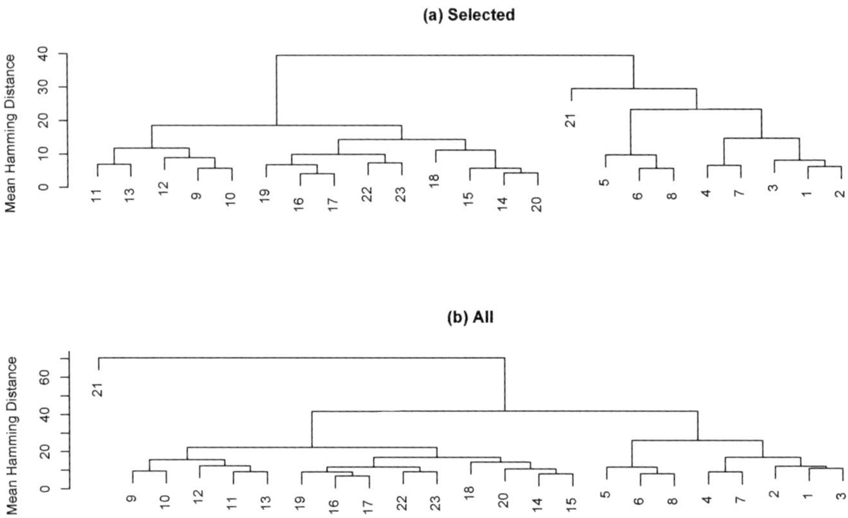

Figure 1. Dendrograms of clusters by mean Hamming distance. This plot is drawn using hierarchical cluster analysis with complete linkage. The top plot uses Hamming distance based only on the 62 highest entropy sites, whereas the bottom plot uses all 566 sites to calculate the Hamming distance.

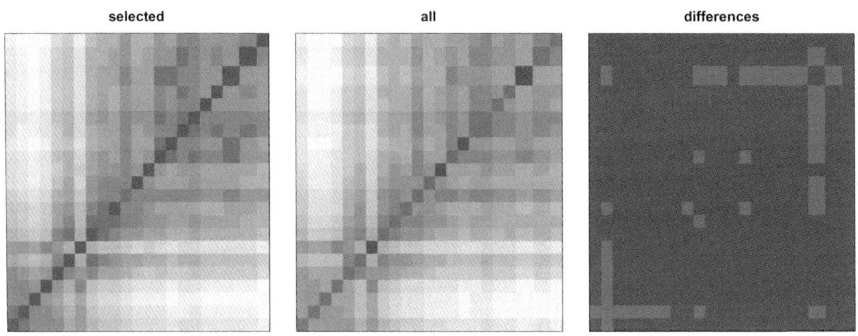

Figure 2. From left to right: mean HD matrix of 62 selected most varied sites, mean HD matrix of the whole sequence with 566 sites, absolute differences of the two matrices. In these plots cluster 21 is excluded. Both matrices (left and centre) have been standardized by dividing by their corresponding maximum values.

have the largest variability. Of the 62 sites, 52 lie within the HA1 domain. The remaining 10 sites lie within the HA2 domain.

3.2. *Sequence Clusters*

Using the method described in the previous section, the 1960 viral sequences were partitioned into 23 clusters. Figure 1 shows two dendrograms of the resulting clusters. The top dendrogram is based on the mean Hamming distance calculated only for the sites of maximal entropy, whereas in the bottom dendrogram the Hamming distance is calculated for all sites. Since the sites of highest entropy cover the mutation hot spots in the amino acid sequence it is expected that the dendrograms should be similar. Indeed, the dendrograms are largely consistent with regard to tree locations. Clusters 1 - 8 are grouped into a clade, while the remaining clusters except 21 are grouped into another clade. Although there exist subtle discrepancies in the specific clade location of some clusters, the mean distances are fairly small.

For a further comparison of the clusters using the entire amino acid sequence and the sites of maximal entropy we compare the Hamming distance matrices. Figure 2 (left, centre) shows the corresponding heat maps of the mean Hamming distance matrices, excluding cluster 21. Figure 2 (right) displays the absolute difference between these two distance matrices. We investigate further by considering also the ratio of the two Hamming distance matrices. The results are shown in Figure 3, and we can see that once cluster 21 is excluded, there appear only spurious inconsistencies off the main diagonal. Hence, the ratio is relatively constant (again, excluding

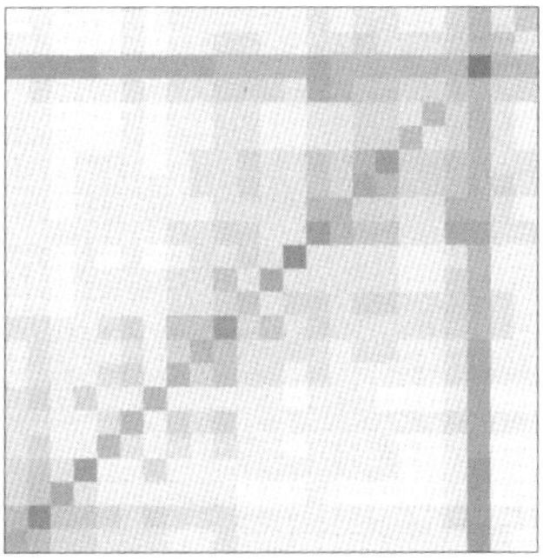

Figure 3. Ratio of distance matrices using the 62 most varied sites and all 566 sites in the Hamming distance calculations.

cluster 21), which implies that the two distances (using the 62 sites and using all 566 sites) are approximately equivalent. Therefore, the plots indicate that, once cluster 21 is removed, the 62 most varied sites can stand for the whole sequence effectively with regard to the distances between sequences. In what follows, we proceed with our analysis of the results, keeping cluster 21.

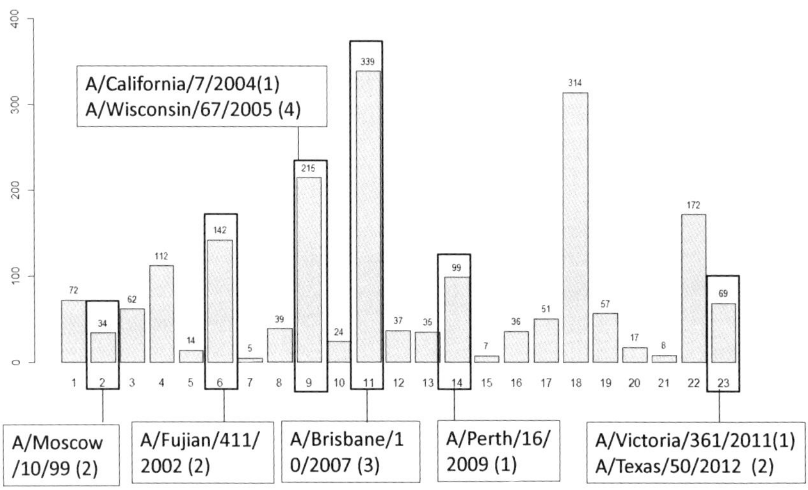

Figure 4. Histogram of cluster size and vaccine location. The clusters have been reordered by earliest year of isolation.

3.3. *Spatio-Temporal Evolution of the Clusters*

Figure 4 shows the number of sequences in each cluster sorted by the first year of isolation. The clusters that house the vaccine strains are also indicated. The vaccine locations are consistent with the calendar year according to their strain name (and year closeby) eg. "A/California/7/2004" and "A/Wisconsin/67/2005", "A/Victoria/361/2011" and "A/Texas/50/2012" are clustered together. This helps to verify the validity of the cluster pattern.

We can observe from Figure 4 that large clusters are generally surrounded by clusters of much smaller size. Thus, the dominant clusters over many years can be identified. We can also observe that more small clusters are generated in recent years. This may be due to higher reporting rates, as rapid sequencing technologies have become increasingly available.

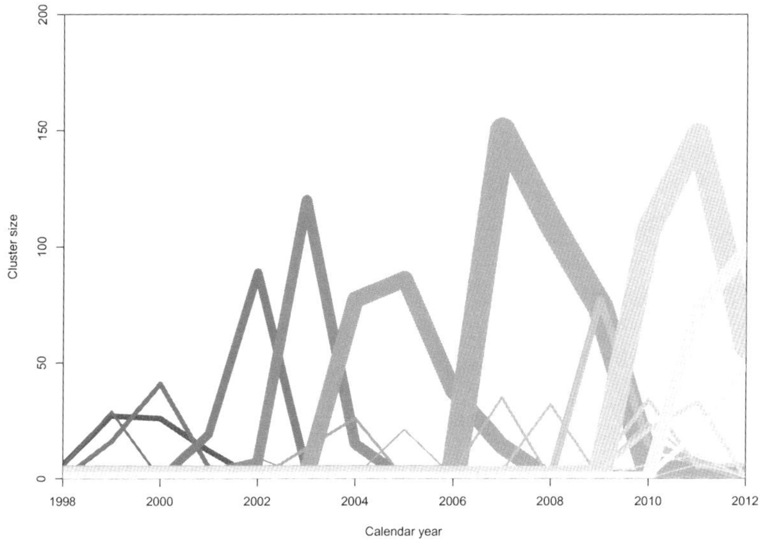

Figure 5. The number of HA protein sequences within each cluster plotted versus calendar year of isolation. Each cluster is indicated by a different colour, and the line width reflects the cluster size. The dominant sequence clusters tend to replace each other every 2-5 years. (The number of sequences each year is only for the unique sequences and does not reflect the severity of infections.)

Each cluster houses strains that exist over one or more influenza seasons. In Figure 5, we plot the number of sequences in each cluster as a function of their isolation year. The size of each cluster is indicated by line thickness (with large clusters indicated with thicker lines). It is observed that some clusters are significantly more long-lived than others, but that no cluster spans more than seven years. It is also observed that clusters first increase and then decrease in size over their lifespan. Dominant clusters of viral sequences replace one another every 2-5 years, but once HA evolves away from a given region of the sequence space, it does not later revisit that region. This agrees with previous studies of influenza evolution[2,9].

3.4. *Evaluation of recommended vaccines for each season*

In Table 3, we identify which clusters contain which vaccines, focusing on the clusters that exist over the calendar years 2000-2012. The dominant cluster, vaccine strain, and cluster that houses the vaccine strain for each year are provided for each vaccine. Ideally, the strain used as the basis for a vaccine each year would correspond to the dominant cluster. Considering the time lag that exists between the disease outbreak and time of isolation, the cluster housing the vaccine strain should be as close to the dominant cluster as possible.

Table 3. Vaccines and Clusters by Year. Dominant clusters are identifed in bold. Vaccine clusters are identified by circles.

Season	Vaccine	Clusters	Dominant cluster	Vaccine cluster
2000-2001	A/Moscow/10/99	**1** 3 4 5	1	2
2001-2002	A/Moscow/10/99	1 **4** 5 6	4	2
2002-2003	A/Moscow/10/99	4 5 **6** 7 8	6	2
2003-2004	A/Moscow/10/99	4 **6** 8 9	6	2
2004-2005	A/Fujian/411/2002	(6) 8 **9** 10	9	6
2005-2006	A/California/7/2004	1 (**9**) 10	9	9
2006-2007	A/Wisconsin/67/2005	(9) 10 **11** 12 13	11	9
2007-2008	A/Wisconsin/67/2005	(9) **11** 13	11	9
2008-2009	A/Brisbane/10/2007	(**11**) 14 15 16 17 19 20 21	11	11
2009-2010	A/Brisbane/10/2007	(11) 14 16 17 **18** 19 20	18	11
2010-2011	A/Perth/16/2009	11 (14) 16 17 **18** 19 20 21 22 23	18	14
2011-2012	A/Perth/16/2009	(14) 16 17 18 21 **22** 23	22	14

A common observation over all of the years shown is that the relationship between the vaccine strain cluster and the cluster housing the dominant strain diverges (that is, is further apart on the dendrogram tree). For example, from 2000-2004 the same vaccine strain "A/Moscow/10/99" was used, however, the dominant cluster changes each year in this time period, moving from a mean distance of 6.15 amino acids (aa) from the vaccine sequence, to 8.37aa, and then 18.68aa in 2002-2004.

From Table 3 we can also see that cluster extinction often coincides with the existence or introduction of a well matched vaccine strain. In particular, the extinction of clusters 6 and 9 coincides with the introduction of vaccines housed in the same cluster. Something similar can be seen on clusters 1 and 11. Ultimate extinction of a cluster, however, is a result of a combination of various factors, including vaccine strain and competition between strains (that may have higher fitness). An exploration of strain fitness is a course for future work.

4. Discussion

In this paper, we have presented a new method for clustering protein sequences, and have applied the method to clustering of HA sequences of seasonal influenza A H3N2. Including vaccine sequences in the analysis, allows us to present important relationships between the vaccine and dominant influenza strain evolution.

In our approach, 62 sites of highest entropic variability were used in the clustering in lieu of the full sequence, greatly reducing the computational complexity of the problem. These 62 sites lie within the HA1 *and* HA2 regions of the HA genome. Our clustering methodology separated the HA dataset into 23 clusters. Analysis of these clusters found that HA generally clusters by year. Upon further analysis, we found that clusters replace one another every 2-5 years, that the evolution of the dominant cluster diverges from that of the cluster housing the vaccine strain, and that extinction of a dominant cluster often coincide with the existence or introduction of a well-matched vaccine.

Our results are highly consistent with previous studies of HA evolution (i.e., see Plotkin et al.[9] and Nelson and Holmes[26] for a review). Previously, Plotkin et al.[9] found that the persistence of clusters can be used to predict the next season's influenza sequences. In their analysis, however, only the HA1 component of the HA protein was considered. Therefore, some sites of high variability were neglected. Through choosing those most varied sites of the whole sequence, all the potential evolutionary hot spots can be taken into account. This can better help to find more diversified strains and evaluate the outbreak and epidemics of each cluster.

In a recent study, Luksza and Lassig[10] employed a new method using an ensemble of trees to infer the genealogy of influenza strains over time, trace the evolution of strain clades, and predict vaccine strains from year to year. A careful comparison of their results to that presented here is currently underway. It is important to note, however, that the computational complexity of their method is higher than the method we employ. Also, a key feature of our algorithm is that a statistical test is employed to determine whether a suspected cluster structure is justified. Such a test is not included in the methodology of Luksza and Lassig[10].

Our method can be applied to other components of the influenza virus genome. A similar study on the NA glycoprotein is currently underway.

Our methodology can be improved in several ways. The discussion in Section 3.2 indicates that cluster 21 behaves differently from the remaining clusters. The within and between mean Hamming distances based on the 62 high entropy sites of cluster 21 is similar to, or even smaller, than for other clusters. However, when the Hamming distance is calculated for all sites, the situation is considerably different. Here, cluster 21 has the greatest within and between distance values. This difference may be the result of numerous factors. It is possible that in the analysis, we have missed some hot spots or some diversifying mutations when choosing the amino acid sites using the entropy approach. Alternatively, this could be the result of some inherent behaviour of cluster 21, which has yet to be understood. Note that cluster 21 is made up of only 8 sequences. It is important to investigate this discrepancy further, and is part of our ongoing research.

Secondly, we use the term "dominant" to denote clusters with the greatest number of sequences in a given season. However, this definition only indicates that the cluster contains the largest number of unique sequences. Therefore, the definition does not account for (a) the actual number of sequences reported in a season, or (b) their relationship with the frequency of the strain within the population. Although the first issue is relatively straightforward to fix, the second is more problematic. The influenza sequences available via IRD are based on voluntary contributions, and are therefore not the result of random sampling. It is thus possible that systematic biases exist in the data set, including yearly and regional variations[10]. Translating the observed sequences on IRD into an appropriate representation of population level frequencies is an important statistical problem which requires careful consideration in our future work.

Lastly, we point out that our analysis is based on the Hamming distance (1). This means that sequences close in Hamming distance (in amino acids) can be regarded as close in lineage evolution history. If we infer the genealogy of all the strains by ensemble of trees, those with small Hamming distance will be grouped into a clade, i.e. in the same trunk of the phylogenic tree. This approach is "purely mathematical" in that it does not include any potential information on the level of importance of specific amino acid differences, or their locations. Incorporating such additional information, will improve the quality of our analysis, and will therefore be included in future analysis.

References

1. World Health Organization, http://www.who.int/influenza/en/.
2. D.M. Knipe and P.M. Howley, *Philadelphia: Lippincott Williams and Wilkins.* (2007).
3. F. Carrat, A. Flahault, *Vaccine* **25**, 6852-6862 (2007).
4. W.M. Fitch, R.M. Bush, C.A. Bender and N.J. Cox, *Proc. Natl. Acad. Sci.* **94**, 7712-7718 (1997).
5. W.M. Fitch, R.M. Bush, C.A. Bender, K. Subbarao and N.J. Cox, *J. Hered.* **91**, 183-185 (2000).
6. T.T.-Y. Lam, Y.L. Chong, M. Shi, et al., *Infection, Genetics and Evolution* **18**, 367-378 (2013).
7. K.B. Westgeest, C.A. Russell, X. Lin, et al., *J. Virol.* **88** 2844-2857 (2014).
8. M. Escalera-Zamudio, M.I. Nelson, A.G.C Guemes, et al., *PLoS One* **9** e102453 (2014).
9. Joshua B. Plotkin, Jonathan Dushoff, Simon A. Levin, *Proc. Natl. Acad. Sci.* **99**(9), 6263-6268 (2002).
10. Marta Luksza and Michael Lassig, *Nature* **507**, 57-61 (2014).
11. Influenza Research Database, www.fludb.org.
12. World Health Organization, http://www.who.int/influenza/vaccines/virus/recommendations/en/.
13. K. Tamura, D. Peterson, N. Peterson, G. Stecher, M. Nei and S. Kumar, *Mol. Bio. Evo.* **28**, 2731-2739 (2011).
14. Robert C. Edgar, *Nucleic Acids Res.* **32**(5), 1792 - 1797 (2004)
15. G. Forney, *Information Theory* **12**(2), 125-131 (1966).
16. Zuzana Staneková, Eva Varečková, *Virology Journal* **7**, 351 (2010).
17. B.S. Everitt and D.J. Hand. *New York: Springer.* (1981).
18. Chris Fraley and Adrian E Raftery, *J. Amer. Statist. Assoc.* **97**, 611-631 (2002).
19. Elodie Ghedin, Naomi A. Sengamalay, Martin Shumway, et al., *Nature* **437**, 1162-1166 (2005).
20. Peng Zhang, Xiaogang Wang and Peter X.-K. Song, *J. Amer. Statist. Assoc.* **101**(473), 355-367 (2006).
21. Arthur Chun-Chieh Shih, Tzu-Chang Hsiao, Mei-Shang Ho and Wen-Hsiung Li, *Proc. Natl. Acad. Sci.* **104**, 6283-6288 (2007).
22. R Development Core Team, *R Foundation for Statistical Computing*, Vienna, Austria, http://www.R-project.org. (2008).
23. Chris Fraley and Adrian E Raftery, *MCLUST Version 3 for R: Normal Mixture Modeling and Model-based Clustering*, Technical Report No. 504, Department of Statistics, University of Washington, (revised 2009).
24. S. Bhatt, E.C. Holmes and O.G. Pybus, *Mol. Biol. Evol.* **28**, 2443-2451 (2011).
25. Christopher J.R. Illingworth and Ville Mustonen, *PLoS Pathog.* **8**(12), e1003091 (2012).
26. M.I. Nelson and E.C. Holmes, *Nature Rev.* **8** 196-205 (2007).

ON A MULTISCALE ANALYSIS OF A MICRO-MODEL OF HEAT TRANSFER IN BIOLOGICAL TISSUES*

ABDELHAMID AINOUZ

Department of mathematics, University of Sciences and Technology Houari Boumedienne
Po Box 32 El-Alia, Bab-Ezzouar 16111, Algiers, Algeria
E-mail: ainouz@usthb.dz

A bio-heat transfer model for biological tissues in a micro-scale and periodical settings is investigated. It is assumed that the model is a two-component system consisting of solid particles representing tissue cells and interconnected pores containing either arterial or venous blood. This tissue-blood system is described by two energy equations, one equation for the solid tissues and the other for the surrounding blood. On the interface between them, it is assumed that the heat transfer is governed by Newtons cooling law. Using homogenization techniques, it is shown that the obtained macro-model presents some extra-terms and it can be seen as a new mathematical model for human thermotherapy and human thermoregulation system.

1. Introduction

Heat transfer in human body is an important field in biomedical engineering such as laser-induced thermotherapy, thermoregulation system, wound healing, cryosurgery, skin burns, etc... See for instance J.C. Chato[11], M. Gautherie[14], K. Khanafer *et al.*[18], M. Miyikawa and J.C. Bolomey (eds)[19], C. Sturesson[23] and the references therein. Many mathematical models were proposed to predict the distribution of the temperaure in biological tissues. One of the most widely used model is the bioheat equation after the pioneering work of H.H. Pennes[20]. It is based on the well-known Fourier law with the concept of blood perfusion. It reads as follows:

$$\rho c \partial_t T - \operatorname{div}(\kappa \nabla T) + \omega_b \rho_b c_b \rho (T - T_b) = f \tag{1}$$

where T, ρ, c and κ are respectively the temperature, the density, the specific heat and the heat conductivity coefficient of the tissue, ω_b is the blood

*Work supported by grant B00220120074 of the C.N.E.P.R.U Project.

perfusion, ρ_b and c_b are the density and the specific heat of the blood and T_b is the temperature of the arterial flow. Finally f is some external source of heating and it is generally written as the sum of sources due to absorbed laser light and metabolic activity. Many scientists have attempted to justify the Helmoltz term of (1). We mention for example M.M. Chen, K.R. Holmes[12], P. Wust *et al.* [24], R. Hochmuth and P. Deuflhard[15]. In the latter, a homogenization technique was developped on a microvascular model consisting of tissues (solid) surrounded by blood (fluid). More precisely they study the following system

$$-\Delta T^\varepsilon = S^\varepsilon \text{ in } \Omega^\varepsilon, \tag{2}$$

$$T^\varepsilon = 0 \text{ on } \partial\Omega, \tag{3}$$

$$\frac{\partial T^\varepsilon}{\partial n^\varepsilon} = \varepsilon\alpha\left(T_b^\varepsilon - T^\varepsilon\right) \text{ on } \partial Q^\varepsilon \tag{4}$$

where T^ε is the temperature and S^ε the source term in the tissues represented by Ω^ε whereas T_b^ε is the temperature of the blood. The domain Ω^ε is obtained by removing from Ω, a bounded and regular domain, a set of holes Q^ε where blood flows. In (4), n^ε is the unit normal of ∂Q^ε outward to Ω^ε and $\alpha > 0$ a physiological parameter. They obtained the following homogenized model:

$$-\text{div}\left(\mathcal{A}\nabla T^*\right) + \alpha^*\left(T^* - T_b\right) = S$$

where $\mathcal{A}$ is the homogenized tensor, T^* is the weak limit in $H^1(\Omega)$ of some extended temperature $P^\varepsilon(T^\varepsilon)$, S^ε is the weak limit in $L^2(\Omega)$ of the source term S^ε and finally α^* is the effective Helmoltz term. For more details, see R. Hochmuth and P. Deuflhard[15]. In fact, biological tissues can be seen as porous media where cells (matrix) are separated by voids or pores which are filled with blood. This system of cells-blood can be interpreted as a two-constituent medium. In connection with binary composites presenting thermal barriers at the interfacial contact, we mention especially the work by J.L. Auriault and H. Ene[8] where they study heat transfer in a two-component composite with conductivities of the same order of magnitude. The macroscopic model is shown to belong to two main types of field models: one-temperature and two-temperature, depending on the order of magnitude of the interfacial thermal conductance. In the present paper, we shall be concerned with a micro-model for the heat transfer in a biological tissue made of two interacting systems (cells tissues and blood regions) where the conductivities are assumed to be of different order of magnitude. We also assume that the transition between these two regions on the interface is governed by Newton's cooling law. That is the heat flow through the interface is proportional, by the thermal conductance of the layer, to the jump of the temperature field, see for example H.S. Carslaw and J.C. Jaeger[10]. We mention that this kind of boundary transmission condition was used for the homogenization in porous media, see for e.g. A. Ainouz[2,3,4]. In fact, there are many works showing how transport theories in porous media enhance the understanding of flow and heat transfer in biological tissues. For more details, we refer the reader to the survey paper by A.-R.A. Khaled and K. Vafai[17].

The paper is organized as follows: In Section 2, the geomety of the domain and the micro-model are set. In Section 3, we prove an existence and uniqueness result for a weak solution to our micro-model (6a)-(6g). Finally in Section 4, a formal expansion technique is used to derive the homogenized model. Note that in this paper we shall only derive formally the homogenized system (see (10)-(13) below) of our micro-model. A rigorous justification of this derivation is done in A. Ainouz[5].

2. Setting of the Problem

We start by introducing the notation used throughout this paper. We consider Ω a bounded domain in $\mathbb{R}^d$ $(d \geq 2)$ with smooth boundary $\partial\Omega$. Let $Y \overset{\text{def}}{=}]0,1[^d$ be the generic cell of periodicity divided as $Y = Y_1 \cup Y_2 \cup \Sigma$ where Y_1, Y_2 are two connected, open disjoint subsets of Y and $\Sigma \overset{\text{def}}{=} \partial Y_1 \cap \partial Y_2$ is a smooth $(d-1)$-dimensional manifold. As in G. Allaire and F. Murat[7], we assume that the $Y-$periodic continuation of Y_1 to the whole space $\mathbb{R}^d$, namely $\widetilde{Y_1} = \cup_{k \in \mathbb{Z}^d} (k + Y_1)$ is smooth and connected. Note that no connectedness assumption is made on the part $\cup_{k \in \mathbb{Z}^d} (k + Y_2)$.

Let χ_1 (resp. χ_2) denote the Y-periodic characteristic function of Y_1 (resp. Y_2). Denoting $\varepsilon > 0$ a sufficiently small parameter, we set

$$\Omega_1^\varepsilon \overset{\text{def}}{=} \{x \in \Omega : \chi_1(\frac{x}{\varepsilon}) = 1\}, \; \Omega_2^\varepsilon \overset{\text{def}}{=} \{x \in \Omega : \chi_2(\frac{x}{\varepsilon}) = 1\},$$

and let $\Sigma^\varepsilon \overset{\text{def}}{=} \overline{\Omega_1^\varepsilon} \cap \overline{\Omega_2^\varepsilon}$. Without no loss of generality, we assume that the region Ω_2^ε is strictly embedded in the region Ω_1^ε, in the sense that $\overline{\Omega_2^\varepsilon} \subset \Omega$. In this connection, Ω_1^ε is referred as the cellular domain and Ω_2^ε as the voids filled with blood. We see that the boundary of Ω_2^ε is the interface Σ^ε and the boundary of Ω_1^ε consists then of two parts: Σ^ε and the exterior boundary Γ. We can write that

$$\partial\Omega_2^\varepsilon = \Sigma^\varepsilon \text{ and } \partial\Omega_1^\varepsilon = \partial\Omega \cup \Sigma^\varepsilon.$$

Thanks to the connectedness of $\widetilde{Y_1}$, we see that Ω_1^ε is connected while Ω_2^ε may or may not be connected.

Let us denote ρ, c and κ the density, the specific heat and the heat conductivity coefficient of the tissue, respectively. Let ω_b denote the blood perfusion, ρ_b and c_b the density and the specific heat of the blood. Let κ_b the heat conductivity coefficient of the blood. We shall assume that the phenomenological parameters: $\rho, c, \kappa, \omega_b, \rho_b, c_b$ and κ_b are positive constant and independent of ε.

Let $(0,T)$ be the time interval. Put

$$Q \overset{\text{def}}{=} (0,T)\times\Omega,\ \ \Gamma \overset{\text{def}}{=} (0,T)\times\partial\Omega,\ \ S \overset{\text{def}}{=} (0,T)\times\Sigma,$$
$$Q_1^\varepsilon \overset{\text{def}}{=} (0,T)\times\Omega_1^\varepsilon,\ Q_2^\varepsilon \overset{\text{def}}{=} (0,T)\times\Omega_2^\varepsilon,\ S^\varepsilon \overset{\text{def}}{=} (0,T)\times\Sigma^\varepsilon$$

Let also

$$\alpha = \frac{\kappa}{\rho c},\ \alpha_b = \frac{\kappa}{\rho_b c_b},\ \alpha_b^\varepsilon = \varepsilon^2\alpha_b,\ \gamma = \omega_b\rho_b\frac{c_b}{c}. \tag{5}$$

The micro-model that we shall study in this paper is as follows:

$$\partial_t T^\varepsilon - \alpha\Delta T^\varepsilon = f \quad \text{in } Q_1^\varepsilon, \tag{6a}$$
$$\partial_t T_b^\varepsilon - \alpha_b^\varepsilon\Delta T_b^\varepsilon = f_b \quad \text{in } Q_2^\varepsilon, \tag{6b}$$
$$\alpha\nabla T^\varepsilon\cdot\nu^\varepsilon = \alpha_b^\varepsilon\nabla T_b^\varepsilon\cdot\nu^\varepsilon \quad \text{on } S^\varepsilon, \tag{6c}$$
$$\alpha\nabla T^\varepsilon\cdot\nu^\varepsilon = -\varepsilon\gamma\left(T^\varepsilon - T_b^\varepsilon\right) \quad \text{on } S^\varepsilon, \tag{6d}$$
$$T^\varepsilon = 0 \quad \text{on } \Gamma, \tag{6e}$$
$$T^\varepsilon(0,\cdot) = h(\cdot) \quad \text{in } \Omega_1^\varepsilon, \tag{6f}$$
$$T_b^\varepsilon(0,\cdot) = h_b(\cdot) \quad \text{in } \Omega_2^\varepsilon \tag{6g}$$

where f (resp. f_b) be some external source of heating in cells (resp. blood), ν^ε stands for the unit normal of Σ^ε outward to Ω_1^ε and h, h_b are the initial temperature field in Ω_1^ε, Ω_2^ε respectively. Without no loss of generality, we shall assume

$$f,\ f_b,\ h,\ h_b \in L^2(\Omega). \tag{7}$$

Our system actually models heat flow in a porous medium (biological tissue). In this connection, Ω_1^ε represents the matrix-cells space region and Ω_2^ε the pores which are filled with blood. The thin layer Σ^ε is an interfacial flow barrier with heat conductance given by $\gamma^\varepsilon = \varepsilon\omega_b\rho_b c_b\rho$. The unknowns T^ε and T_b^ε are the temperatures in Q_1^ε and Q_2^ε respectively. The first equation describes the heat flow in the cells with large conductivity and the second describes the heat flow in the blood region with low conductivity. Condition

(6c) expresses flux continuity across the interface. However, the temperature may present in general jumps across Σ^ε. Here, we have employed Newton's cooling law described by (6d), see for instance H.S. Carslaw and J.C. Jaeger[10]. We can say that Ω_1^ε can be considered as a good conductor, while Ω_2^ε a poor one. The interface Σ^ε can be seen as a heat exchanger. The condition (6e) is the standard homogeneous Dirichlet condition on the exterior boundary. Finally equations (6f)-(6g) are the initial conditions, which close the system under consideration. Note that in (5), the coefficient α_b^ε is scaled by ε^2 to provide the correct scaling for the heat flow in the block regions. Indeed, this scaling is the unique choice that makes every term of the porous medium equation in the block cell reappears in the leading order asymptotic expansion, so that the form of the equation is preserved on the small scale independently of ε.

Let us now establish a variational framework of our problem. To this end, we first introduce some notations. Let

$$H^\varepsilon = L^2\left(\Omega_1^\varepsilon\right) \times L^2\left(\Omega_2^\varepsilon\right) \text{ and } V^\varepsilon = \left(H^1\left(\Omega_1^\varepsilon\right) \cap H_0^1\left(\Omega\right)\right) \times H^1\left(\Omega_2^\varepsilon\right).$$

We shall consider on H^ε and V^ε the following inner products:

$$(\varphi, \psi)_{H^\varepsilon} = \int_{\Omega_1^\varepsilon} \varphi_1 \psi_1 \, \mathrm{d}x + \int_{\Omega_2^\varepsilon} \varphi_2 \psi_2 \, \mathrm{d}x, \quad \varphi = (\varphi_1, \varphi_2), \quad \psi = (\psi_1, \psi_2),$$

$$(\varphi, \psi)_{V^\varepsilon} = \int_{\Omega_1^\varepsilon} \nabla\varphi_1 \nabla\psi_1 \, \mathrm{d}x + \varepsilon^2 \int_{\Omega_2^\varepsilon} \nabla\varphi_2 \nabla\psi_2 \, \mathrm{d}x + \varepsilon \int_{\Sigma^\varepsilon} (\varphi_1 - \varphi_2)(\psi_1 - \psi_2) \, \mathrm{d}\sigma^\varepsilon$$

where $\mathrm{d}x$ denotes the Lebesgue measure on $\mathbb{R}^d$ and $\mathrm{d}s^\varepsilon$ the Hausdorff measure on Σ^ε. The norms induced in H^ε and V^ε are denoted by $\|\cdot\|_{H^\varepsilon}$ and $\|\cdot\|_{V^\varepsilon}$, respectively. Clearly, H^ε and V^ε are Hilbert spaces when equipped with their respective norms. Moreover, it can easily be shown that V^ε is separable, dense and continuously embedded in H^ε. Let us introduce the bilinear form $a^\varepsilon(\cdot,\cdot) : V^\varepsilon \times V^\varepsilon \longrightarrow \mathbb{R}$ defined by

$$a^\varepsilon(\varphi, \psi) = \int_{\Omega_1^\varepsilon} \alpha \nabla\varphi_1 \nabla\psi_1 \, \mathrm{d}x + \int_{\Omega_2^\varepsilon} \alpha_b^\varepsilon \nabla\varphi_2 \nabla\psi_2 \, \mathrm{d}x + \int_{\Sigma^\varepsilon} \gamma^\varepsilon (\varphi_1 - \varphi_2)(\psi_1 - \psi_2) \, \mathrm{d}s^\varepsilon$$

where $\varphi = (\varphi_1, \varphi_2)$, $\psi = (\psi_1, \psi_2) \in V^\varepsilon$. We see that $a^\varepsilon(\cdot,\cdot)$ is continuous and uniformly coercive on $V^\varepsilon \times V^\varepsilon$. Let $(V^\varepsilon)'$ denote the dual space of V^ε.

Let $\mathcal{A}^\varepsilon \in \mathcal{L}\left(V^\varepsilon, (V^\varepsilon)'\right)$ be given by

$$\mathcal{A}^\varepsilon (\varphi) \psi = a^\varepsilon (\varphi, \psi), \qquad \varphi, \psi \in V^\varepsilon.$$

For convenience we shall denote $w^\varepsilon = (T^\varepsilon, T_b^\varepsilon)$, $\quad g = (h, h_b)$ and

$$f^\varepsilon = f\chi_1^\varepsilon + f_b\chi_2^\varepsilon, \quad \chi_m^\varepsilon (x) = \chi_m \left(\frac{x}{\varepsilon}\right), m = 1, 2.$$

Let

$$W^{1,2}(0, T; H^\varepsilon) = \left\{ w \in L^2(0, T; H^\varepsilon) : w' = \frac{dw}{dt} \in L^2(0, T; H^\varepsilon) \right\}.$$

The variational formulation for (6a)-(6g) reads as follows: find $w^\varepsilon \in L^2(0, T; V^\varepsilon)$ such that for every $\varphi \in W^{1,2}(0, T; H^\varepsilon) \cap L^2(0, T; V^\varepsilon)$ with $\varphi(T) = 0$, we have

$$\begin{aligned} &-\int_0^T (w^\varepsilon (t), \partial_t \varphi (t))_{H^\varepsilon} \,\mathrm{d}t + \int_0^T \mathcal{A}^\varepsilon (w^\varepsilon (t)) \varphi (t) \,\mathrm{d}t \\ &= \int_0^T (f^\varepsilon, \varphi (t))_{H^\varepsilon} \,\mathrm{d}t + (g, \varphi (0))_{H^\varepsilon}. \end{aligned} \tag{8}$$

where $\mathrm{d}t$ denotes the Lebesgue measure on $(0, T)$.

Next, we state the existence and uniqueness result for (8) the proof of which is given in the next section.

Theorem 2.1. *Let $\varepsilon > 0$ be a sufficiently small parameter. Then, there exists a unique weak solution $w^\varepsilon \in L^2(0, T; V^\varepsilon)$ of Problem (8) and the following energy estimate holds:*

$$\|w^\varepsilon\|_{L^\infty(0,T;H^\varepsilon)} + \|w^\varepsilon\|_{L^2(0,T;V^\varepsilon)} \leq C. \tag{9}$$

We define the overall temperature in the biological tissue region $\Omega_1^\varepsilon \cup \Omega_2^\varepsilon$ by

$$u^\varepsilon (t, x) = \chi_1 \left(\frac{x}{\varepsilon}\right) T^\varepsilon (t, x) + \chi_2 \left(\frac{x}{\varepsilon}\right) T_b^\varepsilon (t, x), \text{ a.e. } (t, x) \in Q.$$

Using the two-scale convergence technique (see G. Allaire[6]), it is shown the following convergence result.

Theorem 2.2. *There exists a subsequence of (u^ε), still denoted (u^ε) such that there exist $T \in L^2\left(0, T; H_0^1(\Omega)\right)$ and $T_b \in L^2\left(Q; H_\#^1(Y)\right)$ with*

(1) $\chi_1\left(\frac{x}{\varepsilon}\right) u^\varepsilon$ weakly converges to $|Y_1|\, T$ in $L^2(Q)$;
(2) $\chi_2\left(\frac{x}{\varepsilon}\right) u^\varepsilon$ weakly converges to $\int_{Y_2} T_b \mathrm{d}y$ in $L^2(Q)$;

(3) T is a solution to the homogenized problem:

$$\partial_t T - \int_0^t \mathcal{H}(t-\tau) T(\tau)\, d\tau - \operatorname{div}\left(\widetilde{A}\nabla T\right) + \widetilde{\gamma} T = \mathcal{F} \text{ in } Q, \tag{10}$$

$$T = 0 \text{ on } S, \tag{11}$$

$$T(0,x) = |Y_1|\, h(x), \; x \in \Omega \tag{12}$$

where $\mathcal{H}$, $\widetilde{A}$, $\widetilde{\gamma}$ and $\mathcal{F}$ are respectively given by (56), (41), (55) and (57);

(4) The temperature T_b is related to T by:

$$\begin{aligned} T_b(\tau,x,y) = T_{bi}(\tau,x,y) &+ \int_0^\tau \partial_\tau \omega(\tau - t, y)\, T(t,x)\, \mathrm{d}t \\ &+ \int_0^\tau \partial_\tau \sigma(\tau - t, y)\, f_b(t,x)\, \mathrm{d}t \end{aligned} \tag{13}$$

where T_{bi}, ω and μ are respectively given by (46), (47)-(50) and (51)-(54).

Proof. See A. Ainouz[5]. □

Finally, we end this section by noticing that the equation (10) is an integro-differential equation of Barbashin type. It contains in particular an extra-term of integral type : the convolution between $\mathcal{H}$ and the temperature T. This derived macro-model (10)-(13) can be seen as a new mathematical model for human thermotherapy and human thermoregulation system. As it was pointed out in the introduction, we shall only derive formally the homogenized system (10)-(13) by using the two-scale asymptotic technique, see A. Bensoussan *et al.*[9] or E. Sanchez-Palencia[21].

3. Some auxiliary lemmas and proof of Theorem 2.1

We begin this section with some standard lemmas needed for proving the existence and uniqueness result and also for establishing uniform a priori estimates that are specifically important when using compactness techniques.

Lemma 3.1. *There exists a constant $C > 0$, independent of ε such that for all $\varphi_1 \in H_0^1(\Omega) \cap H^1(\Omega_1^\varepsilon)$ we have*

$$\|\varphi_1\|_{0,\Omega_1^\varepsilon} \leq C \|\nabla \varphi_1\|_{0,\Omega_1^\varepsilon}. \tag{14}$$

Proof. See for instance G. Allaire and F. Murat[7] . □

Lemma 3.2. *There exists a constant $C > 0$, independent of ε such that for all $\varphi_2 \in H^1(\Omega_2^\varepsilon)$ we have*

$$\|\varphi_2\|_{0,\Omega_2^\varepsilon}^2 \leq C\left(\varepsilon^2 \|\nabla\varphi_2\|_{0,\Omega_2^\varepsilon}^2 + \varepsilon \|\varphi_2\|_{0,\Sigma^\varepsilon}^2\right). \tag{15}$$

Proof. See C. Conca[13] . □

Lemma 3.3. *There exists a constant $C > 0$, independent of ε such that for all $\varphi \in H^1(\Omega_1^\varepsilon)$ we have*

$$\varepsilon \|\varphi\|_{0,\Sigma^\varepsilon}^2 \leq C\left(\varepsilon^2 \|\nabla\varphi\|_{0,\Omega_1^\varepsilon}^2 + \|\varphi\|_{0,\Omega_1^\varepsilon}^2\right), \tag{16}$$

and

$$\varepsilon \|\varphi\|_{0,\Sigma^\varepsilon}^2 \leq C\left(\|\nabla\varphi\|_{0,\Omega_1^\varepsilon}^2\right). \tag{17}$$

Proof. Using the trace theorem on Y_1 (see for e.g. R. A. Adams and J. F. Fournier[1]), we know that there exists a constant $C(Y_1) > 0$ such that for every $\psi \in H^1(Y_1)$

$$\int_\Sigma |\psi|^2 \,\mathrm{d}\sigma \leq C\left(\int_{Y_1} |\nabla\psi|^2 \,\mathrm{d}y + \int_{Y_1} |\psi|^2 \,\mathrm{d}y\right).$$

Then, we change y by x/ε and we get

$$\varepsilon \int_{\Sigma^{\epsilon k}} |\varphi|^2 \,\mathrm{d}s^\varepsilon \leq C\left(\varepsilon^2 \int_{Y_1^{\epsilon k}} |\nabla\varphi|^2 \,\mathrm{d}x + \int_{Y_1^{\epsilon k}} |\varphi|^2 \,\mathrm{d}x\right) \tag{18}$$

for every $\varphi \in H^1\left(Y_1^{\epsilon k}\right)$, where

$$Y_1^{\epsilon k} = \varepsilon(k + \Omega_1^\varepsilon), \quad \Gamma^{\epsilon k} = \varepsilon(k + \Gamma) \quad k \in \mathbb{Z}^d.$$

Note that the constant C appearing in (18) is the same for all $\varepsilon > 0$ and for all $k \in \mathbb{Z}^d$. Now, by taking the sum of the inequalities (18) over all the cells $Y_1^{\epsilon k}$ contained in Ω, we obtain (16). In fact, on the part of the cells which contain a portion of the exterior boundary, the estimate (16) still holds true, since those cells lie at a distance $O(\varepsilon)$. As ε is sufficiently small, say $\varepsilon < 1$, we have from (16) that for all $\varphi \in H^1(\Omega_1^\varepsilon)$

$$\varepsilon \int_{\Gamma^\epsilon} |\varphi|^2 \,\mathrm{d}s^\varepsilon \leq C\left(\int_{\Omega_1^\varepsilon} |\nabla\varphi|^2 \,\mathrm{d}x + \int_{\Omega_1^\varepsilon} |\varphi|^2 \,\mathrm{d}x\right) \tag{19}$$

and using the Friedrich inequality (14) in (19), we finally get (17). The Lemma is then proved. □

Proof. [Proof of Theorem 2.1] We shall use the Lions Lemma (see for instance R. Showalter[22]). Since $a^{\varepsilon}(\cdot,\cdot)$ is coercive and continuous, it only remains to prove the continuity of the form:

$$\varphi = (\varphi_1, \varphi_2) \mapsto L^{\varepsilon}((\varphi_1, \varphi_2)) = \int_0^T (h^{\varepsilon}, \varphi(t))_{H^{\varepsilon}} \, \mathrm{d}t$$

on $L^2\left(0, T; (V^{\varepsilon})'\right)$. First, using Cauchy-Schwarz inequality and (7), we see that for all $\varphi = (\varphi_1, \varphi_2) \in V^{\varepsilon}$,

$$\begin{aligned} |L^{\varepsilon}((\varphi_1, \varphi_2))| &= \left| \int_0^T \left(\int_{\Omega_1^{\epsilon}} f_1 \varphi_1 \, \mathrm{d}x + \int_{\Omega_2^{\epsilon}} f_2 \varphi_2 \, \mathrm{d}x \right) \right| \\ &\le M(f_1, f_2) \left(||\varphi_1||_{0,\Omega_T^{\varepsilon}} + ||\varphi_2||_{0,\Omega_T^{\varepsilon}} \right), \end{aligned} \tag{20}$$

where

$$M(f_1, f_2) = \max\left(||f_1||_{0,\Omega_T}, ||f_2||_{0,\Omega_T}\right) < +\infty$$

is a constant independent of ε. Observe that $f \in L^2(0, T; H^{\varepsilon})$. Next, from (15), we get

$$\begin{aligned} \int_{\Omega_2^{\epsilon}} |\varphi_2|^2 \, \mathrm{d}x \le C \Bigg(& \varepsilon^2 \int_{\Omega_2^{\epsilon}} |\nabla \varphi_2|^2 \, \mathrm{d}x + \varepsilon \int_{\Gamma^{\epsilon}} |\varphi_1 - \varphi_2|^2 \, \mathrm{d}s^{\varepsilon} \\ & + \varepsilon \int_{\Gamma^{\epsilon}} |\varphi_1|^2 \, \mathrm{d}s^{\varepsilon} \Bigg). \end{aligned} \tag{21}$$

Now, combining (17) and (21) we have

$$\begin{aligned} \int_{\Omega_2^{\epsilon}} |\varphi_2|^2 \, \mathrm{d}x \le C \Bigg(& \int_{\Omega_1^{\epsilon}} |\nabla \varphi_1|^2 \, \mathrm{d}x + \varepsilon^2 \int_{\Omega_2^{\epsilon}} |\nabla \varphi_2|^2 \, \mathrm{d}x \\ & + \varepsilon \int_{\Gamma^{\epsilon}} |\varphi_1 - \varphi_2|^2 \, \mathrm{d}s^{\varepsilon} \Bigg) \end{aligned}$$

which means that

$$\int_{\Omega_2^{\epsilon}} |\varphi_2|^2 \, \mathrm{d}x \le C \, \|(\varphi_1, \varphi_2)\|_{V^{\epsilon}}^2 . \tag{22}$$

Observe that (15) yields

$$\int_{\Omega_1^{\epsilon}} |\varphi_1|^2 \, \mathrm{d}x \le C \, \|(\varphi_1, \varphi_2)\|_{V^{\epsilon}}^2 . \tag{23}$$

Using (20), (22) and (23) we deduce that

$$|L^{\varepsilon}((\varphi_1,\varphi_2))| \le C\,\|(\varphi_1,\varphi_2)\|_{V^{\epsilon}}\,. \tag{24}$$

Thus, L^{ε} is continuous on $L^2(0,T;V^{\varepsilon})$. Note that the constant C appearing in (20) is independent of ε. By Lions Lemma, we conclude that there exists a unique solution $(T^{\varepsilon},T_b^{\varepsilon}) \in L^2(0,T;V^{\varepsilon})$ to the weak formulation of (6a)-(6g). Finally, putting $(\varphi_1,\varphi_2)=(T^{\varepsilon},T_b^{\varepsilon})$ in (8), using the uniform coerciveness of $a^{\varepsilon}(\cdot,\cdot)$, the continuity of L^{ε} and the Gronwall inequality we obtain the uniform estimate (9). This concludes the proof of Theorem 2.1.

□

Remark 3.1. If $h=(h,h_b)$ is given in V^{ε} then one can easily see that $w^{\varepsilon}=(T^{\varepsilon},T_b^{\varepsilon}) \in W^{1,2}(0,T;H^{\varepsilon})$ and therefore $w^{\varepsilon} \in C(0,T;V^{\varepsilon})$.

4. The Homogenization procedure

We shall use the formal two-scale method to derive the homogenized system of (6a)-(6g). For more details on this technique we refer the reader to the standard books by Bensoussan *et al.*[9] and E. Sanchez-Palencia[21]).

Let us then assume the following formal expansions for the two temperatures:

$$T^{\varepsilon}(t,x) = T_0(t,x,y) + \varepsilon T_1(t,x,y) + \varepsilon^2 T_2(t,x,y) + \ldots \tag{25}$$

$$T_b^{\varepsilon}(t,x) = T_b(t,x,y) + \varepsilon T_{b1}(t,x,y) + \varepsilon^2 T_{b2}(t,x,y) + \ldots \tag{26}$$

where $y = x/\varepsilon$ is the microscopic variable and $T_0(\cdot,\cdot,y)$, $T_b(\cdot,\cdot,y)$, $T_k(\cdot,\cdot,y)$, $T_{bk}(\cdot,\cdot,y),\ldots$ $(k=1,2,\cdots)$ are smooth unknown functions that are Y-periodic in the third variable y. The idea of the two-scale method is to plug the above asymptotic expansions (25)-(26) into the set of equations (6a)–(6g) and to identify powers of ε. This yields a hierarchy of (initial) boundary value problems for the successive terms T_0, T_b, T_k and T_{bk}.

Notation 4.1. In what follows, the subscript x, y on a differential operator denotes the derivative with respect to x, y respectively.

At the first step, Equation (6a) at ε^{-2} order and Equation (6c) at ε^{-1} order give respectively

$$-\mathrm{div}_y(\alpha\nabla_y T_0) = 0 \text{ in } Q\times Y_1 \tag{27}$$

and

$$\alpha \nabla_y T_0 \cdot \nu = 0 \text{ on } Q \times \Sigma, \tag{28}$$

where ν is the unit outward normal to Σ. Testing (27) by $\zeta \in H^1_\# (Y_1)/\mathbb{R}$, integrating by parts on Y_1, taking into account (28) and the Y-periodicity of $\nabla_y T_0$, we get the following weak formulation:

$$\begin{cases} a_{Y_1}(T_0, \zeta) \overset{\text{def}}{=} \int_{Y_1} \alpha \nabla_y T_0 \cdot \nabla_y \zeta \, \mathrm{d}y = 0 \;\; \text{for every } \zeta \in H^1_\# (Y_1)/\mathbb{R}, \\ T_0 \in H^1_\# (Y_1)/\mathbb{R}. \end{cases}$$

The bilinear form a_{Y_1} is clearly continuous and coercive on $H^1_\# (Y_1)/\mathbb{R}$ and therefore standard results on uniformly elliptic equations in periodic domain (A. Bensoussan *et al.*[9]) yields that T is independent of the periodic variable y, namely there exists $T(t,x)$

$$T_0(t,x,y) = T(t,x), \qquad \text{for a.e. } t \in (0,T) \text{ and } \quad x \in \Omega.$$

Next, in the second step, Equation (6a) at ε^{-1}, Equation (6c) at ε^0 orders lead to the following corrector problem:

$$-\alpha \Delta_y T_1 = 0 \text{ in } Q \times Y_1, \tag{29}$$

$$(\alpha \nabla_y T_1) \cdot \nu = -(\alpha \nabla T) \cdot \nu \text{ on } Q \times \Sigma, \tag{30}$$

$$y \longmapsto T_1(x,y) \;\; Y - \text{periodic}, \;\; (t,x) \in Q. \tag{31}$$

The corresponding weak formulation is given by

$$(\mathcal{P}) \begin{cases} a_{Y_1}(T_1, \zeta) = -a_{Y_1}(T, \zeta), \;\; \forall \zeta \in H^1_\# (Y_1)/\mathbb{R}, \\ T_1 \in H^1_\# (Y_1)/\mathbb{R}. \end{cases}$$

As before, there exists a unique solution $T_1(t,x,\cdot) \in H^1_\# (Y_1)/\mathbb{R}$ of problem $(\mathcal{P})$ which can be computed as follows. Let us consider for $1 \leq i \leq d$, the following cell problems:

$$(\mathcal{P}_i) \begin{cases} a_{Y_1}(\omega_i, \zeta) = -a_{Y_1}(e_i, \zeta), \;\; \forall \zeta \in H^1_\# (Y_1)/\mathbb{R}, \\ \omega_i \in H^1_\# (Y_1)/\mathbb{R} \end{cases}$$

where $(e_i)_{1 \leq i \leq d}$ is the canonical basis of $\mathbb{R}^d$. These problems $(\mathcal{P}_i)$ are obtained from $(\mathcal{P})$ by replacing ∇T with the vector e_i, $i = 1, 2, \cdots, d$. It follows that for each i Problem $(\mathcal{P}_i)$ admits a unique solution $\omega_i \in H^1_\# (Y_1)/\mathbb{R}$.

Furtheremore, thanks to the linearity of $(\mathcal{P})$, we may write that:

$$T_1(t,x,y) = \sum_{i=1}^{d} \frac{\partial T}{\partial x_i}(t,x)\,\omega_i(y) + \tilde{u}(t,x) \tag{32}$$

for a.e. $(t,x,y) \in Q \times Y_1$ and where $\tilde{u}(t,x)$ is any additive constant.

Furthermore, Equation (6a)-(6b) at ε^0, Equations (6c)-(6d) at ε^1 orders yield the following initial boundary-value problem:

$$\begin{aligned} -\alpha\Delta_y T_2 = f - \partial_t T + \alpha \mathrm{div}_y(\nabla_x T_1) + \\ \alpha \mathrm{div}_x((\nabla_y T_1 + \nabla_x T)) \text{ in } Q \times Y_1, \end{aligned} \tag{33}$$

$$\partial_t T_b - \alpha_b \Delta_y T_b = f_b \text{ in } Q \times Y_2, \tag{34}$$

$$(\alpha\nabla_y T_2)\cdot\nu = -(\alpha\nabla_x T_1)\cdot\nu + (\alpha_b\nabla_y T_b)\cdot\nu \text{ on } Q\times\Sigma, \tag{35}$$

$$\alpha\nabla_y T_2\cdot\nu = -\alpha\nabla_x T_1\cdot\nu + \gamma(T - T_b) \text{ on } Q\times\Sigma, \tag{36}$$

$$y \mapsto T_2(t,x,y)\ \ Y - \text{periodic}, \tag{37}$$

$$y \mapsto T_b(t,x,y)\ \ Y - \text{periodic}. \tag{38}$$

The weak formulation of Equations (33)–(38) is

$$(\mathcal{P})\begin{cases} a_{Y_1}(T_2,\zeta) = \langle F,\zeta\rangle \quad \text{for every } \zeta \in H^1_{\#}(Y_1)/\mathbb{R}, \\ T_2 \in H^1_{\#}(Y_1)/\mathbb{R} \end{cases}$$

where

$$\begin{aligned} \langle F,\psi\rangle = \left(\int_{Y_1} -\zeta\,\mathrm{d}y\right)\partial_t T + \int_{Y_1} \mathrm{div}_x(\alpha(\nabla_y T_1 + \nabla T))\,\zeta\,\mathrm{d}y \\ -\int_{Y_1} \alpha\nabla_x T_1\cdot\nabla_y\zeta\,\mathrm{d}y + \int_{Y_1} f\zeta\ \mathrm{d}y + \int_{\Sigma}\gamma(T - T_b)\,\zeta\,\mathrm{d}s, \end{aligned}$$

where $\mathrm{d}s$ denotes the Hausdorff measure on Σ. Now, using the divergence Theorem (as in E. Sanchez-Palencia[21]), a necessary and sufficient condition for the existence and uniqueness of T_2 is that $\langle F,1\rangle = 0$, namely

$$\begin{aligned} \int_{Y_1} (-\partial_t T + \mathrm{div}_x(\alpha(\nabla_y T_1 + \nabla T)))\,\mathrm{d}y \\ + \int_{\Sigma}\gamma(T - T_b)\,\mathrm{d}\sigma = |Y_1|\,f. \end{aligned} \tag{39}$$

By (32), equation (39) becomes

$$|Y_1|\,\partial_t T - \operatorname{div}\left(\widetilde{A}\nabla T\right) + \int_\Sigma \gamma\,(T - T_b)\,\mathrm{d}s = |Y_1|\,f, \tag{40}$$

where $|Y_1|$ stands for the volume of Y_1, The matrix $\widetilde{A}$ is given by

$$\widetilde{A} = (\tilde{a}_{ij})_{1\le i,j\le d}\,, \quad \tilde{a}_{ij} = \int_{Y_1} \alpha\,(\nabla_y\omega_i + e_i)\cdot(\nabla_y\omega_j + e_j)\,\mathrm{d}y. \tag{41}$$

Equation (40) is the so-called macroscopic equation for the temperature T. The boundary condition for T is obtained from (6e) at ε^0 order and it reads

$$T = 0 \text{ on } \partial\Omega. \tag{42}$$

Similarly, Equations (6f)-(6g) give the initial conditions for T and T_b:

$$T\,(0,x) = |Y_1|\,h\,(x)\,,\ x\in\Omega, \tag{43}$$

$$T_{b0}\,(0,x,y) = \chi_2\,(y)\,h_b\,(x)\,,\ x\in\Omega,\quad y\in Y_2 \tag{44}$$

Next, we proceed further and focus our attention on fluid temperature $T_b\,(t,x,y)$. From (35) and (36) it is easily seen that

$$\alpha_b\nabla_y T_b\cdot\nu = \gamma\,(T - T_b) \text{ on } Q\times\Sigma. \tag{45}$$

Now (34), (45) and (44) is a classical parabolic problem and therefore admits a unique weak solution $T_b \in H^1_\#\,(Y_2)$. Moreover, denoting

$$b\,(\zeta,\eta) = \int_{Y_2}\alpha_b\nabla_y\zeta\cdot\nabla_y\eta\,\mathrm{d}y + \int_\Sigma \gamma\zeta\eta\,\mathrm{d}s, \quad \zeta,\eta\in H^1\,(Y_2)$$

and $\mathcal{B}\,:\,H^1\,(Y_2)\longrightarrow H^1\,(Y_2)$ defined by $\langle\mathcal{B}\,(\zeta)\,,\eta\rangle = b\,(\zeta,\eta)$, applying (as in U. Hornung[16]) the Duhamel's principle to equations (34), (45) and (44), the leading term T_b ca be thus decomposed as the sum of three terms (see (13)):

$$\begin{aligned} T_b\,(\tau,x,y) = T_{bi}\,(\tau,x,y) &+ \int_0^\tau \partial_\tau\omega\,(\tau - t,y)\,T\,(t,x)\,\mathrm{d}t \\ &+ \int_0^\tau \partial_\tau\sigma\,(\tau - t,y)\,f_b\,(t,x)\,\mathrm{d}t \end{aligned}$$

where T_{bi} is the evolution of the initial temperature h_b. It is given by

$$T_{bi}\,(t,x,y) = \mathrm{e}^{-t\mathcal{B}}h_b\,(x,y) = \theta\,(t,y)\,h_b\,(x) \tag{46}$$

where $\theta(t,y)$ is the unique of the weak solution of the cell problem:

$$\begin{aligned}
&\partial_t\theta - \alpha_b\Delta_y\theta = 0 \text{ in } (0,T)\times Y_2,\\
&\alpha_b\nabla_y\theta\cdot\nu + \gamma\theta = 0 \text{ on } (0,T)\times\Sigma,\\
&y \longmapsto \theta(t,y)\ \ Y-\text{periodic},\\
&\theta(0,y) = 1 \text{ in } \Omega\times Y_2.
\end{aligned}$$

On the other hand ω, σ are respectively the unique weak solutions of the following cell problems:

$$\partial_t\omega - \alpha_b\Delta_y\omega = 0 \text{ in } (0,T)\times Y_2, \tag{47}$$

$$\alpha_b\nabla_y\omega\cdot\nu + \gamma\omega = \gamma \text{ on } (0,T)\times\Sigma, \tag{48}$$

$$y \longmapsto \omega(t,y)\ \ Y-\text{periodic}, \tag{49}$$

$$\omega(0,y) = 0 \text{ in } \Omega\times Y_2, \tag{50}$$

and

$$\partial_t\sigma - \alpha_b\Delta_y\sigma = 1 \text{ in } (0,T)\times Y_2, \tag{51}$$

$$\alpha_b\nabla_y\sigma\cdot\nu + \gamma\sigma = 0 \text{ on } (0,T)\times\Sigma, \tag{52}$$

$$y \longmapsto \sigma(t,y)\ \ Y-\text{periodic}, \tag{53}$$

$$\sigma(0,y) = 0 \text{ in } \Omega\times Y_2. \tag{54}$$

Inserting (13) into (40) we get the homogenized integro-differential equation of Barbashin type for the temperature T (see (10)-(12)):

$$\begin{aligned}
&\partial_t T - \int_0^t \mathcal{H}(t-\tau)\,T(\tau)\,d\tau - \operatorname{div}\left(\widetilde{A}\nabla T\right) + \widetilde{\gamma}T = \mathcal{F} \text{ in } Q,\\
&T = \ 0 \text{ on } S,\\
&T(0,x) = |Y_1|\,h(x),\ \ x\in\Omega
\end{aligned}$$

where

$$\widetilde{\gamma} = \frac{1}{|Y_1|}\int_\Sigma \gamma\ \mathrm{d}s \tag{55}$$

and, where $\mathcal{H}$ and $\mathcal{F}$ are given by

$$\mathcal{H}(\tau,x) = \frac{1}{|Y_1|}\int_\Sigma \gamma(y)\partial_t\omega(\tau,y)\,\mathrm{d}s, \tag{56}$$

$$\mathcal{F}(\tau,x) = f(x) + \frac{1}{|Y_1|}\left(\int_\Sigma \gamma(y)\,T_{bi}(\tau,x,y)\,\mathrm{d}s + \int_0^\tau\int_\Sigma \partial_\tau\sigma(\tau-t,y)\,f_b(\tau,y)\,\mathrm{d}s\mathrm{d}t\right). \tag{57}$$

References

1. R.A. Adams and J.F. Fournier, Sobolev Spaces, 2nd edition, pure and applied mathematics series, Elsevier Ltd., 2003.
2. A. Ainouz, *Mathematica Bohemica* **136**, 357 (2011).
3. A. Ainouz, *Electr. J. Diff. Equ.* **2013**, 1 (2013).
4. A. Ainouz, *To appear in Applications of mathematics.*
5. A. Ainouz, *In preparation.*
6. G. Allaire, *SIAM J. Math. Anal.* **23**, 1482 (1992).
7. G. Allaire and F. Murat, *Asymptotic Analysis* **7**, 81 (1993).
8. J.L. Auriault and H.I. Ene, *Int. J. Heat and Mass Transfer* **37**, 18 (1994).
9. A. Bensoussan, J.L. Lions and G. Papanicolaou, Asymptotic analysis for periodic structure North Holland, Amsterdam, 1978.
10. H.S. Carslaw and J.C. Jaeger, Conduction of heat in solids, Oxford university press, 1959.
11. J.C. Chato, Fundamentals of Bioheat Transfer, Springer-Verlag, Berlin, 1989.
12. M.M. Chen and K.R. Holmes, *Ann. N. Y. Acad. Sci.* **335**, 137 (1980).
13. C. Conca, *J. Math. Pures Appl.* **64**, 31 (1985).
14. M. Gautherie, Clinical Thermology: Thermotherapy, Vols. 1–4, Springer-Verlag, Heidelberg, 1990.
15. R. Hochmuth and P. Deuflhard, *Math. Methods. Appl. Sci.* **27**, 971 (2004).
16. U. Hornung ed., Homogenization and Porous Media, interdisciplinary applied mathematics **6**, Springer-Verlag, 1997.
17. A.-R.A. Khaled and K. Vafai, *Inter. J. of Heat and Mass Transfer* **46**, 4989 (2003).
18. K. Khanafer, A. AlAmiri, I. Pop and J.L. Bull, *Emerging topics in heat and mass transfer in porous media, P. Vadász editor, Theory and applications of transport in porous media* **22**, 327 (2008).
19. M. Miyakawa and J.C. Bolomey (Eds.), Non-invasive Thermometry of Human Body, CRC Press, Boca Raton, 1996.
20. H.H. Pennes, *J. Appl. Physiol.* **1**, 93 (1948).
21. E. Sanchez-Palencia, Nonhomogeneous media and vibrating theory, Lecture Notes in Physics, **127**, Springer-Verlag, Berlin, 1980.
22. R. Showalter, Monotone Operators in Banach Space and Nonlinear Partial Differential Equations, Mathematical Surveys and Monographs, **49**, American Mathematical Society, 1997.
23. C. Sturesson, Medical Laser-Induced Thermotherapy: Models and Applications, Lund Reports on Atomic Physics, LRAP-235, Doctoral Thesis, Lund Institute of Technology, October 1998.
24. P. Wust, J. Nadobny, R. Felix, P. Deuflhard, W. John, and A. Louis, *Strahlenther. und Onkol.* **165**, 751 (1989).

MULTI-SCALE MODELLING IN CELL DYNAMICS

M. BANERJEE
Department of Mathematics and Statistics, Indian Institute of Technology Kanpur, Kanpur 208016, India

M. BENMIR
Faculté des Sciences, University Hassan II of Casablanca, Casablanca 20100 Maroc

V. VOLPERT
Institut Camille Jordan, UMR 5208 CNRS, University Lyon 1 69622 Villeurbanne, France
E-mail: volpert@math.univ-lyon1.fr

The paper is devoted to multi-scale modelling in cell population dynamics with intracellular and local or global extracellular regulation. In the case of global extracellular regulation we study the role of random initial conditions of intracellular proteins on cell fate and on stable production of a given number of differentiated cells. In the case of local extracellular regulation, we observe travelling waves of cell differentiation and discuss emerging patterns of differentiated cells.

1. Multi-scale models in biology

Multi-scale models in biology actively develop in order to describe physiological processes. There are various approaches to multi-scale modelling and numerous applications (see [1,3,4,11,15] and the references therein). Cell populations in multi-scale models can be described by discrete or continuous methods. Cellular automata, lattice Boltzmann method and various particles methods are among them [1,5,8,12,13,14]. These approaches allow a detailed description of cell behavior, cell-cell interaction and other aspects of complex biological media. On the other hand, they are applicable for a relatively small number of cells and they do not admit analytical investigation. Continuous models are represented by ordinary and partial differential equations for cell concentrations. In particular, these can be reaction-diffusion equations which take into account random motion of cells and their birth and death; Navier-Stokes equations and Darcy equations describe convective motion of the medium.

Multi-scale models include intracellular and extracellular regulations of biological cells. Intracellular regulation is particularly important because it determines the cell fate, that is the choice between its self-renewal, differentiation and apoptosis. It can be described by ordinary differential equations for intracellular concentrations, by Boolean approach or by probabilistic methods if a small number of molecules participate in this regulation and their concentrations cannot be considered.

Extracellular regulation can be effectuated by various local mechanisms in the given tissue (growth factors, cytokines) or by means of global control from other organs and tissues through endocrine signaling. Extracellular substances diffuse in the tissue and influence intracellular regulation of cells. Distribution of these substances can be described by reaction-diffusion equations.

In this work we will study multi-scale models with intracellular regulation of cell fate and with two different types of extracellular regulation. First we will consider global extracellular regulation by means of endocrine signaling and then local extracellular regulation through cell interaction in the same tissue. We will restrict ourselves to two model examples of such regulation which will allow us to explain some important features of these processes. In particular the role of stochasticity and emergence of spatial patterns.

2. Global extracellular regulation

2.1. *Model*

We will consider cells of two types, A and B. Every given time interval τ a new cell A appears due to differentiation of less mature cells into cells of type A. Cells of the type A can either differentiate into cells of the type B or, otherwise, they die by apoptosis. Hence we restrict ourselves here to the cell choice between differentiation and apoptosis. We do not consider here self-renewal of cells A in order to study first the functioning of the system in this simpler case.

Cells of the type B have a finite and fixed life span T. After that they die. The total number $N(t)$ of cells B at time t is determined by the number of cells A differentiated during the time interval $[t-T, t]$. Suppose that the system should possess N_0 cells B for the normal functioning of the organism. Then the rate of differentiation of cells A should depend on $N(t)$

and N_0.

Let us begin with the simplest model which describes dynamics of this cell population. If we introduce concentrations of cells A and B and use for them the same notation, then we can write the system of two ordinary differential equations:

$$\frac{dA}{dt} = a - (k_d + k_a)A, \quad \frac{dB}{dt} = k_d A - \sigma B, \tag{1}$$

where a is the influx of cells A, k_a and k_d are the rates of their differentiation and apoptosis, σ is the mortality coefficient of cells B. Assuming that the coefficients k_d and k_a depend on B, $k_d(B)$ is a decreasing function and $k_a(B)$ increases, then this system of equations has a unique stationary point (A_0, B_0), where

$$B_0 = \frac{a}{\sigma} \frac{k_d(B_0)}{k_a(B_0) + k_d(B_0)}, \quad A_0 = \frac{\sigma B_0}{k_d(B_0)}.$$

This stationary point is globally asymptotically stable. Let us note that differentiation of cells A is not instantaneous. If we take this into account, we obtain delay differential equations.

In general, the rates of differentiation and apoptosis cannot be considered as given functions of B. Indeed, cell response on the extracellular regulation can depend on its intracellular regulation. Therefore we need to introduce this regulation in the model.

Intracellular regulation. We consider here the simplest regulation with only two intracellular proteins p_1 and p_2 and two equations for their concentrations:

$$\frac{dp_1}{dt} = F_1(p_1, p_2), \quad \frac{dp_2}{dt} = F_2(p_1, p_2) . \tag{2}$$

Each cell has its own values $p_1(t)$ and $p_2(t)$ described by this system of equations. Obviously, these concentrations can be different in different cells.

When a new cell A appears, we prescribe it some initial values p_1^0 and p_2^0. Hence we can determine the evolution of these concentrations inside each cell. We suppose that if the concentration $p_1(t)$ reaches some critical level p_1^*, then the cell dies by apoptosis. If the concentration $p_2(t)$ reaches some critical level p_2^*, then the cell differentiates.

Evolution of the intracellular concentrations depends on the functions F_1 and F_2. We will consider them in the following form:

$$F_1(p_1,p_2) = k_1p_1(1-a_{11}p_1-a_{12}p_2), \quad F_2(p_1,p_2) = k_2p_2(1-a_{21}p_1-a_{22}p_2), \tag{3}$$

where k_1, k_2 and a_{ij} are positive parameters. It is similar to the system of competition of species and it describes bistable dynamics. It can have up to four stationary points with non-negative coordinates: $P_0 = (0,0)$, $P_1 = (1/a_{11}, 0)$, $P_2 = (0, 1/a_{22})$, $P_3 = (\tilde{p}_1, \tilde{p}_2) > 0$, where $\tilde{p}_1$ and $\tilde{p}_2$ satisfy the system of equations

$$a_{11}p_1 + a_{12}p_2 = 1, \quad a_{21}p_1 + a_{22}p_2 = 1.$$

We suppose that it has a positive solution.

We will consider the case where the point P_3 is unstable, the points P_1 and P_2 are stable. This condition can be easily expressed in terms of the coefficients a_{ij}. The separatrix of the point P_3 separates the basins of attraction of the points P_1 and P_2. If the initial condition is below this curve, then the trajectory converges to the point P_1, if the initial condition is above this curve, then the trajectory converges to the point P_2. If the trajectory approaches the point P_1, then the concentration p_1 becomes large, the concentration p_2 small. In this case the cell will die by apoptosis. In the second case, p_2 becomes large, p_1 small, the cell differentiates.

Cell number and extracellular regulation. The number of cells B at time t is denoted by $N(t)$. It equals the number of cells A differentiated during the time interval from $t - T$ to t. Suppose that the system should produce N_0 cells. If $N(t)$ is different from N_0, then there is a feedback control which acts on the rate of differentiation. Since the choice between differentiation and apoptosis of cells A is determined by the concentrations p_1 and p_2 described by system (2), we will suppose that its coefficients depend on the difference $N(t) - N_0$. Namely, we set

$$a_{21}(N) = a_{21}^0 + \alpha(N - N_0), \tag{4}$$

where α is a positive constant. All other coefficients are independent of N.

If $N(t) > N_0$, then the coefficient a_{21} increases in comparison with a_{21}^0. Therefore the function $F_2(p_1, p_2)$ in (3) also decreases. Hence growth of the concentration p_2 decelerates and, consequently, the rate of differentiation also decreases. Thus the dependence of the coefficients of the intracellular regulation on the cell number provides a global control over the system with the purpose to get the required number of differentiated cells.

Stochasticity in the initial conditions. Let us recall that there is a constant influx of cells A. This means that every time interval τ a new cell enters the compartment with cells A. We need to prescribe the initial concentrations p_1^0 and p_2^0 for each new cell. Since these values are usually not known and they can hardly be measured experimentally, this becomes an additional unknown parameter which can influence behavior of the system. We will choose the initial conditions from the square domains

$$D = \{a - k \leq p_1^0 \leq a + k, \quad b - k \leq p_2^0 \leq b + k\}, \tag{5}$$

where a, b and k are chosen in such a way that the whole domain D has non-negative coordinates. The initial condition will be chosen from the domain D as a random variable with a uniform distribution.

2.2. *Results*

We present results of numerical simulations of the model described in Section 2.1. We set $\tau = 0.1$, $T = 20$, $a_{11} = a_{22} = 0.5$, $a_{12} = a_{21}^0 = 1$, $p_1^* = p_2^* = 1.8$ and we will vary other parameters.

Deterministic initial conditions. We begin with the case where the initial conditions for the concentrations p_1^0, p_2^0 are fixed. Therefore all cells A have exactly the same initial conditions in the intracellular regulation. Then this problem does not have solutions with a constant value $N(t) \equiv N$. In this sense we can say that the problem with deterministic initial conditions does not have stationary solution. Indeed, suppose that such solution exists and consider the phase plane of the system (2). Since $a_{21}(N)$ is constant (time independent) then the separatrix which separates basins of attraction of the points P_1 and P_2 is also fixed. If the initial condition (p_1^0, p_2^0) belongs to the basin of attraction of the point P_1, then all cells will die. If it belongs to the basin of attraction of the point P_2, then all cells will differentiate. Finally, if the initial condition is exactly at the sepratrix, the corresponding trajectory will converge to the unstable point P_3 and it will never reach the critical values p_1^* or p_2^*. Hence in all three cases the number of differentiated cells cannot be equal N if it is different from 0 or T/τ.

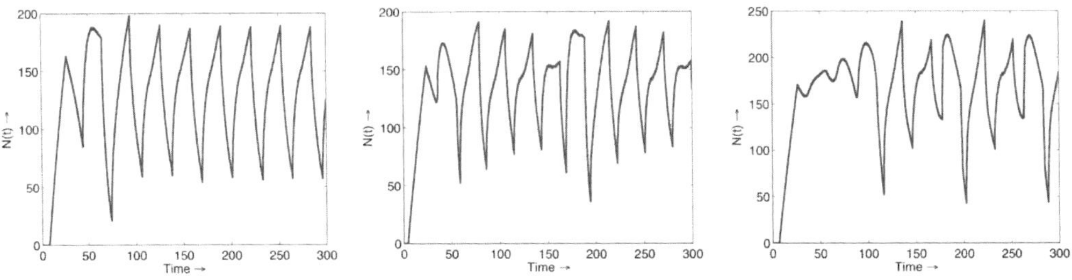

Figure 1. The number of differentiated cells $N(t)$ for a constant initial condition $p_1^0 = p_2^0 = 0.1$. $N_0 = 150, \alpha = 0.003$ (left), $N_0 = 150, \alpha = 0.009$ (middle), $N_0 = 200, \alpha = 0.003$ (right).

Figure 1 shows numerical simulations in the case of a fixed (deterministic) initial condition. The number of differentiated cells oscillates around the target value N_0. Therefore the separatrix oscillates and the initial condition (p_1^0, p_2^0) belongs to different basins of attraction during different time intervals. These oscillations do not result from the instability of a stationary solution since the stationary solution does not exist.

Random initial conditions. In this section the initial conditions p_1^0 and p_2^0 will be taken from the square domain $0 \leq p_1^0, p_2^0 \leq k$ as a random variable with a uniform distribution. During the time interval T there are T/τ new cells A. In the case of symmetric parameters and initial conditions, a half of them will differentiate. For the values of parameters considered here, we obtain $N = 100$ differentiated cells B. Figure 2 shows examples of numerical simulations for $N_0 = 100$ and different values of α. If α is sufficiently large, then there is an overshoot in cell number. Its value for large time converges to N_0 after small decaying oscillations.

Figure 2 shows the evolution of the number of differentiated cells $N(t)$ in time for different values of parameters. In all cases it converges to some limiting value N_∞. It is interesting to note that the function $N(t)$ is practically independent of the value of k which determines the choice of initial conditions. In the wide range of variation of this parameter, $0.1 \leq k \leq 1.5$, the curves coincide. A slight difference appears for $k = 2$.

Let us recall that in the case of deterministic initial conditions, $N(t)$

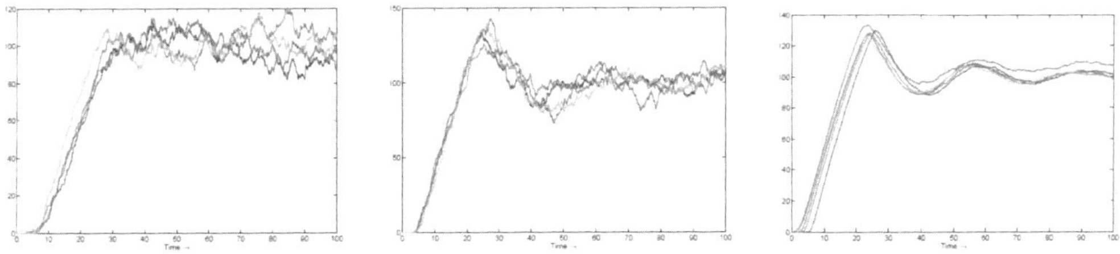

Figure 2. The number of cells $N(t)$ as a function of time for $N_0 = 100$, $\alpha = 0$ (left) and $\alpha = 0.009$ (middle). Four curves show different simulations with the same values of parameters. The curves in the right figure correspond to different values of k (from 0.1 to 2). Average value of $N(t)$ for 50 simulations, $N_0 = 100$, $\alpha = 0.009$.

does not converge to a stationary value. Hence we can expect appearance of oscillations for very small values of k. We will return to this question below. Therefore convergence to a stationary solution is determined by randomness in the initial conditions.

The domain D of initial conditions is split by the separatrix into two subdomains which belong to the basins of attraction of the points P_1 and P_2. Depending on the position of the initial condition, the corresponding trajectory will converge to one of these stationary points. The areas of these two subdomains determine the proportion of differentiated cells and the value N_∞. Let us note that system of equations (2)-(4) has time dependent coefficients. However for t sufficiently large, $N(t)$ is close to a constant, and this system is close to the autonomous system.

The global regulation by the number of differentiated cells $N(t)$ through the coefficients of the system (see (4)) influences the limiting value N_∞. It increases with the increase of N_0 but in general it is different from N_0. Clearly, if $N(t) \equiv N_0$, then from (4) we have $a_{21} = a_{21}^0$. In this case, the system is symmetric and exactly half of the total cell number differentiate. For the values of parameters under consideration $N_\infty = 100$. Therefore, if $N_0 \neq 100$, then $N_\infty < N_0$. The value of N_∞ increases and approaches N_0 when we take greater values of α.

The simulations presented above are carried out for a fixed value a_{21}^0 in (4). We can also introduce its dependence on N. Then we can obtain a better approximation of N_0 by N_∞. We varied a_{21}^0 in large limits and obtained convergence to a stationary value N_∞ which depends on the value of a_{21}^0 (not shown here).

Thus global regulation allows us to obtain a given stationary value of the number of cells for a large interval of variation of the coefficients and of the range of random initial conditions.

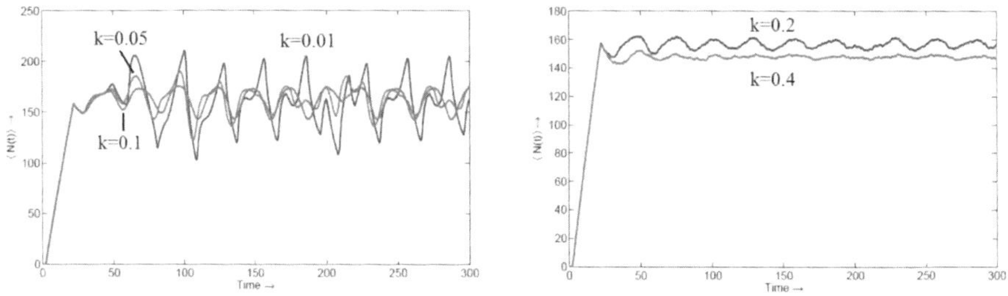

Figure 3. The number of differentiated cells $N(t)$ for $N_0 = 150$, $\alpha = 0.009$, $a = 0.4, b = 0.6$ and different values of k. When k increases, the amplitude of oscillations decay.

Oscillating solutions. We considered deterministic initial conditions and observed periodic oscillations of the number of differentiated cells. Next, we studied the case with random initial conditions from a sufficiently large domain. In this case, the number of differentiated cells converges to a constant value. Let us now consider transition between these two cases. Initial conditions will be uniformly distributed in the domain $[a-k, a+k] \times [b-k, b+k]$. Figure 3 shows the number of differentiated cells in time for different values of k and all other parameters fixed. The amplitude of oscillations decreases when k increases, and they practically disappear for $k = 0.4$. Hence randomness of initial conditions can suppress oscillations.

If we now take other values of a and b, then the oscillations are practically independent of the value of k (in the considered range). Persistence of the oscillations can be related to large time delay for cells to differentiate. Indeed, the time which the trajectory needs to reach the critical value p_2^* depends on the initial condition.

System (2) has time dependent coefficients (4). Therefore the extracellular regulation through $N(t)$ acts on the cell during its all life time. Indeed, some of the curves $(p_1(t), p_2(t))$ cross each other because the system is not autonomous. This extracellular regulation can have more influence on the intracellular concentrations than the choice of initial conditions. This is why in some cases the results weakly depend on the range of initial conditions.

3. Local extracellular regulation

In this section we consider the problem of lineage choice where undifferentiated cells differentiate into one of two types of differentiated cells. There are three cell types, undifferentiated cells A, differentiated cells B_1 and B_2. Cells A contain two intracellular proteins, p_1 and p_2. Their concentrations are described by ordinary differential equations:

$$\frac{dp_1}{dt} = F_1(p_1, p_2) + b_{11}u_1 + b_{12}u_2, \tag{6}$$

$$\frac{dp_2}{dt} = F_2(p_1, p_2) + b_{21}u_1 + b_{22}u_2. \tag{7}$$

The functions F_1 and F_2 will be specified below. Extracellular concentration u_1 is produced by differentiated cells B_1, and u_2 is produced by cells B_2. Concentrations u_1 and u_2 are governed by the equations

$$\frac{\partial u_1}{\partial t} = d_1 \frac{\partial^2 u_1}{\partial x^2} + W_1, \tag{8}$$

$$\frac{\partial u_2}{\partial t} = d_2 \frac{\partial^2 u_2}{\partial x^2} + W_2. \tag{9}$$

Cells of the type A are located at $x > \xi(t)$, cells B_1 and/or B_2 at $x < \xi(t)$. The value of $\xi(t)$ increases in time, that is the wave speed is positive.

If the intracellular concentration p_1 is greater than a critical value, $p_1 \geq p_1^*$, then the cell A changes its type to B_1, if $p_2 \geq p_2^*$, then it changes its type to B_2. Once a cell is differentiated, it cannot change any more.

We introduce the concentration c_A of cells A, the concentration c_1 of cells B_1 and the concentration c_2 of cells B_2. At each space point x these three concentrations can have only two values, 0 or 1. Consider some interval filled by cells A. Therefore, in this interval $c_A = 1$, $c_1 = c_2 = 0$. If at some point x of this interval p_1 becomes greater than p_1^*, then the cell A changes type to B_1. Hence, for this x, $c_A = 0$, $c_1 = 1$, $c_2 = 0$. Similarly, if p_2 becomes greater than p_2^*, then $c_A = 0$, $c_1 = 0$, $c_2 = 1$.

We set

$$W_1 = k_1 c_1, \quad W_2 = k_2 c_2.$$

This means that cells B_1 produce u_1, cells B_2 produce u_2. The rates of production vanish if the concentrations of the corresponding cells equal zero.

In this section we consider the intracellular kinetic functions in the form

$$\begin{aligned} F_1(p_1, p_2) &= k_1 p_1^2 (1 - a_{11} p_1 - a_{12} p_2) - s_1 p_1, \\ F_2(p_1, p_2) &= k_2 p_2^2 (1 - a_{21} p_1 - a_{22} p_2) - s_2 p_2. \end{aligned} \tag{10}$$

The stationary points of the corresponding system

$$\frac{dp_1}{dt} = F_1(p_1, p_2), \quad \frac{dp_2}{dt} = F_2(p_1, p_2) \tag{11}$$

are as follows: $P_0 = (0,0)$, $P_{10} = (p_1^{(1)}, 0)$, $P_{20} = (p_1^{(2)}, 0)$, where $p_1^{(1)}$ and $p_1^{(2)}$ are solutions of the equation

$$p_1(1 - a_{11} p_1) = \frac{s_1}{k_1} ,$$

$P_{01} = (0, p_2^{(1)})$, $P_{02} = (0, p_2^{(2)})$, where $p_2^{(1)}$ and $p_2^{(2)}$ are solutions of the equation

$$p_2(1 - a_{22} p_2) = \frac{s_2}{k_2} ,$$

and also up to four stationary points with positive coordinates which can be found as solutions of the system of equations

$$p_2 = \frac{1 - a_{11} p_1}{a_{12}} - \frac{s_1}{a_{12} k_1} \frac{1}{p_1} , \quad p_1 = \frac{1 - a_{22} p_2}{a_{21}} - \frac{s_2}{a_{21} k_2} \frac{1}{p_2} .$$

It can have from zero to four positive solutions depending on the values of parameters.

It can be easily verified that the point P_0 is stable. Indeed, the corresponding matrix has negative eigenvalues. Let us assume that $0 < p_1^{(1)} < p_1^{(2)}$ and $0 < p_2^{(1)} < p_2^{(2)}$. Then the points P_{20} and P_{02} are also stable. In the case where there are four stationary points with positive coordinates, one of them is stable.

There are two different patterns of solutions depending on parameters b_{ij}. Let us consider two specific examples. If $b_{11}, b_{22} > 0$ and $b_{12} = b_{21} = 0$, then the substances $u_i, i = 1, 2$ produced by cells B_i stimulate production of intracellular substances p_i. In their turn, they lead to differentiation of cells

A into cells B_i. Therefore we observe here a positive feedback between intracellular regulation, extracellular regulation and cell differentiation. One of the cell lineages B_1 or B_2 dominates another one. It expands on the whole space interval. All cells differentiate into only one cell lineage.

In the second example, $b_{11} = b_{22} = 0$ and $b_{12}, b_{21} > 0$. This means that the substance u_1 produced by cells B_1 stimulate production of the intracellular substance p_2, while u_2 stimulates production of p_1. Hence there is a negative feedback, and cells B_1 upregulate production of B_2, while cells B_2 promote productions of cells B_1. Behavior of the system is qualitatively different in this case compared with the previous one. Both types of differentiated cells can be observed here.

Figure 4 shows propagation of the wave of cell differentiation for the case $b_{11} = b_{22} = 0$ and $b_{12} = b_{21} = 0.1$. The distribution of the concentration of cells c behind the wave represents a periodic patterns with the values $c = 1$ (cells B_1) and $c = 2$ (cells B_2). Distributions of p_1 and p_2, u_1 and u_2 have also space oscillations.

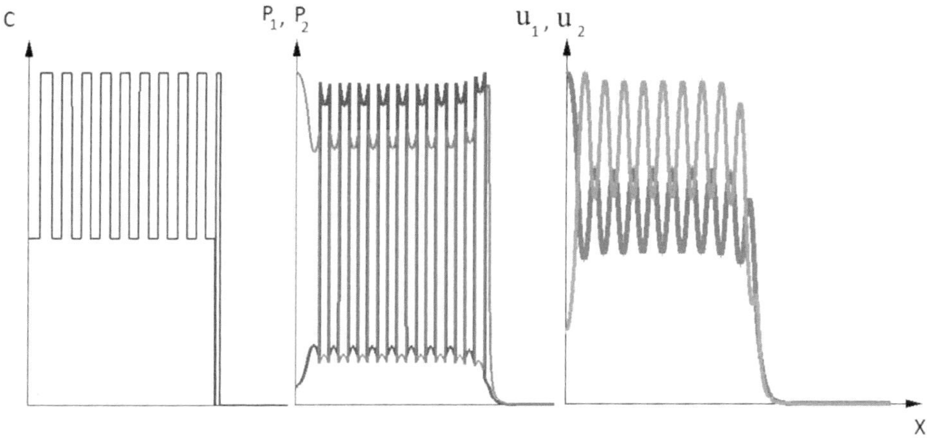

Figure 4. A snapshot of solution. A travelling wave of cell differentiation propagates from the left to the right. The distribution of cells B_1 ($c = 1$) and B_2 ($c = 2$) behind the wave is periodic (left image). Distributions of p_1, p_2 (middle) and of u_1, u_2 (right) oscillate in space.

Let us note that the patterns with two lineages of differentiated cells can be obtained only in the case where the stationary point P_0 of system (10) is stable. Otherwise, if it is unstable, then only one cell lineage will be obtained even in the case of negative feedback between cell differentiation and production of intracellular proteins. We can give the following explanation. When the two cell lineages appear, they produce the extracellular substances u_1 and u_2 which diffuse and influence undifferentiated cells where production of intracellular proteins p_1 and p_2 begins. Even if the concentrations of u_1 and u_2 are small, they are sufficient to initiate intracellular reactions since the point P_0 is unstable. Therefore even small extracellular concentrations determine the future choice of undifferentiated cells between two cell lineages far ahead the front of differentiated cells. One of them will finally win this competition, and all differentiated cells will belong to the same type. If the point P_0 is stable, then small concentrations u_1 and u_2 will not be sufficient to start intracellular reactions. They will begin only when differentiated cells are sufficiently close. If these are cells B_1, then they will stimulate production of p_2 and vice versa. This negative feedback results in the coexistence of two cell lineages.

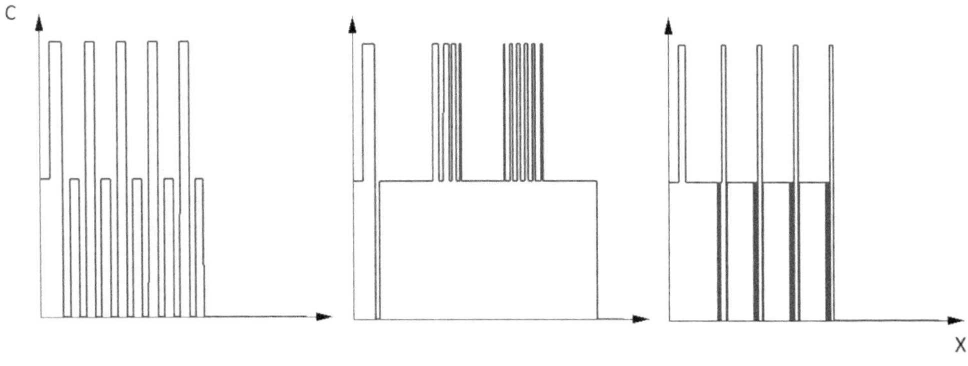

Figure 5. Distributions of differentiated cells for three different sets of parameters: $q_1 = q_2 = 0.01, D_1 = D_2 = 0.1$ (left), $q_1 = q_2 = 0.01$, $D_1 = 0.1, D_2 = 0.5$ (middle), $q_1 = 0.1, q_2 = 0.01, D_1 = D_2 = 0.1$ (right).

Different patterns are shown in Figure 5. Undifferentiated cells can coexist with both types of differentiated cells (left and right images) and distribution of differentiated cells can be different in comparison with the previous figure (Figure 5, middle).

Thus we obtain coexistence of cell lineages in the case where undifferentiated cells are stable from the point of view of intracellular regulation and where the feedback between cell differentiation and intracellular regulation is negative.

4. Discussion

In this work we study multi-scale models of cell dynamics with intracellular regulation and global or local extracellular regulation. We show how the interaction of these regulations determine the cell fate and control the number of cells and their distribution.

In the case of global regulation, cells make a choice between differentiation and apoptosis. This choice is determined by the intracellular regulation (bistable dynamics) and controlled by the extracellular regulation through the total number of differentiated cells. Such regulation occurs in various physiological processes, in particular in erythropoiesis, production of red blood cells in the bone marrow [6,9,10]. The choice between differentiation and apoptosis is possible in this model because new cells have random initial conditions of intracellular proteins. We analyze the role of this randomness and show that it allows the system to ensure stable production of a given number of cells required for normal functioning.

In real physiological systems, global extracellular regulation is usually combined with local regulation. So we illustrate how local extracellular regulation interacts with the intracellular regulation in the case of lineage choice. We show that cell differentiation propagates as a reaction-diffusion wave [15,16]. One of the important questions here is how to preserve both lineages of differentiated cells. We show that it is possible if undifferentiated state is stable and if there is a negative feedback of differentiated cells on lineage choice.

Thus, we study local and global extracellular regulation separately on two model examples. An interesting question for further analysis is how they work together.

References

1. A. R. A. Anderson, M. Chaplain, K.A. Rejniak. Single cell based models in biology and medicine. Birkhäuser, Basel, 2007.
2. M. Banerjee, V. Volpert. On the role of regulation and stochasticity in cell fate. Submitted.
3. S. Bernard. Modélisation multi-échelles en biologie. In: Le vivant discret et continu. N. Glade, A. Stephanou, Editeurs, Editions Materiologiques, 2013, 65-89.
4. V. Cristini, J. Lowengrub. Multiscale modeling of cancer: an integrated experimental and mathematical modeling approach. Cambridge University Press, Cambridge, 2010.
5. A. Deutsch, S. Dormann. Cellular automaton modeling of biological pattern formation. Birkhäuser, Boston, 2005.

6. N. Eymard, N. Bessonov, O. Gandrillon, M.J. Koury, V. Volpert, *The role of spatial organisation of cells in erythropoiesis.* J. Math. Biol., (2014), in press.
7. N. Glade, A. Stephanou, Editeurs. Le vivant discret et continu. Editions Materiologiques, 2013.
8. M. Karttunen, I. Vattulainen, A. Lukkarinen. A novel methods in soft matter simulations. Springer, Berlin, 2004.
9. P. Kurbatova, S. Bernard, N. Bessonov, F. Crauste, I. Demin, C. Dumontet, S. Fischer, V. Volpert. *Hybrid Model of Erythropoiesis and Leukemia Treatment with Cytosine Arabinoside.* 2011, SIAM J. Appl. Math., 71 (2011), no. 6, 2246–2268.
10. P. Kurbatova, N. Eymard, V. Volpert. *Hybrid model of erythropoiesis.* Acta Biotheoretica, 61 (2013), no. 3, 305-315.
11. J.M. Osborne, A. Walter, S.K. Kershaw, G.R. Mirams, A.G. Fletcher, P. Pathmanathan, D. Gavaghan, O.E. Jensen, P.K. Maini, H.M. Byrne. *A hybrid approach to multi-scale modelling of cancer.* Phil. Trans. R. Soc. A, 368 (2010), 5013–5028.
12. A.A. Patel, E.T. Gawlinsky, S.K. Lemieux, R.A. Gatenby. *A Cellular Automaton Model of Early Tumor Growth and Invasion: The Effects of Native Tissue Vascularity and Increased Anaerobic Tumor Metabolism.* J. Theor. Biol., 213 (2001), 315–331.
13. A. Satoh. Introduction to practice of molecular simulation. Elsevier, 2011.
14. S.Y. Trofimov. Thermodynamic consistency in dissipative particle dynamics. Eindhoven University Press, 2003.
15. V. Volpert. Elliptic partial differential equations. Volume 2. Reaction-diffusion equations. Birkhäuser, 2014.
16. A. Volpert, Vit. Volpert, Vl. Volpert. Traveling wave solutions of parabolic systems. Translation of Mathematical Monographs, Vol. 140, Amer. Math. Society, Providence, 1994.

TOPOLOGY OF CELL MEMBRANES

E. I. KATS

Landau Institute for Theoretical Physics, RAS,
142432, Chernogolovka, Moscow region, Russia, and
Moscow Institute of Physics and Technology,
141700, Dolgoprudny, Moscow region, Russia

M. I. MONASTYRSKY

Institute of Theoretical and Experimental Physics, Moscow 117218, Russia
Email:michael.monastyrsky@gmail.com

Physical and biological observation methods provide a variety of bilayer membranes' shapes and their transformations. Besides, the topological and geometrical methods allow us to deduce a classification of all possible membrane surfaces. This double-sided approach leads to a deeper insight into membranes properties. Our goal is to apply an appropriate mathematical technique for classifying vesicles (closed surfaces in mathematical terminology) and for their transformation ways. We exhibit that all vesicles can be decomposed in a small number of universality classes generated by a few "bricks": a torus, a screwed torus, and the real projective plane. We consider several ways of transforming membrane surfaces, bearing in mind that they possess an additional extremal property. Our method exploits different constructions of minimal surfaces in S^3. We estimate energetic barrier for transformation of minimal membrane surfaces using the so-called doubling procedure. This problem is far from being a pure theoretical exercise. For instance, almost all cells' biological functions, or tumor progression, are accompanied by apparently singular cell membrane transformations.

1. Introduction

Biological (and also artificial, single lipid layer) cell membranes are rather complex bilayer sheets, separating (and simultaneously connecting via pore openings) the cell's interior and its surrounding medium. Of course this is not the only role of membranes. In biological systems the membranes also perform many other amazingly sophisticated functions, participating in fission, fusion and signaling processes. While performing all these functions membrane conformations can be changed considerably, passing through singular shapes with very narrow (molecular size) necks, conical points, cusps or self-intersections (see e.g. pictures in [1]). Moreover, different vesicle topologies are observed even when the same preparation procedure is used, suggesting that not only energetics, but sometimes kinetics of self-assembly plays a role in the selection of vesicle morphologies. One of our goals is to describe possible stable, metastable or even kinetically stabilized membrane structures.

However, despite this complexity, on a scale larger than bilayer thickness, membranes can be treated merely as a two dimensional surfaces M^2 living in three dimensional space R^3 and determined by the extremum of the so-called Helfrich-Willmore functional (see section 4). Naturally, for such scales the mathematical methods of classifying surfaces and describing their transformations can be very useful. This approach became especially demanded recently due to the program intended to design topologically controlled materials [2]. Usually, in mathematical studies of membranes or vesicles, only embedded surfaces were considered. Our claim, motivated by observations of strongly deformed membranes (e.g., by proteins' insertion or during cell fusion and fission processes), is to show that the class of embeddings is too narrow to describe all possible transformations of membrane surfaces. One could also think about immersions rationalizing geometrically stage of local pore formations in membranes. Note that in liquid membranes, individual molecules on a membrane can diffuse quite efficiently. On the contrary, global motions which are mandatorily required by surface transformation, also require coherent movement of all lipids on a vesicle and on all solvent molecules inside of it. Usually this process is considerably slow. For such a long time scale we can consider very fast finger-like molecular protrusions as local singular immersion points. Budding of vesicles (nucleation of a new small vesicle from a larger one) is another process of this type. Indeed the membranes' transformations are intertwined with the collective behavior of proteins. It is believed that the stability of proteins to generate and to sense the membrane curvature are crucial in driving budding of vesicles. Here again we have to deal with a slow collective membrane transformations combined with formation of singular surface points.

We do believe that the immersion and the cobordism theories supplied by some energetic and kinetic arguments represent appropriate mathematical tools to classify singular membranes and to rationalize physical scenarios behind each of such transformations. To avoid any confusion we should emphasize that we don't discuss geometric shape transformations governed merely by surface metric characteristics (volume V or excess (at a given V) area A_{ex}). This problem was intensively studied in 1990s (see e.g., [3,4]). Our general classification of membranes into a finite number of immersion classes (see section 3) is coarse grained, and disregard metric degrees of freedom, tacitly assuming that V and A_{ex} somehow adapted to allow immersed surfaces of a given class.

The structure of immersed manifolds has been studied by mathematicians during the past 70 years but as far as we know this technique was never applied to membranes and vesicles. The main goal of our paper is to point out that under natural physical assumptions, the theory of immersions is well adapted to analyze transformations of membranes with different topological genus g.

Our paper is organized as follows. In the next section 2 we introduce the concept of immersion and develop the appropriate mathematical description. Section 3 is devoted to classification of immersed surfaces in R^3. In section 4 we explain the relation between minimal surfaces immersed in S^3 and Willmore surfaces in R^3. In section 5 we describe different method of constructing minimal surfaces embedded in S^3, and also discuss the relation of our construction with the general classification of physical states based on cobordism theory. We perform (section 6) some energetic estimations for transformations of the minimal surfaces in the same cobordant class, and summarize our results in conclusion section 7.

2. Immersed surfaces

In this section we shall treat the problem of classifying immersed surfaces in R^3 and recall some basic notions and definitions.

- Let M^2 be a compact surface, possibly with a boundary, and f is a smooth map of $M^2 \to R^3$. We say that M^2 is immersed in R^3 if the Jacobian of the map f is of rank 2 at each point x in M^2, or equivalently that a mapping of $T_x M^2 \to T_{f(x)} f(M^2)$ is one to one. Here $T_x M^2$ denotes the tangent space of M^2 at the point x. If the map is also one to one mapping of $M^2 \to R^3$, we say that M^2 is embedded in R^3. It follows from the definition, that intersecting points or lines may occur in the immersed surfaces.
- A regular homotopy is a smooth homotopy that is an immersion at each stage. It is natural to classify immersions of manifolds up to a regular homotopy. In the case of immersions of two-dimensional manifolds in R^3 the following results are known (for all definitions of general topological notions we refer to [5,6]) :

(i) if M^2 is a compact surface and $h \equiv dim H_1(M^2, Z_2)$, then the space of immersions f $M^2 \to R^3$ has 2^h connected components [7]. Here $H_1(M^2, Z_2)$ denotes a one dimensional homology group with coefficients in the cycling group

To formulate the following result (see (ii) below) we recall the important notion of modern topology so-called cobordism. Let us consider two compact manifolds M_1^n and M_2^n. We say that M_1^n and M_2^n are cobordant if there exists a manifold W^{n+1} with boundaries $\partial W^{n+1} = M_1^n \cup M_2^n$. Here $\cup$ means the operation of taking the non-connected sum. The manifold W calls cobordism of the manifolds M^1 and M^2. The cobordant relation is an equivalence relation on the space of n-dimensional compact manifolds. The set of all n-dimensional cobordant classes– $\Omega(M^n)$ makes up a group with respect to the operation $\cup$.

(ii) The cobordism group $\Omega(M^2)$ of compact surfaces immersed in R^3 is Z_8 [8].

It is possible that two immersed surfaces f and g are not regularly homotopic but f is regularly homotopic to $g \circ \phi$ where ϕ is diffeomorphism $M^2 \to M^2$. Here $\circ$ denotes the operation of composition. It means that the surfaces are geometrically equivalent, but differently parameterized. From physical point of view these surfaces are the same. So we shall classify immersed surfaces $[f]$ and $[g]$ up to a regular homotopy. Therefore using immersions to classify membranes as physical surfaces one has to care about such physically equivalent surfaces. The classification by cobordism groups is more coarse since two cobordant surfaces can have different genus.

3. Classification of Immersed Surfaces

In this section we shall follow to papers [9,10]. To formulate the key result we recall some classical notions. Let M^2 be a compact surface and $H_1(M^2, Z_2)$ is one-dimensional homology group with coefficients in the cycling group Z_2. Let us determine a quadratic form $q : H_1(M^2, Z_2) \to Z_2$ by the formula:

$$q(x + y) = q(x) + q(y) + x \cdot y (mod 2)$$

where $x \cdot y$ denotes the $mod2$ intersection number of cycles x and y. It is possible to get the classification of immersed surfaces in R^3 up to regular homotopy in pure algebraic terms using quadratic forms q, To do this, first of all we relate to quadratic form q some topological invariant of immersed surfaces. Let M^2 be a compact surface, $f : M^2 \mapsto R^3$. Then

- At first step we define a function $q_f : \Gamma \to Z_4$ where Γ is the set of all smooth embedding $\gamma : S^1 \to M^2$, $\gamma \in \Gamma_f$. Let $\Gamma_f \subset \Gamma$ denote the set of all embedding $\gamma : S^1 \to M^2$ such that $f \circ \gamma$ is also an embedding. Let us consider a tubular neighborhood N_γ of $\gamma(S^1)$ in M^2. We define $q_f(\gamma)$ as a linking coefficient lk [11].

$$q_f(\gamma) = lk(f \circ \gamma, f(\partial(N_\gamma)))$$

and $q_f \in Z_4$.
- We borrow from [9] the following statement. Let f and g are two immersion of M^2 in R^3. Then

a) f and g are regularly homotopic if and only if (iff) $q_f = q_g$.

b) For every quadratic form $q : H_1(M^2, Z_2) \to Z_4$ there exists an immersion $fM^2 \to R^3$ such that $q = q_f$

c) The immersed surfaces $[f]$ and $[g]$ are regular homotopic iff there is a linear automorphism $\alpha : H_1(M^2, Z_2) \to H_1(M^2, Z_2)$ such that $q_f = q_g \alpha$.

Said above allows us to define the following characteristics, relevant for the classification of immersed surfaces, Arf invariant of a quadratic form

$$Q = Arf\,(q) = \frac{1}{\sqrt{2^h}} \sum_{\alpha \in H} (-1)^{q(\alpha)}\,, \tag{1}$$

where $h = dim H_1(M^2, Z_2)$. For this quantity we can use the known result [9] which states that if q and q_1 are two quadratic forms on $H_1(M^2, Z_2)$ then there exists an automorphism $\alpha : H_1(M^2, Z_2) \to H_1(M^2, Z_2)$, such that $q_1 = q \circ \alpha$ if and only if (iff) the corresponding Arf invariants Q and Q_1 are equal.

The usefulness of the introduction of Arf invariant based on the following proposition, the key ingredient to our classification of the immersed surfaces:

The surfaces $f(M)$ and $g(M)$ are regularly homotopic iff (iff) $Arf\,(q_1) = Arf\,(q_2)$. The proposition means that in particular, an immersed surface is regularly homotopic to its embedding counterpart iff Arf invariant equals $+1$. Important advantage of our classification of the immersions by Arf invariant ensures in clear distinction between the internal properties of M^2 and embedding in different spaces, e.g. S^3 or R^3.

Now we are going over to the classification of immersed surfaces. This is in parallel to well-known classification of two-dimensional manifolds M^2 up to a homeomorphism but is more subtle. The basic bricks in the classification are:

- M^2 is orientated. Every M^2 is homeomorphic either to a sphere or a connected sum of tori.
- M^2 is non-orientated. Every M^2 is homeomorphic to a connected sum of projective planes P^2.
- The set of classes of homeomorphic surfaces converted to a semi-group with the connected sum as a group operation. The unit element is a sphere S^2.

In the case of the immersed surfaces we have to add some new features. We consider also two types of surfaces. Namely for the orientable M^2 there exist two types of "elementary" bricks - tori depending of the value of Arf invariant: S and T, where S is a standard torus: $S^1 \times S^1$ with $Arf = 1$ and T is an immersed torus with $Arf = -1$, obtained by rotating a plane lemniscate around a vertical axis while rotation is also in its plane. We call T as a screw torus. However if M^2 is nonorientable, then the elementary brick is the real projective plane P^2. As it is known P^2 is impossible to embed in R^3 or S^3 as a compact surface (i.e. without self intersections). Furthermore there exist different ways of immersions. The less singular immersion of P^2 without singular points and continuous curvature is so-called Boy surface B [12] (W. Boy 1902). B surface has an axis of symmetry and we can consider also the $\bar{B}$ surface obtained by reflection in a plane containing this axis. B and $\bar{B}$ lead to different values of Arf invariant and they are not regularly homotopic.

The following theorem [9] describes the relations in the semi-group I of classes of immersed surfaces:

- Every compact orientable immersed surface $M^2 \subset R^3$ that is not sphere is regularly homotopic to a connected sum of S and T.
- Every compact non-orientable immersed surface $M^2 \subset R^3$ is regularly homotopic to a connected sum of several copies of B and $\bar{B}$.
- The semigroup I is generated by S, T, B and $\bar{B}$. The zero element of I is represented by any immersed sphere in R^3.

Let us recall the classical notion,so called connected sum operation making possible to construct a new manifold from two preassigned manifolds in regular way. We consider two compact manifolds ${M_1}^n$ and ${M_2}^n$ and cut two discs ${D_1}^n$ ${D_2}^n$, $D_i \subset M_i, i = 1, 2$ correspondingly. Then we glue our manifolds M_i over the disc's boundaries ∂D_i and obtain a new manifold relative to the operation of the connected sum $N = M_1 \# M_2$. Thus we arrive at the following relations :

$$S\#S = T\#T$$

$$B\#B\#B\#B = \bar{B}\#\bar{B}\#\bar{B}\#\bar{B}\#\bar{B}$$

$$S\#\bar{B} = B\#\bar{B}\#\bar{B}$$

$$T\#\bar{B} = B\#B\#B$$

$$T\#B = \bar{B}\#\bar{B}\#\bar{B}$$

Two results, proved by U. Pinkall [9] are relevant to our work:

- Two compact immersed surfaces in R^3 of the same topological type are regularly homotopic iff they are cobordant. This statement is false if we omit the restriction of equality of topological types. For example, a circle l_1 is cobordant to two circles, l_2 and l_3 but it is evident that it is impossible to get the homotopy l_1 to $l_2 \cup l_3$. In case of embedded surfaces more stronger statement is valid.
- Two compact embedded surfaces of the same genus (topological type) are regularly homotopic.

It follows from the classification of the immersed surfaces up to regular homotopy mentioned in this section that the group I is generated by a few elementary bricks:

- In case of orientable surfaces by S^2, S and T with $Arf \in Z_2$,
- For non-orientable case by $B, \bar{B}$, $B\#\bar{B}$, $B\#B$, $\bar{B}\#\bar{B}$, $B\#B\#B$, $B\#\bar{B}\#T$, $\bar{B}\#\bar{B}\#\bar{B}$ with $Arf \in Z_8$.

4. Willmore surfaces and Minimal surfaces in S^3

Our aim in this section is to apply the general and universal (i.e., valid for any kind of surfaces) classification presented in the previous section to the specific case, relevant for the membrane vesicles, Willmore surfaces. First we recall basic notions concerning the Helfrich-Willmore (HW) functional which determines a broad class of physical membranes.

Let M^2 be a surface (closed, with boundary or non-compact) embedded in R^3. A surface M^2 is called Willmore surface if it is an extremum of the HW functional:

$$F = \kappa \int_{M^2} H^2 \, dA + \bar{\kappa} \int_{M^2} G \, dA \tag{2}$$

where κ and $\bar{\kappa}$ are phenomenological elastic moduli, respectively bending and the Gaussian rigidity coefficients, H is the mean curvature, G is the Gauss curvature, and dA is an area element. The bending modulus κ has to be positive, to provide elastic stability, and $\bar{\kappa}$ should be negative to provide topological stability. According to the Gauss-Bonnet theorem, the second term in (2) is a topological invariant, independent of membrane shape. Then for a given topological genus the HW functional is equivalent to:

$$F = \kappa \int_{M^2} (k_1^2 + k_2^2) \, dA$$

where k_1 and k_2 are the principal curvatures of M and

$$H = \frac{k_1 + k_2}{2}$$

Therefore the extremum of functional (2) determines the most flat surface of fixed topology.

The following result due to J. Weiner [13] is important for future use. Let M^2 be a minimal surface in S^3 and γ is a stereographic projection $S^3 \mapsto R^3$. Then

$$\gamma(\hat{M}^2) = M^2 \qquad \text{and} \qquad F(M^2) = \sigma(\hat{M}^2) \tag{3}$$

where $\sigma(\hat{M}^2)$ is the area of the minimal surface. This statement is valid both for closed surfaces and for those having boundaries. Willmore surfaces are not exhausted by the projection of the minimal surfaces. For example, there exists an infinite set of Willmore tori which are not equivalent to minimal torus in S^3 [16]. These tori are extrema of (2) but not minima. This result, apparently, is valid also for higher genus surfaces. Willmore surfaces of higher genus generated by minimal surfaces in S^3 were constructed in [14]. It is important to emphasize that Willmore surfaces embedded in R^3 satisfy the following condition [15]:

$$\int_{M^2} H^2 \, dA \geq 4\pi n$$

where n is the number of intersection points (n–different pre-images under projection). From this and the square estimation of minimal surfaces embedded in S^3 follows that all Willmore surfaces for $g \geq 2$ have at least one point of intersections. Note that the mentioned above equivalence between special Willmore surfaces and minimal surfaces leads to a deep analogy with the string theory. Minimal surfaces $M^2 \subset S^3$ are the "world sheets" of the string determined in S^3 with Nambu-Goto action. The surfaces M^2 also play the role of instantons in the Yang-Mills theory determined in the space S^3.

5. Minimal Immersions in S^3

In this section we consider several examples of minimal immersions of two dimensional surfaces in S^3. There exist three general approaches to construct minimal immersed and embedded surfaces.

Reflection Method.

We illustrate the method applying it to 2 types of the surfaces

- Lawson surfaces.

  In his remarkable paper [14] H. B. Lawson Jr. constructed series of embedded and immersed minimal surfaces in S^3 of an arbitrary genus g. Here we give a sketch of his construction.

  Let Γ be a polygon in S^3 and G_Γ is the finite subgroup of $O(4)$ generated by reflections across the geodesic sub–arcs of Γ. If Γ satisfies some regularity conditions, then there exists the fundamental domain D_Γ which has a boundary Γ and generates minimal compact submanifold M_Γ in S^3. To illustrate how it works we represent $S^3 \subset C^2 \ni z, w$ by equation: $|z|^2 + |w|^2 = 1$. Then we consider two great circles: $C_1 = (0, w) \in C^2$, $C_2 = (z, 0) \in C^2$. Let k and m be non-negative integers and choose points $P_1, P_2, \in C_1$ and $Q_1, Q_2, \in C_2$ such that distance (in spherical metric) $(P_1, P_2) = \pi/(k+1)$, and distance $(Q_1, Q_2) = \pi/(m+1)$. We define $\Gamma_{m,k}$ to be the polygon $P_1P_2Q_1Q_2$. If we act on the polygon $P_1P_2Q_1Q_2$ by the group $\Gamma_{m,k}$ we obtain the surface $\xi_{m,k}$ (in Lawson notations). He has proved the following theorem:

For any pair of positive integers m and k, there exists an embedded compact minimal surfaces ξ_{mk} in S^3 of genus mk. In particular there exists at least one embedded surface of any given genus, and there are at least two such surfaces (unless g is a prime number). If we change the ordering in the fundamental polygon and use the same construction as in the above theorem we can produce various families of minimal surfaces. In his influential paper[14] Lawson found two other series of minimal surfaces in S^3. The geodesic polygons is $P_1P_2Q_2Q_1$. The corresponding family of surfaces $\tau_{m,k}$ has Euler characteristic zero. Iff mk is devisable by 2, $\tau_{m,k}$ is an immersed Klein bottle.

- Another family of non-orientable surfaces can be obtained if we choose the fundamental polygon as $P_2Q_2P_1(-Q_1)$ (here $m, k \geq 1$, and see more details in [14]).

- Karcher, Pinkall and Sterling surfaces [17]. They utilized similar ideas and studied various tessellations of S^3 into cells. They have found a new but finite number of minimal surfaces embedded in S^3 with g: 3, 5, 6, 7, 11, 17, 19, 73, 601. These surfaces glue from fundamental domains of Platonic solids in R^3 .

Doubling Construction.

Let us consider two minimal embedded surfaces, e. g. two Clifford tori, and cut a number of holes in each of them and connect them by minimal surfaces like catenoid bridges. The theorem proven by Kapouleas and Yang claims [18]:

There exists a closed minimal surface M_m of genus m^2+1, embedded in the unit sphere $S^3 \subset C^2$, which can be obtained by a "doubling" construction from the Clifford torus, by using m^2 catenoidal bridges centered at the points of an $m \times m$ lattice $L \subset T$, and connect two nearby copies of T It is worth to note that Kapouleas' and Yang's method of proof is applicable for sufficiently large m without explicit estimation of a value of m.

In the section 6 we supplement this procedure by some energetic estimations.

Desingularization.

To define the procedure of desingularization we consider two two-dimensional manifolds (M_1 and M_2) immersed in R^3 with nonempty intersection L. We call the minimal surface M' desingularizing surface of M_1 and M_2 if we blow up the singularity set L and obtain a smooth minimal surface. For example, for two great spheres orthogonal to each other in S^3

by desingularization procedure we obtain Lawson's minimal surfaces ξ_{mk}. A natural question related to the desingularization procedure is whether starting from Clifford torus one can get by desingularization procedure a new minimal surface? The answer follows from the statement proven by Choe and Soret [19]. Let $T_1, \ldots, T_m, \subset S^3$ be the Clifford tori intersecting each other along a great circle C_1 at an angle of π/m. Then there exits a compact minimal surface $T^0_{m,k}$, desingularizing $T_1 \cup, \ldots, \cup T_m$ for each $k = 2ml$ with integer $l \geq 1$:

- $T^0_{m,k}$ is embedded and has genus $1 + 2k(m-1)$
- $T^0_{m,k}$ is invariant under a group of screw motions,
- $T^0_{m,k}$ has no reflection symmetry across a great sphere,
- Area (A), $A(T^0_{m,k}) \leq [2m\pi]^2$.

It is worth to note that there exists also another family of embedded surfaces $T^e_{m,k}$ with the same properties itemized above but with the area $A(T^0_{m,k}) < A(T^e_{m,k})$. It concerns with the fact that there are two ways of desingularizing the tori along the great circle C_2.

5.1. *Regular homotopy of different minimal surfaces*

One of the interesting consequences of our construction is the following observation. Let us consider two surfaces M_1 and M_2 embedded in S^3 of the same genus g. Let M_L belongs to a series of Lawson (L) and M_{CS} to a series of Choe-Soret (CS) surfaces. It is possible to find M_1 and M_2 belonging to different series L and CS but with the same g. It follows from the theorem mentioned already in the end of section 3 that two compact embedded surfaces of the same genus are regularly homotopic. Then two surfaces M_L and M_{CS} are regularly homotopic. The minimal genus of such surfaces when it occurs is 9. The similar comparison exists in other series of minimal surfaces, e.g. in Kapouleas and Yang M_KY and M_CS surfaces.

From this consideration it is clear that to study transformations in real membranes it is necessary to estimate the energy of creation "holes", "handles" and other "defects" in minimal surfaces. This very respectable aim is beyond the scope of our work, and hardly doable without heavy numerics. In the section 6 we present only simple semi-qualitative estimations for the doubling procedure.

5.2. *Cobordisms and Vacuum Manifolds*

In this short subsection we shall show that our approach to classify different types of membrane surfaces and their transformations is related very naturally to the known construction in field theories with the broken symmetry, based on the ideas of the cobordism theory [21]. The theory deals with situations when there exist solutions with degenerate ground states. Such situation appears in many interesting physical phenomena, e.g., in the field theory well-known examples are gauge monopoles and instantons. In condensed matter physics the examples include coexistence of A and B phases in superfluid He^3 or interacting point and line defects in nematic and cholesteric liquid crystals.

We recall the basic ideas of the paper [21]. Let us consider a manifold M^k embedded in N^r. In physical applications k is usually equal to 3 (stationary case) or 4 in time dependent problem. M^k has a boundary ∂M^k consisting in several pieces $({P_1}^{k-1}, \ldots, {P_n}^{k-1})$. The field Φ defined at M^k yields a continuous map f of M^k to isotopical space J. The components $({P_1}^{k-1}, \ldots, {P_n}^{k-1})$ mapped by the same f to degenerate vacuums of J. It means that we have a bordism mapping

$$f : (M^k, {P_1}^{k-1} \ldots, {P_n}^{k-1}) \to (J, V_1, \ldots, V_n)$$

Topological invariants of this mapping encoded all possible configurations of a system. In this paper we are found in a different physical situation but there is an exact analogy with the general construction. There are no topological obstacles to transfer one surface to another cobordant surface. This is the case for the reflective method:gluing procedure for the Lawson's surfaces. One can expect that transformations of cobordant surfaces are energetically not very expensive, since it requires to pay only a surface matching energy. On the contrary, Kapouleaus– Yang doubling procedure requires to overcome bulk energy barrier. However, energetic estimations exhibit that this process realized at the room temperature conditions .

6. Physical Consequences of Immersions

As it was already said our aim is to classify all possible closed membranes(vesicles) in R^3, minimizing the HW energy functional. We show that if one allows immersions, the classification problem can be reduced to that for the minimal surfaces in S^3. Moreover, we find a low expensive energetically algorithm to generate such immersed minimal surfaces. Below we obtain energetic estimations to produce a minimal surface starting from a given more simple minimal surface using the doubling procedure.

To illustrate how it works we follow Kapouleas and Yang recipe [18]. We start from the Clifford torus $g = 1$ and its energy is chosen as a reference state $E_0 = 0$. Then by the doubling procedure we can produce a surface with $g = m^2 + 1$, where $m \gg 1$ is a number of catenoid bridges, connecting the initial Clifford torus with its counterpart. To estimate energetic cost of the doubling procedure we proceed in the spirit of the work [23] for smectic-like fluid membrane phases and our work [24], where similar procedure has been performed for small pores in the minimal surfaces. Since the catenoid is the minimal surface, its own bending energy is zero ($H = 0$ in the HW functional (2). The only shape-dependent energy to pay for creation of a single small size catenoid is the deformation energy of a locally flat membrane where the catenoid is created. The energy can be estimated if we replace the HW-functional H (2) by an inverse size of the deformed catenoid bridge region. The deformed region can be approximately estimated as the catenoid neck. The energy cost of a single catenoid passage is

$$E_1 \simeq \kappa(\xi) \left(\frac{r}{\xi} \right)^2 , \tag{4}$$

where $\kappa(\xi)$ is scale dependent (due to thermal fluctuations, [25]) bending elastic modulus, ξ is membrane surface deformation range, produced by a single passage, and r is the catenoid neck size (on the order of the membrane thickness δ). We neglect in (4) the topological Gaussian term, which is independent of the bridge shape. If the deformation regions of the neighboring catenoids are overlapping, the catenoids repel each other. Therefore we can roughly estimate ξ as the lattice constant for $m \times m$ lattice of catenoids. Since the entropy of the non-interacting catenoids distribution $\propto k_B T \ln \frac{A(X/\xi)^2}{m^2}$ (with $X \le \xi$ being the average distance between catenoids) is maximized (like in ideal gas) for the uniform over the whole surface (area A) catenoids distribution (i.e., for $X \simeq \xi$), we conclude that

$$\xi \simeq \sqrt{\frac{A}{m^2}} . \tag{5}$$

It follows from (4) that the energetic barrier to overcome for the doubling procedure is

$$E_b \simeq \kappa \left(\frac{r^2 m^2}{A} \right) g . \tag{6}$$

Such transformations (from $g = 1$ to $g = m^2 + 1$) can be realized by thermal fluctuations if the energetic barrier (6) is on the order of $k_B T$ (k_B

is Boltzmann constant and T is temperature):

$$m \leq \left(\frac{k_B T A}{\kappa r^2}\right)^{1/4}, \tag{7}$$

However, because typically $A/r^2 > 10^6$, and $k_B T/\kappa \simeq 10^{-2}$ it is not a very restrictive condition and one can indeed produce very high genus surfaces by the doubling procedure.

The estimate of the energetic barrier is only a part of the story. If one thinks about membrane's transformations, the important question is whether this surface with high genus $g = m^2 + 1$ is energetically favorable with respect to the initial state (Clifford torus). The answer depends on the sign of the Gaussian rigidity modulus $\bar{\kappa}$. Only if $\bar{\kappa} > 0$, the high genus surfaces gain the energy $\bar{\kappa} g$. If this gain is larger than the deformation energy penalty (6), the high genus surface indeed is energetically favorable. Thus we get the second physical restriction on the maximal genus attainable by the doubling procedure

$$g \leq \left(\frac{\bar{\kappa} A}{\kappa r^2}\right). \tag{8}$$

7. Conclusion

We claim in this paper that many relevant membranes' transformations can be described by immersions. Mathematics offers a universal classification of all possible immersions, which can be decomposed in a few elementary surfaces - "building bricks", determined by the value of so-called Arf invariant. Furthermore there are three procedures (reflection, doubling, and desingularization) enable to construct more complicated surfaces from these elementary bricks. Although all three procedures are feasible topologically, physical transformation requires to overcome certain energetic barrier. We presented some estimation of the energy barrier only for the doubling process, but a natural question, should be answered. Why at all presented above elegant mathematical classification of the immersed surfaces should have something to do with biological membranes? The matter is that the cell membranes are intrinsically dynamic objects, whose conformations are associated with various biological functions. Membrane fluidity allows the membrane to rearrange very rapidly (characteristic molecular protrusion time scale), its local shape and composition to perform needed functions. Visual observations [20] and numeric simulations [22] confirm our guess that dynamically singular (looking as self-intersections) immersion points are not forbidden. The stabilization of such immersed surfaces is not just a question of topology. Kinetics and energetic (see arguments below) also come into play. Another situation, when locally (in time and in space) immersion points are not forbidden is a crumpling or cracking of a thin membrane sheet under action of applied pressure. At crumpling the deformation goes to a strongly localized regions occurring in an arbitrarily small fraction of the sample. We report here mainly a brief summary of our results and leave the systematic analysis and applications of described mathematical methods to physical and biological membranes for future work.

Acknowledgments

We would like to thank A. Gabibov and R. Kusner for stimulating discussions and R. Kerner for valuable remarks. E.I.Kats contribution into this work was funded by the Russian Science Foundation (grant 14-12-00475). M. Monastyrsky thanks Institut des Hautes Études Scientifiques (Bures sur Yvette, France), where part of this work was done for stimulated atmosphere and support. His work was partially supported by the grant RFBR 13-01-00314 and by the government grant of the Russian Federation for support of research projects implemented by leading scientists, Lomonosov Moscow State University under the agreement No. 11.G34.31.0054.

References

1. L.S.Hirst, A.Ossowski, M.Fraser, J.Geng, J.V.Selinger, PNAS, **110**, 3242 (2013).
2. B.Senyuk, O.Liu, S.He, R.D.Kamien, R.B.Kusner, T.C.Lubensky, I.I.Smalyukh, Nature, **493**, 200 (2013).
3. F.Julicher, U.Seifert, R.Lipowsky, Phys. Rev. Lett., **71**, 452 (1993).
4. X.Michalet, D.Bensimon, Science, **269**, 666 (1995).
5. B. Dubrovin, A. Fomenko, S. Novikov, Modern Geometry, v.1-3(2ed) Springer, Berlin, 1990, 1992
6. M. Monastyrsky, Topology of Gauge Fields and Condensed matter, Plenum, N.Y. 1993
7. I. James, E. Thomas, Topology 4, 351-359 (1966).
8. R. Wells, Topology 5, 281-294 (1966)
9. U.Pinkall,Topology, 24 ,421-434 (1985)
10. R. Kusner, N. Schmitt, ArXiv :dg-ga/9512003
11. D. Rolfsen, Knots and Links, Publish or Perish, Wilmington, 1976
12. D.Hilbert, S. Cohn-Fossen, Geometry and the Imagination, AMS Chelsea Pbl., Providence, 1952.
13. J.L. Weiner, Indiana Math. J. 27,19-35 (1978)
14. H. Lawson Jr., Ann.Math,92 335-374 (1970)

15. P.Li, S.T. Yau, Invent.Math. 69,269-291 (1982)
16. U.Pinkall, Invent.math. 81, 379-386 (1985)
17. H. Karcher, U. Pinkall, I. Sterling, J. Diff. Geom 28, 169-185 (1988)
18. N. Kapouleas, S. D. Yang, Amer. J. Math. 132, 257-295 (2010)
19. J.Choe, M. Sorret, ArXiv:1304.3184
20. T.Hawkins, L.R.Stephens, T.Bretschneider, BMC Bioinformatics, **14**, 296 (2013).
21. V. Golo, M. Monastyrsky, Ann. Inst. Henri Poincaré, 28, 75-89 (1978)
22. Y.Chelm, K.Azzag, P.Canadns, J.Averseng, S.Baghdiguian, B.Maurin, J. of Biomechanics, **46**, 1075 (2013).
23. L.Golubovic, Phys. Rev. E, **50**, 1279 (2000).
24. E.I.Kats, M.I.Monastyrsky, JETP, **91**, 1279 (2000).
25. W.Helfrich, J. Phys. (Paris), **46**, 1263 (1985).

GEOMETRY OF MORPHOGENESIS

NADYA MOROZOVA

CNRS FRE 3377 "Epigenetique et Cancer", CEA Saclay, 91191,
Gif-sur-Yvette France
and Insitut des Hautes Études Scientifiques
38 route des Chartres, 91440 Bures-sur-Yvette France
E-mail: morozova@ihes.fr, nadezda.morozova@cea.fr

ROBERT PENNER

Insitut des Hautes Études Scientifiques
38 route des Chartres, 91440 Bures-sur-Yvette France
and Math and Physics Departments, Caltech
Pasadena, CA 91125 USA
E-mail: rpenner@ihes.fr, rpenner@caltech.edu

We introduce a formalism for the geometry of eukaryotic cells and organisms. Cells are taken to be star-convex with good biological reason. This allows for a convenient description of their extent in space as well as all manner of cell surface gradients. We assume that a spectrum of such cell surface markers determines an epigenetic code for organism shape. The union of cells in space at a moment in time is by definition the organism taken as a metric subspace of Euclidean space, which can be further equipped with an arbitrary measure. Each cell determines a point in space thus assigning a finite configuration of distinct points in space to an organism, and a bundle over this configuration space is introduced with fiber a Hilbert space recording specific epigenetic data. On this bundle, a Lagrangian formulation of morphogenetic dynamics is proposed based on Gromov-Hausdorff distance which at once describes both embryo development and regenerative growth.

1. Introduction

We seek to posit a geochemical model for morphogenesis of eukaryotic organisms that includes cellular details. Our aim is to describe the main laws underlying the morphogenetic processes during normal development as well as during possible perturbations resulting in creating structure de novo as in the course of regeneration[5]. In this inherently multi-scale context, we must describe the nano-biology of the cell in its environment without unnecessary detail in a manner compatible with macro-biological aspects of the geometrical shape of an organ or entire organism.

Indeed, we shall proceed with a series of ansätze, one after another, the first group of which are effectively axiomatic mathematical statements about the structure of cells and the organisms which they comprise. Some aspects of this are inspired by biology as we shall discuss in detail, and others are nothing more than rather obvious but new mathematical formalizations of elementary biological structure. Indeed, for the mathematician and biologist reader alike, the depth of certain ansätze may range from sublime to ridiculous–though likely quite differently for the two fields as we shall comment in specific cases.

The second group of ansätze governs the "morphogenetic field", a term used by René Thom[10] with the concept carefully articulated[8] by the first-named author with Misha Shubin and discussed in some detail here, which determines the time-evolution of cell states. In fact, our treatment here differs insofar as they[8] consider four discrete cell events (apoptosis, division, movement, growth), whereas in our language, the first two are represented by birth/death events (death alone or death with the birth of two daughters), and the latter two by continuous evolution as a path in a certain bundle (a Hilbert space* as fiber over a configuration of finitely many distinct points in three-space).

One profound contribution[8] is what we shall call the "Epigenetic Hypothesis": *the distribution of oligo sacchyride residues of glycoconjugates on the cell surface determines the morphogenetic field via an as-yet unknown code.* In fact, we are not so absolute as that and relax it here by allowing any other distribution or gradient on the cell surface as well, notably protein or electromagnetic gradients for example. One nice aspect of the mathematical formalism we have found is that such gradients on the cell surface can be explicitly prescribed in terms of real-valued functions on the two-dimensional sphere.

Another notable aspect of the formalism here is the seamless jump in scale from cells to organisms provided by any Gromov-Hausdorff type metric on measured metric spaces. Indeed, in our description, each cell has its explicit extent in three-space, the union of which forms a metric subspace

*A Hilbert space is a vector space with inner product which is complete as a metric space, and a product of such is thus again a Hilbert space.

comprising the organism. Depending upon which features of the organism are to be emphasized, different measures on the cells yield different metrics on the space of organisms.

Our precarious multi-scale endeavor is only confounded by the first ansatz:

Ansatz 1.1. No statement in biology is always true except this one.

2. Cells

We first describe the geometry of a single cell in mathematical detail (based on the biologically sound assumption that cells are star-convex), then the notion of cell state which captures certain epigenetic information (and is succinctly described by a collection of real-valued functions on the two-dimensional sphere $\mathbb{S}^2$ of unit radius centered at the origin in space $\mathbb{R}^3$ owing to star-convexity) and finally discuss cell trajectory, i.e., lifetime from birth to division or to final apoptosis (programmed cell death) or necrosis (death from injury or disease).

The structure of a live cell includes the *Microtubule Organizing Center* (MTOC) or *centrosome*, from which emanates a network of microtubules forming the essential part of the cytoskeleton of the cell and holding in place the cell surface just as a system of poles holds in place the surface of a circus tent. Most animal and some plant MTOC, apart from its main gamma-tubulin content, contain two cylindrical structures called centrioles meeting at a point; see Figure 1.

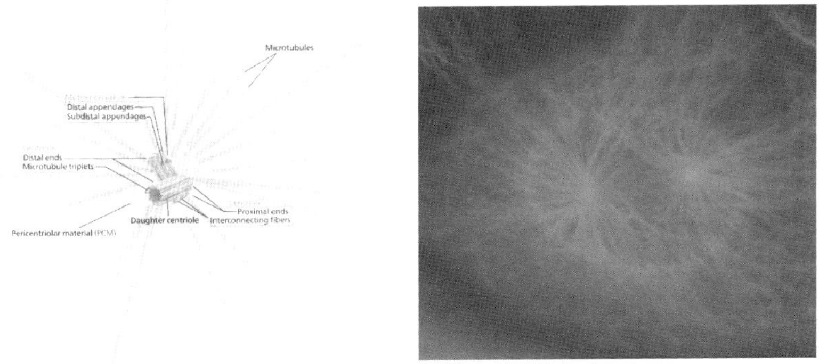

Fig. 1. Centrosome structure (from Wikipedia) on the left and the image of centrosomes in a cell (from wadsworth.org) on the right.

Let us say that a cell is *centriolic* if its centrosome indeed contains such a pair of centrioles, and likewise an organism if its cells are centriolic. These organisms are especially amenable to our methods and analyses here since each cell C has its distinguished *MTOC point* $p(C) \in C \subset \mathbb{R}^3$ given by the point of contact of the two centrioles. This is a mere convenience though, and one may rather make the choice of an appropriate point within the centrosome or even an arbitrary point of star-convexity in general.

2.1. *Shape*

Even comprehending just the geometry of a single cell within an organism already seems hopeless as suggested by Figure 2. One has only to peruse for example protist organisms (which are unicellular eukaryotes) depicted on the left to see a wild geometric diversity of detailed cell shapes. There are furthermore anatomical anomalies such as meter-long human neuro skeletal cells as depicted on the right with their dendritic extremities. Exotic examples abound.

Nevertheless modulo Ansatz 1.1, all cells do share a common geometric feature as follows. Recall that in mathematics, one says that a region $K \subseteq \mathbb{R}^3$ of space $\mathbb{R}^3$ is *star-convex* with respect to a point $p \in K$ provided that for each point $q \in K$ the line segment $\overline{pq}$ connecting p to q also lies in K.

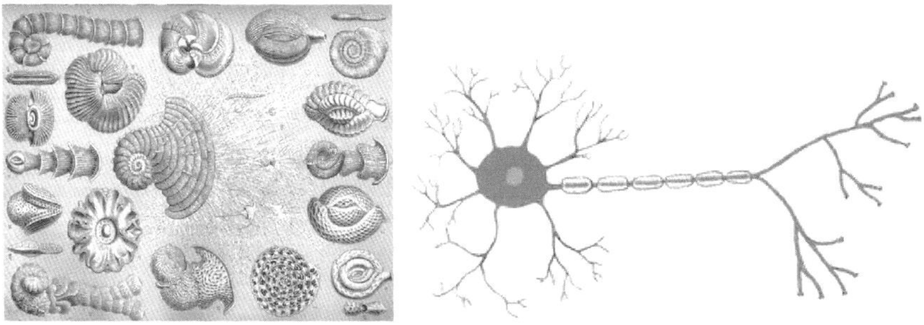

Fig. 2. Cell shapes: collage of protists[2] on the left and a sketch of a multipolar neuron (from studyblue.com) on the right.

Ansatz 2.1. A cell C is star-convex with respect to some point $p(C) \in C$ in its MTOC, and in particular for a centriolic cell, we may take $p(C)$ as the point of contact between centrioles.

It is a biological axiom, to the extent that such there be in light of Ansatz 1.1, that the cytoskeleton of a cell is star-convex with respect to a point in its MTOC. In our idealization of the geometry, we may consider the outer surface of the cytoskeleton as a cell surface. In reality, one would presumably take a small epsilon neighborhood of the star-convex cytoskeleton to describe the cell shape, though many cells are star-convex already with epsilon equal to zero. Indeed, we shall later consider only epsilon neighborhoods of cells anyway, so the geometric idealization of Ansatz 2.1 from cytoskeleton to cell is unimportant.

Remark 2.1. Fix in space the centrosome of a centriolic cell at some instant in time. Let $\vec{u}$ be the unit vector in the direction along the cylindrical axis for the earlier-replicated centriole and $\vec{v}$ the unit vector in the direction of the later-replicated one, with $\vec{w} = \vec{u} \times \vec{v}$. In each case, the direction along the cylindrical axis is determined by the fact that the MTOC point of centriole contact occurs asymmetrically at a preferred starting end of each cylinder. A centriolic cell thus determines a so-called positively oriented orthonormal 3-frame $(\vec{u}, \vec{v}, \vec{w})$ describing the spatial orientation of each cell. We shall not in fact require this further extra structure, but it is worth noting.

Now, fix a cell C with its MTOC point $p = p_C$. Given a unit vector $\vec{\xi} \in \mathbb{S}^2$, where $\mathbb{S}^2$ denotes the usual two-dimensional sphere in $\mathbb{R}^3$ (comprised of endpoints of vectors of unit length based at the origin), there is a corresponding ray from p in the same direction of $\vec{\xi}$, and this ray meets the cell surface at a single point at some distance $\rho(\vec{\xi}) \in \mathbb{R}_{>0}$ from p according to star-convexity. See Figure 3. This assignment $\rho = \rho_C : \mathbb{S}^2 \to \mathbb{R}_{>0}$ of a positive real-valued function on the two-dimensional sphere $\mathbb{S}^2$ completely determines the shape of the cell C.

To recapitulate this section, given a eukaryotic cell C, there is an MTOC point $p = p_C \in C$, which is well-defined without choice provided C is centriolic, with respect to which C is star-convex together with a shape function $\rho = \rho_C : \mathbb{S}^2 \to \mathbb{R}_{>0}$ which completely determines the geometrical cell shape

$$C = \{q \in \mathbb{R}^3 : ||q - p|| \leq \rho(\frac{q - p}{||q - p||})\} \subseteq \mathbb{R}^3,$$

where $||\vec{\eta}||$ denotes the usual length of a vector $\vec{\eta}$ in $\mathbb{R}^3$. There is furthermore a canonical identification of the standard sphere $\mathbb{S}^2$ with the cell surface ∂C given by the mapping $R_C : \vec{\xi} \mapsto p_C + \rho_C(\vec{\xi})\vec{\xi}$. The volume V of C and its

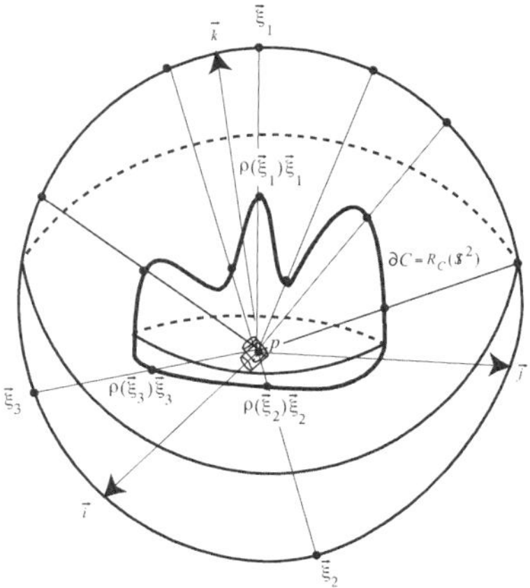

Fig. 3. Centrosome and shape function ρ.

surface area A are respectively given in terms of a sufficiently regular shape function ρ_C as

$$V(C) = \int\int\int_C dV = \frac{1}{3}\int\int_{\mathbb{S}^2} [\rho_C(\vec{\xi})]^3 \; dA,$$

$$A(C) = \int\int_{\mathbb{S}^2} \left|\nabla \rho_C(\vec{\xi}) \cdot \vec{\xi}\right| \; dA.$$

2.2. *Cell state*

What exactly constitutes a cell state for morphogenesis or any other purpose depends upon the implied internal structure of the cell and must reflect the predictions and interactions of the theory that depend upon it. One obvious aspect of cell state is the shape function $\rho_C : \mathbb{S}^2 \to \mathbb{R}_{>0}$ and another is the spatial location of the MTOC point p_C introduced in the previous section. Taken together, these determine the natural parametrization $R_C : \mathbb{S}^2 \to \partial C$ of the cell surface ∂C.

It follows from the existence of the parametrization that any real-valued function $\phi : \partial C \to \mathbb{R}$ can equivalently be specified by simply pulling back to a real-valued function $f : \mathbb{S}^2 \to \mathbb{R}$, i.e., $\phi(q) = f(R_C^{-1}(q))$. In particular, the density of some substance on the cell surface can be conveniently specified by just such a real-valued function on $\mathbb{S}^2$†.

†Of course, a surface density is not simply pulled-back, rather, it is scaled by the Jacobian of R_C to account for area distortion.

Ansatz 2.2. Cell state is determined by the shape function $f_0 = \rho$ together with a collection $f_1, \ldots, f_N$ of cell surface densities $f_i : \mathbb{S}^2 \to \mathbb{R}$, for $i = 1, \ldots, N$.

The collection $(f_1, \ldots, f_N)$ of cell surface densities is called the *epigenetic spectrum* of the the cell. We shall refer to the full $N+1$ tuple $F_C = (f_0, f_1, \ldots, f_N)$ including the shape as the *(internal) cell state.*

In fact, the cell surface ∂C is relatively fluid with sub-regions evolving in time to different shapes within the cell surface. Thus, we should expect the typical distribution of substance on the cell surface to evolve continuously in time.

Of special interest are the distributions of oligo conjugates, small oligosaccharide residues of glycoconjugates‡ on the cell surface. There are up to 12 varieties of monosaccharides and disaccharides in oligosaccharide residues of glycoconjugates (six of hexose type). There have been numerous indications that these oligosaccharide residues are related to cellular morphogenetic pathways as has been formulated as an Epigenetic Hypothesis in[8] and the references therein.

Thus, we propose that the cell state F_C of C includes both the cell shape $f_0 = \rho$ and the epigenetic spectrum of cell densities $f_1, f_2, \ldots, f_{12}$ of the corresponding number of "coding" oligo conjugates, as per the Epigenetic Hypothesis, the nature and number of which should be determined experimentally. Here is an important point of contact between one aspect of our theoretical work and experimental determination. We leave open the possibility of extending this minimal set of functions determining cell state with other molecular compounds or other factors such as protein or electromagnetic gradients defined on the cell surface, where recent works[6] suggest the latter are central to morphogenesis.

Recall[4] that there is a standard distance between two tuples $F = (f_0, f_1, \ldots, f_N)$ and $G = (g_0, g_1, \ldots, g_N)$ of functions $f_i, g_i : \mathbb{S}^2 \to \mathbb{R}$ given by their L^2 *distance*

$$d(F,G) = \sqrt{\int_{\mathbb{S}^2} (f_0 - g_0)^2 + (f_1 - g_1)^2 + \cdots + (f_N - g_N)^2 dA}.$$

Indeed, letting

$$L^2(\mathbb{S}^2) = \{f : \mathbb{S}^2 \to \mathbb{R} : \int \int_{\mathbb{S}^2} f^2 dA < \infty\}$$

‡The term "glycoconjugates" means glyco-residues attached to either lipids or proteins lying inside the cell membrane or cell wall in the case of plants.

denote the so-called square integrable functions on $\mathbb{S}^2$, the distance d gives the collection $\left[L^2(\mathbb{S}^2)\right]^{N+1}$ of all possible cell states F_C the structure of a Hilbert space. Hardly worth mentioning mathematically, the next ansatz gives an explicit concept of distance between cell states.

Ansatz 2.3. The natural distance[§] between cell states $F = (f_0, f_1, \ldots, f_N)$ and $G = (g_0, g_1, \ldots, g_N)$ is the L^2 distance $d(F, G)$.

2.3. *Trajectory*

The sperm-fertilized cell oocyte becomes the zygote, the first cell $C(\emptyset)$ of the organism, which next mitotically divides to ten or more generations in a highly organized paroxysm of cell divisions, no time at all for diffusive or other equilibria to develop, no time for appreciable cell growth or motion and immediately generating all manner of steep gradients across the early embryo.

Recall that the centrosome of a cell C at interphase replicates to create a pair of centrosomes. These centrosomes move to opposite ends of the cell and nucleate microtubules to help form the mitotic spindle separating the chromosomes into two sets, one copy for each daughter cell C', C'' together with one centrosome for each daughter cell. One centrosome of C' or C'' is inherited from the centrosome of C, which we call the mother centrosome, while the other, which we call the daughter centrosome since it is younger than its mother, has been assembled upon the mother centrosome before mitosis.

Insofar as the replicated centrosomes of C are distinguishable, so too are its daughters C' and C''. We can thus label each cell in the organism by its phylogeny: a word in the alphabet m and d, where m indicates the daughter $C(m)$ that inherits the mother centrosome from C and d the daughter $C(d)$ that inherits the daughter centrosome.

Each cell itself is determined by its *phylogeny*, a finite word ω in the two-letter alphabet $\{m, d\}$, including the empty word for the zygote. This labeling of cells by m and d is evidently equivalent to the usual structure of embedded phylogenetic tree: a binary tree with gamete for root and with m on the right (say) and d on the left.

[§]In fact, we mean this in the looser sense that the distance might differently weight different component functions (a "weighted L^2 space") more generally taking $d_{\vec{\alpha}}(F, G) = \sqrt{\int\int_{\mathbb{S}^2} \sum_{i=0}^{N} \alpha_i (f_i - g_i)^2 dA}$, for certain specified parameters $\alpha_i \in \mathbb{R}_{>0}$.

Ansatz 2.4. Each cell C is born at some instant s_C in time and terminates either through death or cell division at a later instant t_C; each cell except for the gamete is born with exactly one distinguishable sister.

Death may be apoptosis, necrosis or it may reflect some crisis for the organism such as amputation or disintegration. A somatic cell is always born with exactly one sister and terminates either alone (by death) or in giving birth to its daughters (by division).

It is worth saying explicitly that Ansatz 2.4 is biologically tautological, no more than a self-evident remark but is a necessary aspect of axiomatic mathematical formulation. Indeed, in other important classical mathematical contexts, births as well as deaths can occur only in pairs. Biologically, the ansatz is trivial and can be ignored, for of course it is tautologically so.

Vertices of our phylogenetic tree other than the root are thus labeled by the cell event of death or cell division. Edges of the tree correspond to the lifetime of cells, and we will assign as a length of the edge labeled by the cell C its lifetime $t_C - s_C$. In this manner, a point in an edge corresponds to the cell $C^t \subseteq \mathbb{R}^3$ at a particular instant t in time, where $s_C \leq t \leq t_C$.

According to our assumptions, two sister cells C' and C'' must share their time of birth $s_{C'} = s_{C''} = t_C$ with the time of death of their mother (but see section 4.2). For an extreme example such as a skeletal muscle cell C or an eye lens cell C which is fully differentiated once and for all time already in the embryo and never divides again, we find s_C at birth and t_C after death of the organism. At another extreme are dermal cells with their profuse and nearly identical divisions.

The potency of a cell to produce progeny with different types of tissues and organs or simply to differentiate into many possible cell types is called a morphogenetic potency or just a potency. Thus in an organism there are totipotent, pluripotent, bipotent and fully differentiated cells.

Let us define the *potency* $\pi(C)$ of a cell C to be zero if it does not divide at all, and if it does divide, say the daughters have respective epigenetic spectra $F', F'' \in [L^2(\mathbb{S}^2)]^N$, then we compare the spectra defining $\pi(C) = d(F', F'')$. Thus, the potency of a fully differentiated cell is nearly zero. In contrast, a bipotent stem cell, for example, can have a large potency if its product cell line differs substantially in epigenetic spectrum from its stem cell line in the L^2 distance.

3. Organisms

We define an organism to be the collection of its constituent cells lying in space as determined by a language over the two-letter alphabet given by following the centrosome. The collection of constituent cell states determines a state of the organism at each instant in time. A convenient mathematical formalism exists for this as a configuration space of distinct points in space which evolve under time either continuously or through specific rules of birth and death at certain event times together with a bundle over this base space with fiber given by an L^2 space of functions on the two-dimensional sphere. All this is quite mathematically routine given what has come before.

At a given moment in time, the cells C underlying an organism comprise a metric subspace which moreover can be further imbued with any function $\mu(C)$ chosen to reflect essential morphological or other parameters of C. For example, we could take mass density of a cell, density of distribution of ion channels or other cell constituent concentrations. We may choose particular functions $\mu(C)$ for specific applications.

A fundamental ansatz is that two organisms each at a given moment in time can be profitably compared using Gromov-Hausdorff (GH) type distances (cf. Appendix A) between corresponding measured metric spaces. This is precisely what allows us to finally pass from the cellular to the capacious scale in our investigations.

It is important to note that the comparison of "two organisms" in this framework should be interpreted first of all as the comparison of the morphology of the same organism at different moments in time reflecting the laws of evolution of its shape. Equally, we consider the comparison of the differences between normal and "crisis" developmental scenarios (that is, strong perturbations of proper morphology) for the same organism at some moment in time.

3.1. *Phylogeny*

The zygote $C(\emptyset)$ comes into being at an instant $s_\emptyset = s_{C(\emptyset)}$ and then undergoes a sequence of binary divisions, the first one occurring at the instant $t_\emptyset = t_{C(\emptyset)}$. At each division, there are distinguishable sister cells produced, and the mother cell producing these progeny subsequently ceases to exist. And so it goes.

The mother of all cells of the organism is the zygote $C(\emptyset)$. For each subsequent offspring produced, i.e., each cell that did, does or will or ever occur in the organism there is a well-defined word ω in the two-letter alphabet $\{m, d\}$ as discussed before, where the length of the word is the number of generations from the zygote.

Not all words occur for a given organism, and the *phylogenetic language* of $\mathcal{O}$ is the collection $\Omega_{\mathcal{O}}$ of words that do in fact occur for the organism $\mathcal{O}$. Let $C(\omega)$ be the cell corresponding to the word $\omega \in \Omega_{\mathcal{O}}$ including the empty word for the zygote. Thus, each cell $C(\omega)$ is born and dies at respective specific times s_ω and $t_\omega > s_\omega$.

We shall let $C_\omega^t \subseteq \mathbb{R}^3$ denote the cell $C(\omega)$ labelled by ω at time t, where $s_\omega \leq t \leq t_\omega$ with time-dependent cell state $F_\omega^t \in [L^2(\mathbb{S}^2)]^{N+1}$, reflecting shape and epigenetic coding, and MTOC point $p_\omega^t \in \mathbb{R}^3$ with auxiliary measure $\mu_\omega^t \in \mathbb{R}_{>0}$.

3.2. *Organism state, configuration and trajectory*

We have the mathematical definitions of organism and state given by the following biological tautology:

Ansatz 3.1. An organism $\mathcal{O}$ is determined by the collection of its constituent cells $\{C(\omega) : \omega \in \Omega_{\mathcal{O}}\}$ and its state by their respective states.

Let us say that a cell C_ω^t *exists* at some time t if $s_\omega \leq t \leq t_\omega$ for some $\omega \in \Omega_{\mathcal{O}}$. It is convenient to suppress the birth/death times of a cell and write simply $C_\omega^t \subset \mathbb{R}^3$ to signify the cell at a time t at which the cell exists and to signify the empty set otherwise. In the same way, let F_ω^t denote the cell state and μ_ω^t the measure at a time t between cell birth and death and to take constant value zero otherwise; finally, p_ω^t denotes the MTOC point and ρ_ω^t the shape function if the cell C_ω^t exists, and these are undefined otherwise.

Let $\Omega^t = \Omega_{\mathcal{O}}^t \subseteq \Omega_{\mathcal{O}}$ be the phylogenetic words of the cells of the organism $\mathcal{O}$ that exist at time t. The organism

$$\mathcal{O}^t = \bigcup \{C_\omega^t : \omega \in \Omega^t\} \subseteq \mathbb{R}^3$$

at a time t is the union of these cells, i.e., the constituent cells of the organism at the instant t. Of course, $\mathcal{O}^t$ or any epsilon neighborhood of it inherits the structure of a metric subspace of $\mathbb{R}^3$, and each cell comes with its measure μ_ω^t which together give the weighted sum

$$\mu^t = \mu_{\mathcal{O}}^t = \sum_\omega \mu_\omega^t \delta_{p_\omega^t}$$

of delta functions as a measure on $\mathcal{O}^t$.

The *state* of the weighted organism $(\mathcal{O}^t, \mu^t)$ at time t consists of the cell states F_ω^t of its constituent cells. The collection of MTOC points

$$\chi^t = \{p_{\omega_1}^t, p_{\omega_2}^t, \ldots, p_{\omega_{\#\Omega^t}}^t\}$$

of cells that exist at time t come in their prescribed lexicographic ordering (where, say $m < d$) which we shall call the *(labelled) configuration* of the organism at time t.

Recall that if X is a metric space, then a *(labelled) configuration* in X of $n \geq 1$ points is a collection of n distinct labelled points in X; see Appendix B for the barest thumbnail discussion tailored to our own needs and[3] for instance for further generalities. It is biologically clear that in fact the MTOC points χ^t do give a configuration in $\mathbb{R}^3$, lying in $\mathcal{O}^t$, i.e., the MTOC points of distinct cells never coincide. Indeed, we have the stronger:

Ansatz 3.2. The cells of an organism that exist at each instant in time have disjoint interiors in space $\mathbb{R}^3$.

Again a biological tautology, this kind of "steric" constraint is commonplace in protein theory for instance. We do not intend to entirely rule out gap or other cell junctions here, rather, let us consider adding that further layer of detail only later.

The number $\#\chi^t$ of MTOC points of the organism $\mathcal{O}^t$ is piecewise constant and jumps at various times of lone death (one point disappears) or birth (two new points appear and an old one disappears) of constituent cells according to Ansatz 2.4. In fact, there may be finitely many coincidentally simultaneous birth or death times, but we make the following finiteness assumption:

Ansatz 3.3. For any fixed t, there is a finite sequence of event times $\tau_0 = s_\emptyset < \tau_1 < \tau_2 < \cdots < \tau_M \leq t$ so that for each time $\tau_{i-1} < s < \tau_i$, for $i \geq 1$, the configurations χ^s are equinumerous and evolve continuously in time.

The discussion of how exactly the cell state before determines the specific cell states after a birth event time is postponed until the more speculative considerations of the next section.

4. Morphogenesis

"The time has come" the Walrus said, "To talk of many things: Of shoes– and ships–and sealing-wax– Of cabbages–and kings– And why the sea is boiling hot– And whether pigs have wings."[¶] This is just to say that now we must discuss things a bit more metaphorically in order to absorb the next ansätze.

4.1. *Organism shape*

As humans, we can perceive the natural world on various levels. For instance, in pond water one finds paramecia, which we might probe in the lab *in vivo* or *in vitro* to discern further structure. Going to an extreme, we might travel to CERN and uncover the constituent elementary particles.

As human observers, we do not even want to try to absorb the molecular let alone the atomic or stringy nature of a paramecium. With what has been[‖] called "willful ignorance" and also[**] termed "twice barbarism", our vision in morphogenesis had better contain the required cellular and other data even if we shall ultimately and barbarically ignore this level of detail in the large.

To handle the need for minute detail in certain biological regards for cells, we have already introduced the notion of cell state. We must somehow pass from this data as we have formalized it, corpuscular by its very nature, to the context of continuous data of substantial spatial extent.

Naturally enough, we might take an epsilon neighborhood of the actual cells at each instant in time, that is, a neighborhood of $\mathcal{O}^t$ in $\mathbb{R}^3$. We furthermore might naturally let epsilon vary over $\mathcal{O}^t$ in order to allow for instance for different packing densities of one type of cell compared to another. An elegant and precise mathematical formalism for this viewpoint is provided by the class of metrics on measured metric spaces inspired by GH distance. For completeness, one such distance (the Sturm L^2 distance[9]) is explicitly defined in Appendix A.

Ansatz 4.1. Given two organisms $\mathcal{O}$ and $\mathcal{O}'$ with their respective measures μ and μ', Gromov-Hausdorff type distances between the measured metric spaces $(\mathcal{O}, \mu)$ and $(\mathcal{O}', \mu')$ such as the Sturm L^2 distance capture the geometry of organism morphology.

Just as the Sturm L^2 distance is one among many sensible choices for comparing measured metric spaces, so too, we have the flexibility to take

¶Lewis Carroll, The Walrus and The Carpenter (1872).
‖René Thom, Semiophysics: a sketch (1989)
**Claude Levi-Strauss, Tristes Tropiques (1955)

different measures $\mu(C)$ on cells C. As an example, the mass density can be one special case of measures involved in building the morphology of an organism. Presumably as a human observer employing it as we have done here is like observing the world with those X-ray glasses that are bought by gullible children; we would detect density differences between cells within organisms when we compare them wearing our mass density glasses. A tadpole and a frog would be easily distinguished by their bone structures for instance. The formalism we have developed is robust enough to handle general examples exchanging X-ray for other glasses by changing the measure μ^t on the fixed metric space that is the organism $\mathcal{O}^t$.

4.2. *Cell division in detail*

During mitotic division of the cell C, the two daughter cells $C(m)$ and $C(d)$ labelled by centrosomes as before are separated by a so-called mitotic plane passing through the interior of C often asymmetrically. This separating plane divides the cell surface ∂C into two inequivalent regions $\Omega_m, \Omega_d \subset \partial C$ forming the cell boundary of respective daughters.

Ansatz 4.2. The epigenetic spectra of daughters $C(m)$ and $C(d)$ are nearly inherited in the L^2 sense on $\mathbb{S}^2$ by inclusion of the cell surfaces Ω_m, Ω_d from the mother.

On the complementary regions to Ω_m and Ω_d in daughter cell surfaces, there are as-yet unspecified rules of how to extend to a function on the entire cell surface. This ansatz partly describes the *morphogenetic birth effect*, that is, the function that determines the cell states of $C(m)^{\sigma_2}$ and $C(d)^{\sigma_2}$ at the moment σ_2 when mitotis is complete from the cell state of C^{σ_1} of a mother cell as mitosis begins at σ_1.

It is not that the biology breaks down during the interval between times σ_1 and σ_2. After all, there is a sensible moment of birth when the cell membranes of the daughters finally separate that we could also take to be the moment of death of the mother C. The surface concentrations in ∂C of the oligo conjugates or other compounds are everywhere defined until this moment and after become the surface concentrations in $\partial C(m)$ and $\partial C(d)$. In this sense, again the ansatz is disappointingly tautological biologically. However, this scenario does *not hold exactly* and in our barbaric and willful ignorance, we posit in Ansatz 4.2 a precise sense in which L^2 distance suitably measures deviation from this idealized biology.

4.3. *Morphogenetic field*

To recapitulate, given a eukaryotic organism $\mathcal{O}$ and a time t, we have the constituent cells $C_\omega^t \subseteq \mathbb{R}^3$ that exist at time t indexed by a finite set Ω^t of words in the alphabet $\{m, d\}$. The union of these star-convex regions is by definition the organism $\mathcal{O}^t \subseteq \mathbb{R}^3$ at time t as a metric subspace of $\mathbb{R}^3$. This region comes equipped with a measure μ^t, and it is the measured metric spaces $(\mathcal{O}^t, \mu^t)$ that reflect the large-scale geometry of the organism. Standard methods, called GH type metrics, can be applied to these spaces to quantify the sense in which two organisms have the same shape. Nevertheless, the organism has a complicated state, namely, we take the state of the organism to be the collection of cell states of each of its constituent cells, i.e., cell states F_ω^t and MTOC points χ_ω^t, for $\omega \in \Omega^t$.

On the cellular scale, we have organized things in Appendix B into a vector bundle $E \to \Gamma$ over the space Γ of configurations of finitely many points in $\mathbb{R}^3$, where the fiber over a configuration with n points is an n-tuple of elements in $[L^2(\mathbb{S}^2)]^{N+1}$, namely, the cell states of the constituent cells. Thus, the entire organism in its entire state is described at an instant in time as a point in the total space of this bundle over the finite configuration of MTOC points in the base, and the measure μ is an auxiliary function defined on the base.

The idea[8] is that the "morphogenetic field" is a function defined on some bundle of states that governs the time evolution of an organism starting from ovum. Their formalism was different with discrete time cell events as was already mentioned. Here we imagine continuous evolution in some $E^{(n)}$ with jumps in n at finitely many event times compatible with Ansatz 3.3. The morphogenetic field of[8] should be realized here as some sort of functional which governs the time evolution of an organism, namely, as a Lagrangian action:

Ansatz 4.3. There is some Lagrangian formulation of organism evolution, i.e., an action functional on paths in E whose stationary paths are the time trajectories of organisms.

It is important to emphasize that the Lagrangian formulation of Ansatz 4.3 is different from that of the underlying physics: our action should be formulated on the space of organism states as we have defined them.

Using our definition of potency (see the end of section 2.3) and the observation that each organism starts to differentiate from the zygote having a huge potency towards its later less-potent progeny, we posit:

Ansatz 4.4. The average *potency* $\pi(C)$ over the set of all cells C in an organism is strictly decreasing in generation during embryo development.

This is of course in keeping with the second law of thermodynamics.

Ansatz 4.5. Nearly identical sisters almost always remain nearly in contact.

This is interesting as it relates the phylogeny to the geometry. Of course, sisters are never truly identical, but if they were, then the morphogenetic field could not distinguish them, so of course they would evolve in the same manner. Nature is nearly like this as an essentially deterministic physical system. This seems to give a nice explanation for organ formation and other aspects as well and really makes good sense. Again, with the L^2 distance between cell states, we can conveniently quantify differences between sisters or indeed any pair of cells viz Ansatz 4.5.

Ansatz 4.6. The morphogenetic field is locally supported in the sense that for any cell C_ω^t that exists, there is a neighborhood $\mathcal{N} \supseteq C_\omega^t$ of it in $\mathbb{R}^3$ so that the forward time evolution of C_ω^t for some fixed interval of time depends only upon the states of the cells in $\mathcal{N}$.

In particular, phylogeny does not matter *a priori*, which implies that there is indeed a physical determinism underlying the morphological determinism we are studying. And this determinism is governed by the laws formulated as a morphogenetic field concept.

Ansatz 4.7. For any sub-organism $\mathcal{O}' \subseteq \mathcal{O}^t$ in any reasonable state there is a stable ideal outcome organism $\mathcal{O}^\infty$ so that the Lagrangian action of an arbtrary path from $\mathcal{O}'$ to a stable outcome $\mathcal{O}^*$ is given by a Gromov-Hausdorff type distance between $\mathcal{O}^\infty$ and $\mathcal{O}^*$.

This GH-deviation from ideal outcome is admittedly a bold hypothesis for the dynamics providing thereby the key mechanism driving morphogenesis ranging from embryo development to the creation of proper morphology during the regeneration processes in the event of crisis. The ansatz is interesting on several levels first of all in its postulate that there are stable ideal outcome organisms at all under admittedly unspecified reasonable conditions. The comparison of stable outcomes using GH techniques is consistent with the earlier Ansatz 4.1. Indeed, perhaps another more traditional differential calculus type expression of action in terms of the GH-metric could be possible rather than relying on stability of outcomes. Secondly, it ties the cellular scale of the morphogenetic field with the large-scale geometry of organisms through the action, and it evidently implies the earlier Ansatz 4.3 in its specificity.

Acknowledgments

Thanks to Christophe Soulé and Sasha Getmanenko for helpful discussions.

Appendix A.

5. Distances between measured metric spaces

This appendix gives a precise example of the sort of distance between measured metric spaces that we propose for use in comparing organisms. For completeness, definitions are given from first principles.

Definition A.1. If X is a set, then a function $\mu : X \times X \to \mathbb{R}$ is a *metric* if and only if for all $x_1, x_2, x_3 \in X$, we have: [Symmetric] $d(x_1, x_2) = d(x_2, x_1)$; [Triangle Inequality] $d(x_1, x_3) \leq d(x_1, x_2) + d(x_2, x_3)$; [Positive] $d(x_1, x_2) \geq 0$ and $d(x_1, x_2) = 0$ if and only if $x_1 = x_2$.

Given metric spaces (X, d_X) and (Y, d_Y), let $\mathcal{D}(d_X, d_Y)$ denote the set of all possible metrics on the disjoint union $X \sqcup Y$ of X and Y extending d_X on X and d_Y on Y, that is, d satisfies Definition A.1 on $X \sqcup Y$ and

$$d(x_1, x_2) = d_X(x_1, x_2) \text{ and } d(y_1, y_2) = d_Y(y_1, y_2)$$

for all $x_i \in X$ and $y_i \in Y$, for $i = 1, 2$.

Definition A.2. A *Borel probability measure* μ on a space X is the specification of a real number $\mu(A)$ to each set A in the sigma algebra[††] $\mathcal{B}(X)$ of subsets generated by the open sets of X satisfying the following properties: [Sigma additive] if $A_i \in \mathcal{B}(X)$ are pairwise disjoint subsets for $i \in I$ with I countable, then $\mu(\cup\{A_i : i \in I\}) = \sum_{i \in I} \mu(A_i)$; [Non-negative] $\mu(A) \geq 0$ for all $A \in \mathcal{B}(X)$; [Null empty set] $\mu(\emptyset) = 0$. [Probability measure] $\mu(X) = 1$.

Given two measured metric spaces (X, d_X, μ_X) and (Y, d_Y, μ_Y), let $\mathcal{M}(\mu_X, \mu_Y)$ denote the collection of all Borel probability measures μ on

[††] A collection of subsets of X is a *sigma algebra* provided it is closed under countably many applications of the set operations complement, union and intersection.

$X \times Y$ extending μ_X on X and μ_Y on Y, that is, μ satisfies Definition A.2 on $\mathcal{B}(X \times Y)$ and

$$\mu(A \times Y) = \mu_X(A) \text{ and } \mu(X \times B) = \mu_Y(B)$$

for all $A \in \mathcal{B}(X)$, $B \in \mathcal{B}(Y)$.

One collection of distances for measured metric spaces inspired by the GH metric is given by:

Definition A.3. The *Sturm L^p-distance*[9] between measured metric spaces (X, d_X, μ_X) and (Y, d_Y, μ_Y) is defined to be

$$S_p(X,Y) = \inf_{d,\mu} \left(\int_{X \times Y} [d(x,y)]^p \mu(dx, dy) \right)^{\frac{1}{p}},$$

where the infimum is over all $d \in \mathcal{D}(d_X, d_Y)$ and $\mu \in \mathcal{M}(\mu_X, \mu_Y)$, and $p \geq 1$ is some fixed natural parameter.

We can take this Sturm distance, say with $p = 2$ for definiteness, as just one of many possibilities. Further GH type metrics on collections of measured metric spaces as well as discussions of their computational implementation in practice for digital shape comparison can be found in the very accessible[7].

6. Configuration spaces

As in the text, we have:

Definition A.1. If X is a metric space then a *(labelled) configuration* in X of $n \geq 1$ points is a collection of n distinct labelled points in X. Let

$$\Gamma_X^{(n)} = \{(p_1, p_2, \dots, p_n) \in X^n : p_i \neq p_j \text{ if } i \neq j\},$$

where the exponent X^n denotes the n-fold Cartesian product. Let $\Gamma_X = \sqcup_{n \geq 1} \Gamma_X^{(n)}$ denote the disjoint union.

Insofar as $\Gamma_X^{(n)}$ is a subset of the Cartesian product X^n, it inherits a natural subspace topology. For example, $\Gamma_X^{(1)} = X$ while a configuration on two points is the complement of the diagonal in the Cartesian square $\Gamma_X^{(2)} = X^2 - \{(x_1, x_2) \in X^2 : x_1 = x_2\}$.

There is a nice theory of analysis and measure on configuration spaces satisfying appropriate boundedness conditions, cf.[1]. For our purposes here, just the underlying topological spaces and vector bundles over arbitrary configurations are enough.

Definition A.2. Given not only a metric space X but also a Hilbert space V, we define

$$E_X^{(n)} = \{((p_1, \varphi_1), (p_2, \varphi_2), \ldots, (p_n, \varphi_n)) \in (X \times V)^n : p_i \neq p_j \text{ if } i \neq j\},$$

and set $E_X = \sqcup_{n \geq 1} E_X^{(n)}$.

Again, as a subset of the Cartesian product $(X \times V)^n$, the new space $E_X^{(n)}$ inherits its natural metric topology, and the forgetful map $E_X^{(n)} \to \Gamma_X^{(n)}$ has the natural structure of a vector bundle with fiber V^n for each $n \geq 1$. We informally refer to E_X itself as a V-bundle over Γ_X.

Taking $X = \mathbb{R}^3$ and $V = [L^2(\mathbb{S}^2)]^{N+1}$ gives a vector bundle $E = E_X$ over $\Gamma = \Gamma_X$, whose fiber is the state of the organism.

References

1. S. Albeverio, Yu. G. Kondratiev, M. Röckner. Analysis and geometry on configuration spaces. J. Funct. Anal. **154** (1998), 444-500; Analysis and geometry on configuration spaces: The Gibbsian case. J. Funct. Anal. **157** (1998), 242-291.
2. B. Alberts, A. Johnson, J. Lewis, M. Raff, K. Roberts, and P. Walter, Molecular Biology of the Cell 4th Edition, Garland (2002).
3. J. Birman, Braids, Links and Mapping Class Groups, Annal of Mathematical Studies , Princeton University Press (1974).
4. Y. Katznelson, An Introduction to Harmonic Analysis, Cambridge University Press (2004).
5. D. Lobo, W. S. Beane, M. Levin, Modeling Planarian Regeneration: A Primer for Reverse- Engineering the Worm, PLoS Computational Biology **8** (2012).
6. M. Levin, Morphogenetic fields in embryogenesis, regeneration, and cancer: Non-local control of complex patterning, BioSystems **109** (2012), 243-261.
7. F. Mémboli, On the use of Gromov-Hausdorff distances for shape comparison, Europgraphics Symposium on Point-Based Graphics (2007), M. Botsch, R. Pajarola (editors).
8. N. Morozova and M. Shubin, The Geometry of Morphogenesis and Morphogenetic Field Concept, In: Pattern Formation in Morphogenesis - Problems and Mathematical Issues. Springer Proceedings in Mathematics, **15**, (2012),255-282; recent version http://arxiv.org/abs/1205.1158v2, 2013.

9. K.-T. Sturm, On the geometry of metric measure spaces. I. Acta Math. **196** (2006), 65-131.
10. R. Thom, Structural stability and morphogenesis, W. A. Benjamin (1972).

COMPARING SHAPE TRAJECTORIES OF BIOLOGICAL SOFT TISSUES IN THE SIZE-AND-SHAPE SPACE

V. VARANO, S. GABRIELE, L. TERESI
LaMS Modelling & Simulation Lab, University RomaTre,
via della madonna dei monti, 30,
00184, Rome, Italy
E-mail: lams@uniroma3.it

I. DRYDEN
School of Mathematical Sciences, University of Nottingham,
Nottingham NG7 2RD, UK
E-mail: Ian.Dryden@nottingham.ac.uk

P. E. PUDDU, C. TORROMEO, P. PIRAS
Deptartement of Cardiovascular, Respiratory,
Nephrological and Geriatric Sciences, Sapienza University,
V.le del Policlinico, 155,
00161 Rome, Italy
E-mail: paolo.piras@uniroma3.it

Biological soft tissues can suffer for very large size and shape changes during their motion. We aim to study the peculiar deformation parameters of this motion, that can be well represented as trajectories in a particular Riemannian manifold named size-and-shape space. Given a sample of different individuals that experience different morphological trajectories it is possible to compare the shape of trajectories themselves. This comparison is not trivial because the Size-and-Shape Space is non-Euclidean. Dealing with soft tissues the differences between individuals can be very large, as a consequence the morphological trajectories can start from highly scattered points on the Size-and-Shape Space. In such a case ordination methods based on the tangent space (as the PCA) should be performed on a different tangent space for each individual to have a good approximation. In order to compare shape data belonging to different tangent spaces is necessary to perform a parallel transport of the data toward a common tangent space. In this paper we propose a data centering technique based on the Levi-Civita parallel transport in the Size-and-Shape Space useful to compare trajectories. Moreover we give a simple criterion to discriminate in which cases the proposed technique can be well approximated by a suitable hierarchical alignment followed by a linear shift. We illustrate this technique by performing analysis on real human hearts whose 2D left ventricular motion data set are considered.

1. Introduction

Modern morphometrics, i.e. Geometric Morphometrics (GM) starts with the seminal contributions of Kendall[1], where it has been proposed to sample the configuration of a body through the m coordinates of k landmarks; a configuration is thus represented with a point in an $m \times k$ space, called the Configuration Space. Then, to effectively represent a shape, it is considered a quotient of the Configuration Space with respect to a Lie Group of transformations, a group that is selected according to specific needs.

At first, we consider the Size-and-Shape Space, a quotient of the Configuration Space with respect to rigid transformations (translations + rotations), which, representing changes in observer, cannot be considered as attributes of the shape. Filtering out translations is rather simple when the bodies under examination are sampled using homologous landmarks: translations are eliminated by centering the centroids of each configuration. Several criteria exist to filter out rotations, via the alignment of homologous landmarks and the choice of a metric whose minimisation yields the optimal superimposition of all configurations. A popular algorithm for the alignment is the Procrustes distance metric, used in the Generalised Procrustes Analysis (GPA). In [1], it is shown how the regular part of the Size-and-Shape Space is a Riemannian manifold of $m \times (k-1) - m(m-1)/2$ dimensions.

A different space, the Shape Space, is obtained as quotient with respect to the Euclidean similarity transformations (rigid transformations + scaling); scaling transformations filter out differences in size, considered as a non-shape information. In such a case, the size of a configuration is measured by the Centroid Size (CS), i.e. the radius of gyration, and all the configurations are scaled to a common CS; the Shape Space is then an $m \times (k-1) - m(m-1)/2 - 1$ dimensional space.

Given a set of shapes, that is, of points belonging to the Shape Space, its average defines a pole, also called the consensus; when the differences with respect to the consensus are small, the geodesic Procrustes distance is approximated by its projection on the tangent space at the consensus. The aligned coordinates are often subjected to ordination methods, such as Principal Component Analysis (PCA) for further analysis.

In this paper, we shall focus on the study of *morphological trajectories* $\mathcal{T}$, defined as sequences of shapes s_i ordered upon a given criterion, i.e. time, size or other: $\mathcal{T} = (s_1, ..., s_n)$. Given a set of trajectories, it is particularly desirable to investigate about the characteristics of the motion undergone by the shapes in each trajectory; an important goal is to recognise whether or not the different shapes spanned by the trajectories under exam are manifestation of a same motion.

A rationale for this Phenotypic Trajectory Analysis (PTA) has been proposed in [2] and [3], where it has been suggested to consider each shape of a given sequence as a landmark which samples a motion; thus, a trajectory can be considered as the "shape of a motion", and analysed via shape analysis, a possibility granted by the fact that the shapes of each trajectory could be considered as homologous: such a crucial request can be time-based or physiology-based in biological applications. More recently, it has been shown in [4] that shape analysis for trajectories should be done only after a proper representation of them is obtained.

As PTA is meant to assess the "shape of a motion", it is of the essence to filter out any inter-group differences; such a task is often achieved in statistics by applying a group-mean centering, optionally followed by the Grand Mean addition. A problem arises if the data are shape data, lying naturally in non Euclidean spaces.

Frequently, the Levi Civita connection on the Shape Space is used to compute the geodesics between two shapes [5,6]; sometimes, it is also used to transport a deformation along geodesics, so that a "same" deformation can be applied to different shapes[7,8].

Biological soft tissues can suffer for very large shape and/or size changes; this is particularly evident for the motion of the human cardiac left ventricle, an example analysed in the following. There, the change in volume is one of the principal features allowing the pumping function, and one could be interested in studying the correlations between changes in shape and in size. By transporting deformations in the Shape Space, much of the information about the change in size is lost. For these reasons we propose the Size-and-Shape, instead of the Shape Space, as the proper space for transporting deformations in the case of soft matter.

In the following, we briefly summarise the geometry of the Size-and-Shape Space in the case of plane configurations, by characterising the Levi Civita parallel transport of deformations along geodesic paths. Then, we use this characterisation in order to develop a procedure able to center trajectories in such a way that differences between individuals at the beginning of each trajectories are filtered out. Moreover we give a simple criterion to discriminate in which cases the proposed technique can be well approximated by a suitable hierarchical alignment followed by a linear shift. Finally, we illustrate the method with an application in cardiology, showing as the PTA is able to distinguish between health and pathological subjects.

2. The Size and Shape Space

In general, when both size and shape data are available, it is possible to follow two strategies in performing statistical analyses: analyse shape and size separately, that is shape in Shape Space Σ_m^k and size on the positive real line; analyse size and shape jointly in Size-and-Shape Space $S\Sigma_m^k$.

Which method is preferable depends on the aim of the study; as example, the first approach yields more robust inference if one is interested in shape alone. However, in many applications in biology, especially in the case of soft tissues, one is interested in investigating joint size and shape information, or in using size as a covariate, and then the Size-and-Shape Space seems to be more natural. Moreover, even if one wants to analyse shape and size separately, it is important to preserve the information about size changes when transporting deformations.

As well known, in the case of plane configurations the Riemannian structure of both Shape Space and Size-and-Shape Space is simpler than in 3D case. In particular, explicit representations are available for the Levi Civita parallel transport of deformations. For this reason in this paper we will focus on the plane case in order to exemplify our procedures.

A planar, centred configuration with k ordered vertices can be represented by a complex vector $z = (z_1, ..., z_{k-1})^T$, with $z_i \in \mathbb{C}$ a complex number. Its size-and-shape, denoted as $[z]_S$, is the orbit of z under the action of the rotation group $SO(2)$, where $SO(2)$ is identified with the unit circle in $\mathbb{C}$ acting on z as $e^{i\theta}z = (e^{i\theta}z_1, ..., e^{i\theta}z_{k-1})^T$, with $\theta \in \mathbb{R}$.

Thus, the Configuration Space is $\mathbb{C}^{k-1}$, and the Size-and-Shape Space $S\Sigma_2^k$ is the quotient space $\mathbb{C}^{k-1}/SO(2)$. Let z and μ be two centred configurations; their complex inner product is given by $z^*\mu$, where z^* denotes the transposed complex conjugate of z. The real part $\mathrm{Re}(z^*\mu)$ corresponds to the Euclidean product between z and μ, and can be used to construct the Euclidean norm (or the CS) of z as $\|z\| = (z^*z)^{1/2}$; the normalised imaginary part $\mathrm{Im}(z^*\mu)/(\|z^*\mu\|)$ represents the procrustean rotation to be used for superimposing μ on z. Then, $\mathrm{Re}(z^*\mu) = 0$ means that the two configurations are orthogonal, and hence their shapes are most different, while $\mathrm{Im}(z^*\mu) = 0$ means that the two configurations are optimally aligned.

Given a configuration $z \in \mathbb{C}^{k-1}$, its tangent space $\mathcal{T}_z\mathbb{C}^{k-1}$ can be identified with $\mathbb{C}^{k-1}$ itself. A tangent space splits, with respect to the quotient map from $\mathbb{C}^{k-1}$ to $S\Sigma_2^k$, into a vertical and a horizontal subspace:

$$\mathcal{T}_z\mathbb{C}^{k-1} = \mathcal{V}_z\mathbb{C}^{k-1} \oplus \mathcal{H}_z\mathbb{C}^{k-1},$$

characterized as follows:

$$\mathcal{V}_z\mathbb{C}^{k-1} = \{v \in \mathbb{C}^{k-1} : v = i\,\lambda\,z\,,\ \lambda \in \mathbb{R}\},$$

$$\mathcal{H}_z\mathbb{C}^{k-1} = \{v \in \mathbb{C}^{k-1} : \mathrm{Im}(z^* v) = 0\}.$$

In practice, a vertical vector at $\mathcal{T}_z\mathbb{C}^{k-1}$ represents an infinitesimal rotation of the configuration z, while a horizontal one is a vector aligned with z. The most important feature of the horizontal subspace $\mathcal{H}_z\mathbb{C}^{k-1}$ is that it is isometric to $\mathcal{T}_{[z]}S\Sigma_2^k$, the tangent space of $S\Sigma_2^k$ at the size-and-shape $[z]_S$; it follows that a tangent vector $v \in \mathcal{T}_{[z]}S\Sigma_2^k$ can be represented through the corresponding horizontal vector in $\mathcal{T}_z\mathbb{C}^{k-1}$, the so-called horizontal lift. Thus, the parallel transport of vectors in $S\Sigma_2^k$ shall be represented as a linear map between horizontal subspaces on $\mathbb{C}^{k-1}$.

3. Transporting Size and Shape

Let μ_a and μ_b be two centred and aligned configurations, that is $\mathrm{Im}(\mu_a^*\,\mu_b) = 0$; $[\mu_a]_S$ and $[\mu_b]_S$ are the corresponding size-and-shapes. The geodetic segment in the Size-and-Shape Space passing for the two points $[\mu_a]_S$, $[\mu_b]_S$, can be represented through a horizontal geodesic in the Configuration Space[1]:

$$\mu(t) = \mu_a + t\,(\mu_b - \mu_a) = \mu_0 + t\,\mu_1\,, \quad \text{with } t \in [0,1]\,; \tag{3.1}$$

$\mu_1 = (\mu_b - \mu_a) \in \mathcal{H}_{\mu_a}\mathbb{C}^{k-1}$, given the alignment of the configurations, is a horizontal vector. Given any horizontal vector w_a at μ_a, its parallel-transported vector w_b at μ_b[9,10] is:

$$w_b = w_a - i\,\frac{\mathrm{Im}\,(\bar{\mu}_b^*\,w_a)}{1 + (\bar{\mu}_a^*\,\bar{\mu}_b)}\,(\bar{\mu}_a + \bar{\mu}_b)\,, \tag{3.2}$$

where the unit vectors $\bar{\mu}_a := \mu_a/\|\mu_a\|$ and $\bar{\mu}_b := \mu_b/\|\mu_b\|$ are the pre-shapes of μ_a and μ_b. The formula (3.2) has an interesting structure: the first summand is a simple Euclidean translation (also called linear shift); the second one is the correction due to the quotient map which eliminates the infinitesimal rotations. To quantify the error made when approximating the parallel-transported vector w_b with the Euclidean translation w_a, we define the following error measure:

$$\varepsilon^2 = \frac{\|w_b - w_a\|^2}{\|w_a\|^2} = \frac{2\,\mathrm{Im}\,(\bar{\mu}_b^*\,\bar{w}_a)^2}{1 + \bar{\mu}_a^*\,\bar{\mu}_b}\,, \tag{3.3}$$

where the unit vector $\bar{w}_a := w_a/\|w_a\|$ is the pre shape of w_a. As by definitions $0 \leqslant \bar{\mu}_a^* \, \bar{\mu}_b \leqslant 1$ and $0 \leqslant \mathrm{Im}\,(\bar{\mu}_b^* \, \bar{w}_a) \leqslant 1$, the error is intrinsically limited: $0 \leqslant \varepsilon \leqslant \sqrt{2}$. Even though $\sqrt{2} \simeq 1.4$ corresponds to a very large error of 140%, this value has to be considered as an upper bound; as we shall demonstrate, in practical applications the computed ε often results to be small.

Equation (3.3) can be simplified as follows; let δ be the segment joining $\bar{\mu}_a$ to $\bar{\mu}_b$, $\bar{\delta}$ the corresponding unit vector, and ρ the Riemannian Procrustes distance between μ_a and μ_b:

$$\bar{\mu}_b = \bar{\mu}_a + (\bar{\mu}_b - \bar{\mu}_a) = \bar{\mu}_a + \delta = \bar{\mu}_a + \|\delta\|\, \bar{\delta}\,, \quad \|\delta\| = 2\sin(\rho/2)\,. \tag{3.4}$$

Using the previous result, equation (3.3) rewrites:

$$\begin{aligned} \varepsilon^2 &= \frac{2\,\mathrm{Im}\,((\bar{\mu}_a + \delta)^* \, \bar{w}_a)^2}{1 + \bar{\mu}_a^* \, \bar{\mu}_b} = \frac{2\,\|\delta\|^2}{1 + \bar{\mu}_a^* \, \bar{\mu}_b}\, \mathrm{Im}\left(\bar{\delta}^* \, \bar{w}_a\right)^2 \\ &= \frac{2\,(2\sin(\rho/2))^2}{1 + \cos(\rho)}\, \mathrm{Im}\left(\bar{\delta}^* \bar{w}_a\right)^2 = 4\tan^2(\rho/2)\,\mathrm{Im}\left(\bar{\delta}^* \bar{w}_a\right)^2 . \end{aligned} \tag{3.5}$$

The representation (3.5) allows us to calculate the range of the error from the Procrustes distances between $\bar{\mu}_a$ and $\bar{\mu}_b$: $0 \leqslant \varepsilon \leqslant 2\tan(\rho/2)$. Given that for small values of ρ it holds $2\tan(\rho/2) \simeq \rho$, the calculation of the maximum error ε becomes very simple.

To give an example, let us suppose to have a dataset in which the maximum procrustean distance between pairs of shapes is 0.1; then, the maximum error that can be made by approximating the parallel transport with a Euclidean translation is of the order of 10%. These considerations allow us to approximate the parallel transport in Size-and-Shape Space with a Euclidean translation when the procrustean distances are small.

This means that one uses the connection of the Configuration Space in which the Size-and-Shape Space is embedded; it is important to stress that all the previous considerations rely on two hypotheses: 1) μ_a and μ_b are aligned; 2) w_a is horizontal. Then, they do not allow to perform parallel translations *directly* in the configuration space. The whole procedure can be performed only after a *hierarchical alignment* of the configurations has been done.

4. Hierarchical Alignment and Parallel Transport on a Triplet of Configurations

Let us consider two different configurations μ_a and μ_b, very far from each other, and let $\tilde{\mu}_a$ be a third configuration in a neighbourhood of μ_a; our goal is to deform μ_b using the same transformation that maps μ_a onto $\tilde{\mu}_a$.

At first, we perform a hierarchical alignment; by using the same symbols for the elements of the new triplet, the alignment yields $\tilde{\mu}_a$ aligned with μ_a, and μ_a aligned with μ_b, which does not imply the alignment of $\tilde{\mu}_a$ with μ_b. This means to rotate the three configurations in such a way that:

$$\mathrm{Im}\,(\mu_a^*\,\tilde{\mu}_a) = 0 \tag{4.6}$$

$$\mathrm{Im}\,(\mu_b^*\,\mu_a) = 0 \tag{4.7}$$

After the hierarchical alignment has been done, we define the *deformation vector* $w_a = \tilde{\mu}_a - \mu_a$, a horizontal vector at μ_a:

$$\mathrm{Im}\,(\mu_a^*\,w_a) = \mathrm{Im}\,(\mu_a^*\,(\tilde{\mu}_a - \mu_a)) = \mathrm{Im}\,(\mu_a^*\,\tilde{\mu}_a - \mu_a^*\,\mu_a) = \mathrm{Im}\,(\mu_a^*\,\tilde{\mu}_a) = 0 \tag{4.8}$$

Eventually, we can use equation (3.2) to get the parallel-transported vector w_b at μ_b and use it to map μ_b onto $\tilde{\mu}_b = \mu_b + w_b$.

5. Data Centering Based on the Parallel Transport

Once the hierarchical alignment has been defined for a triplet, it is immediate to extend this notion to a data set composed by morphological trajectories. In particular, here we consider the case in which each trajectory spans a neighbourhood of a given reference configuration, as example, the first configuration of the sequence. For a given trajectory, each configuration of the sequence can be represented as a deformation vector starting from the first one; moreover, if each configuration is aligned with the reference, then, every deformation vector is horizontal.

Now, our goal is to investigate the motion undergone by the configurations of each trajectory, using GM tools; as trajectories can be located very far from each other on the Size-and-Shape Space, the differences between reference shapes have to be filtered out.

Our algorithm is as follows: at first, we define the grand mean resulting from a GPA on all the references; then, for any trajectory, we parallel-transport its configurations along the geodesic connecting the reference to the grand mean; in such a way, the (configurations of the) trajectories are displaced so that they have a common reference, see Fig. 1.

After data centering, we apply a standard GPA followed by PCA. This last step allows us to separate the shape information from the size information that remains available after the parallel transport; the whole procedure is summarised in the following:

(1) Hierarchical Alignement:

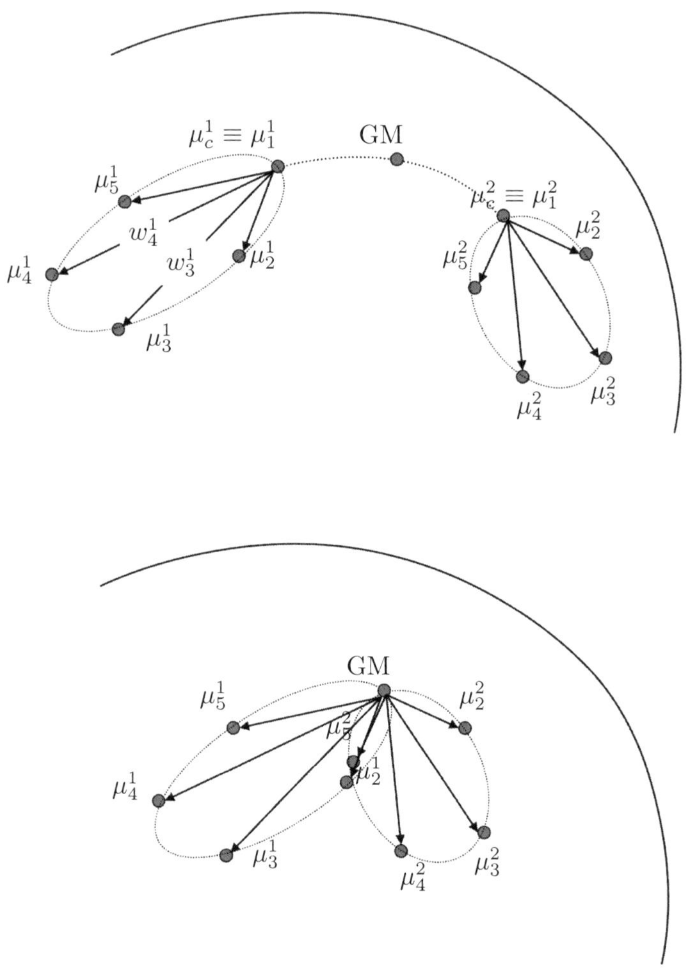

Figure 1. Data centering in curved spaces.

(a) For any trajectory, assign to a configuration of the sequence the role of reference, labelled μ_c^j; this role can be played by the first one of the sequence, or by the local mean, depending on the nature of analysis: motion analysis or phenotypic trajectories analysis. Here we use the first configuration;

(b) Perform a GPA with no scaling, in order to maintain the very nature of the deformation, between all the references μ_c^j and we define the grand mean (M);

(c) Perform loops of OPA with no scaling to unit CS, to align each shape of any cycle with its proper μ_c^j.

(2) Parallel transport in the Size-and-Shape Space:

(a) Define the vectors $w_i^j = \mu_i^j - \mu_c^j$;

(b) Transport the vectors toward the M by using the explicit formula for the parallel transport in the Size-and-Shape Space (3.2);

(c) Add the transported vectors to the M.

(3) Eventually, perform a standard GPA+PCA in the tangent space to M.

When the Procrustes distance between the local references and the M is small enough point (2b) can be performed as an Euclidean translation, by neglecting the second term of the (3.2).

6. Application to Human Cardiac left Ventricular Movement Data

We tested our procedure on actual data, coming from 3D Speckle-Tracking-Echocardiography (3D-STE) of 50 human left ventricles (LV), 48 relative to healthy subjects and 2 to patients with volumetric overload due to aortic insufficiency. These data come from the same research project partially published in[4], and have been collected by means of 3D-STE (PST25SX Artida, Toshiba Medical Systems Corp., Tokyo, Japan).

The whole LV geometry is reconstructed by starting from a set of 6 homologous landmarks, manually detected by the operator for each of the subjects under study. The manual detection of the initial set of landmarks is crucial to fulfil a homology principle, so that comparable anatomical structures of different subjects are detected; completely automated approaches may suffer from error of pattern identification depending on the specific algorithm used for reconstruction.

The data set generated for each subject by 3D-STE consists of a sequence of 3D configurations, acquired every ~ 45 ms, each constituted by 1297 landmark, assumed to be homologous, for both the epicardial and endocardial surfaces, positioned along 36 horizontal circles, each comprised of

36 landmarks, plus the apex (Fig.2a). After the generation of the data set,

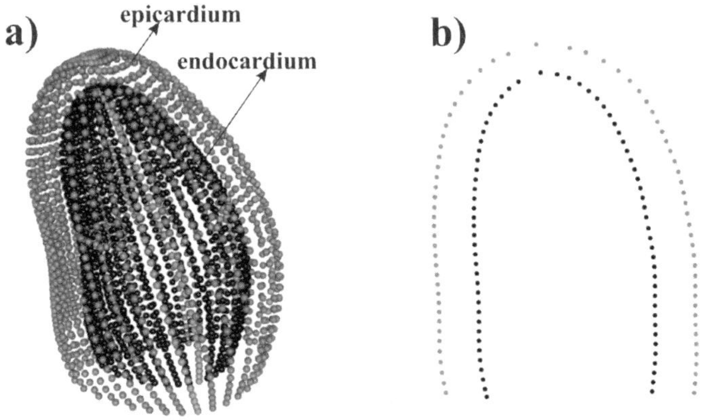

Figure 2. Image from real data. a) Original three-dimensional LV geometry. b) Two-dimensional configuration after projection.

we define 9 time-events, to be used as homologous time-events to sample the heart beating for all the subjects, and we assemble a trajectory for each subject by extracting from the data set a sequence of 9 configurations corresponding to the homologous time-events[4]. Finally, for each trajectory, we project a coronal slice of the 3D configurations on the plane transversal to the LV base, identified by the mitral annulus (Fig.2b); this plane is selected at the first homologous time, i.e. at the R peak.

The first configuration is assumed as reference; once performed a GPA between all the 50 reference configurations, we calculate the Procrustes distance between each reference and the grand mean M, obtaining 0.12 as maximum value, to which there corresponds a maximum 12% error, according to (3.5), if the parallel transport is approximated by an Euclidean translation.

In Fig. 3 the values of the actual errors calculated by means of the (3.3) on the whole dataset are plotted. The values obtained could justify the use of the linear approximation, but could give some appreciable difference in the results, as well. In order to check these differences we perform analysis by using both parallel transport and Euclidean translation. Moreover, we compare these results with that obtained without performing data centering and by performing a *naif* data centering in the configuration space. By summarizing, we perform four analysis:

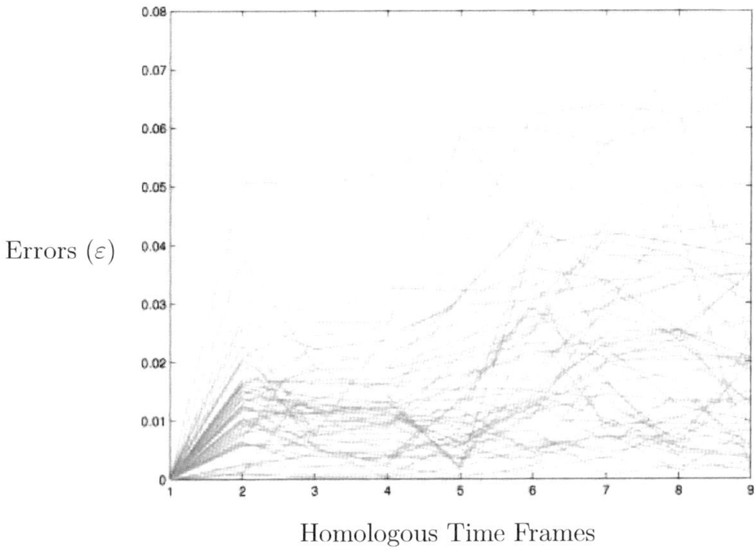

Figure 3. Calculated errors in approximating parallel transports with linear shifts for the case of real data.

a) No Data Centering: Classical GPA+PCA.
b) Naif Data Centering: Euclidean translation in configuration space.
c) Riemannian Data Centering: Parallel transport in size-and-shape space after hierarchical alignment.
d) Approximated Riemannian Data Centering: Euclidean translation after hierarchical alignment.

Fig. 4 illustrates the results of the four procedures. While GPA+PCA still shows evidence of the ambiguity of PCs, that explain concomitantly intra- and inter- individual variation, the naif data centering exhibits more concentrated trajectories but, at the same time, a complete lack of morphological information. On the other hand, the Riemannian (exact or approximated) data centering yields similar results.

The deformation associated to the PCA analyses is shown in Fig. 5; while for classic GPA+PCA there is no hope to observe the sole deformation cleaned up by inter-individual variability, the deformations given by the naif data centering are completely meaningless. On the other side, the deformations illustrated by the Riemannian methods (exact or approximated) are very similar; here, it is possible to appreciate small qualitative

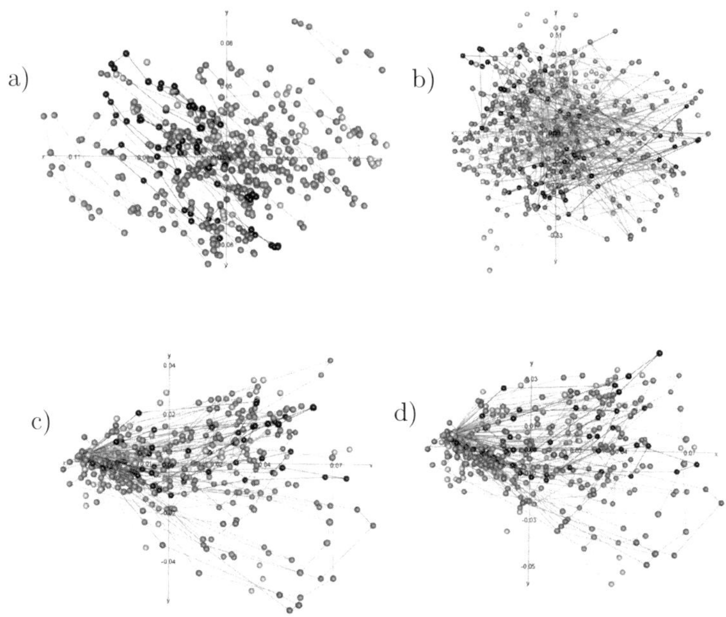

Figure 4. PCA results for the real data. a) Classic GPA+PCA. b) Euclidean translation in the configuration space. c) Parallel transport in size-and-shape space. d) Euclidean translation after hierarchical aligment.

differences between the exact and the approximated approach: in the first one, the deformation grid is more regular and the deformation mode more realistic. In both the Riemannian methods, it is clear that PC1 represents the global contraction and the myocardial thickening, while PC2 is associated to different extent of deformation affecting mitral annulus and ventricular apex.

As explained in Section 3, in motion analysis it is important to correlate PC scores with change in size. In this case, the meaning of PC1 is clearly correlated with the change in size, while for PC2 this correlation is not that obvious. Fig. 6 shows the analysis of the correlations between CS values and PC1, PC2, and PC3, respectively, for each subject; these correlations result to be very significant for the first PC. This indicates that the concentration of allometric change on the first PC allows a meaningful appreciation of those shape changes associated to contraction and those changes that are not.

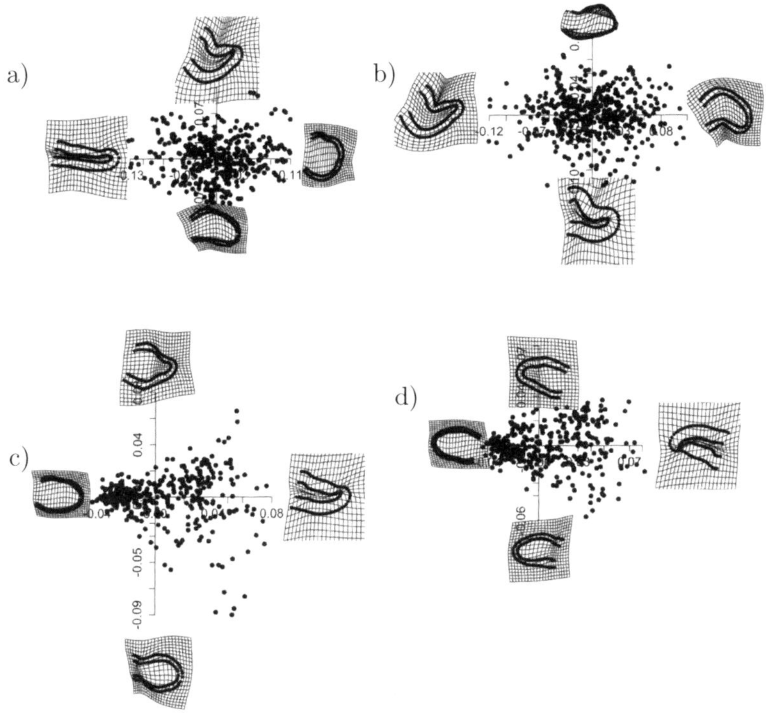

Figure 5. Deformations associated to PCA plots depicted in Fig. 4. a) Classic GPA+PCA. b) Euclidean translation in the configuration space. c) Parallel transport in size-and-shape space. d) Euclidean translation after hierarchical aligment.

Finally, we use the PC scores interpolated at homologous time frames in a GPA followed by a PCA, in order to evaluate the shape of trajectories themselves as done in [2,3,4]. The results of the trajectories analysis are given in Fig. 7, for techniques b), c) and d). It is worth noting that an analysis based on a naif data centering would not able to distinguish between patients (squares) and healthy subjects (dots).

7. Conclusions

In this paper we discussed the problem of comparing morphological trajectories experienced by different specimens of biological soft tissues. A crucial point is to filter inter-individual shape differences. We proposed to solve this problem by performing a data centering in the Size-and-Shape Space, by means of the Riemannian parallel transports. Theoretical considera-

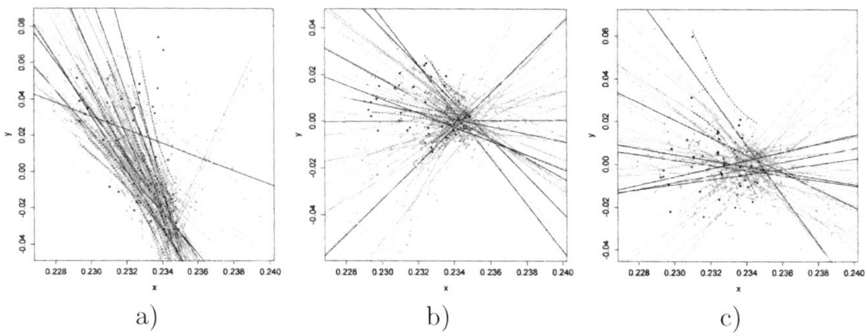

Figure 6. Correlations between the centroid size values and pc-scores. a) First PC b) Second PC c) Third PC. Dotted line represent 95 % confidence interval of regression lines.

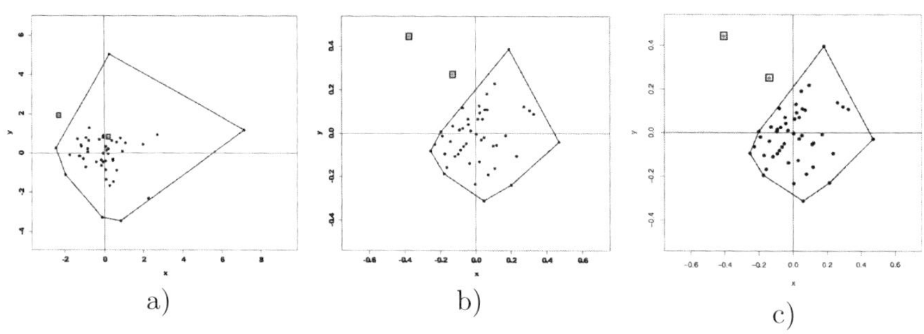

Figure 7. PCA performed on the shapes of the trajectories. Dots represent healthy subjects. Squares represents patients. a) Euclidean translation in the configuration space. b) Parallel transport in size-and-shape space. c) Euclidean translation after hierarchical aligment.

tions, together with analysis on real heart data from human left ventricle suggest that when configurations differ for small Procrustes Distances the parallel transport can be well approximated by a Euclidean translation, after a hierarchical alignment. It is provocative that by this method a quite evident differentiation is obtained of peculiar pathological conditions (where a clear-cut volume increase is evident) as compared to a very large group of normal individuals, at variance with standard methods.

On the other hand, the comparison with a naif data centering performed through a direct Euclidean translations in the Configuration Space shows as an improper treatment of the data before the PCA can lead to completely wrong results together with misunderstanding morphological interpretations. This study gives a more solid theoretical background to the approach proposed in [4] and illustrates as a correct data centering is a crucial point in the analysis of morphological trajectories.

References

1. Kendall, D. G., Barden D., Carne T. K., Le H. 1999. Shape and Shape Theory, Wiley.
2. Adams D.C., Collyer M.L. 2009. A general framework for the analysis of phenotypic trajectories in evolutionary studies. Evolution 63: 11431154.
3. Collyer M.L., Adams D.C. 2013. Phenotypic trajectory analysis: comparison of shape change patterns in evolution and ecology. Hystrix (in press).
4. Piras P., Evangelista A., Gabriele S., Nardinocchi P., Teresi L., Torromeo C., Schiariti M., Varano V., Puddu P. E. 2014. 4D-Analysis of Left Ventricular Heart Cycle Using Procrustes Motion Analysis. PLoS ONE 9(1): e86896.
5. Kume A., Dryden, I.L., Le H. 2007. Shape space smoothing splines for planar landmark data. Biometrika, 94: 513528.
6. Le H. 2003. Unrolling shape curves, J. London Math. Soc. (2) 68: 511526
7. Le H., Kume A. 2000. Detection of shape changes in biological features. Journal of Microscopy, 200 (2): 140147.
8. Huckemann S., Hotz T., Munk A. 2010. Intrinsic MANOVA for Riemannian Manifolds with an Application to Kendall's Spaces of Planar Shapes. IEEE Trans. Patt. Anal. Mach. Intell., 32 (4): 593-603.
9. Varano V., Gabriele S., Teresi L., Dryden I., Puddu P.E., Torromeo C., Piras P. 2014. Comparing Shape Change Trajectories via New Connections in Size and Shape Space. Submitted.
10. Varano V., Dryden I. 2014. Levi Civita Parallel Transport in the Size-and-Shape Space for Plane Configurations. Technical Report.

DYNAMIC SCALING ANALYSIS OF *IN VITRO* AND *IN SILICO* CELL CULTURES*

M. A. C. HUERGO

Instituto de Investigaciones Fisicoquímicas Teóricas y Aplicadas (INIFTA), Universidad Nacional de La Plata, CONICET CCT-La Plata, Sucursal 4, Casilla de Correo 16,(1900) La Plata, Argentina

B. MOGLIA, E.V. ALBANO AND N. GUISONI

Instituto de Física de Líquidos y Sistemas Biológicos (IFLYSIB), Universidad Nacional de La Plata, CONICET CCT-La Plata; Calle 59-789 (1900) La Plata, Argentina

Departamento de Física, Facultad de Ciencias Exactas, Universidad Nacional de La Plata, La Plata, Argentina

The growth of linear cell colony fronts was investigated from both, experimental and theoretical points of view by means of dynamic scaling techniques. Two different cell lines where studied: Vero and HeLa. Experiments show that the front of Vero cell colonies shift at 0.22 ± 0.02 μm min^{-1} average constant linear velocity, while for HeLa cell colonies the speed is of 0.09 ± 0.02 μm min^{-1}. Dynamic scaling analysis of both cultured Vero and HeLa cell colonies shows that the roughness w versus the system width l log-log plots collapse to a single curve and a growth exponent $\beta = 0.33 \pm 0.04$ is derived. From the structure factor analysis a global roughness exponent $\alpha = 0.50 \pm 0.05$ is obtained and a dynamic exponent $z = 1.5 \pm 0.2$. This set of exponents fulfils the Family-Vicsek relationship and is consistent with the predictions of the continuous Kardar-Parisi-Zhang model. On the other hand, the numerical simulations of a minimal discrete model for cell cultures in vitro showed that the growing interface is charactertized by a growth exponent $\beta = 0.32 \pm 0.02$, a roughness exponent $\alpha = 0.48 \pm 0.02$ and a dynamic exponent $z \simeq 1.50 \pm 0.02$. Again, this set of scaling exponents agreed with those predicted by the Kardar-Parisi-Zhang universality class.

1. Introduction

The study and understanding of the processes involved in cell growth are topics of interdisciplinary interest that have attracted the attention of the

*This work was financially supported by the Consejo Nacional de Investigaciones Científicas y Técnicas (CONICET).

scientific community. Of course, the collective behavior of interacting living cells is a complex and intriguing phenomenon that is far from being well understood. In that way, in the last decades the dynamic scaling theory has been applied to investigate the growth mechanism of a variety of biological systems, since it has provided to be a powerful tool for the characterization of growth interfaces [36,21]. In fact, the development of self-affine interfaces [36,21,37,40] can be rationalized in terms of critical exponents that, as in the case of second-order phase transitions [13], allow for the identification of few universality classes. Then, systems belonging to a given universality class share all relevant interfacial properties independently of their intrinsic complexity. In that way, pioneering work in the field of bacterial colony growth [17,18,19,20] has been oriented to investigating the cooperative bacterial motion in relation to the complex structure of colonies and the feasibility of the scaling analysis of colony front profiles. Also, scaling analysis have been applied to *Escherichia coli* and *Bacillus subtilis* linear colonie [20].

Developing interfaces of *in vitro* cell cultures have been employed recently as model systems for the study of cell proliferation under well-controlled experimental conditions [29,41,47]. In that way, both Vero and HeLa cell cultures were analyzed as models of cell growth under two different geometrical regimes: linear and radial growth. Radially spreading systems allow the authors to apply the dynamic scaling techniques to analyze interfaces whose size does not change with time, and only grow in one direction. Instead, radially spreading systems, which imply the spontaneous shape that cell colonies acquire was analyzed with the same techniques and compared with the first geometry to assure whether the validity of extending the mathematical formalism for fixed size colony growth experiments was directly applicable to radially expanding systems. Thus, the comparison of scaling analysis data derived from radial expanding systems to those predicted from computational models proposed for fixed size systems is a fact that has been under several discussions [45,46,48,8,26,41].

From the theoretical point of view, one can gain insight into the complex processes involved in cell growth with the aid of models, also known as *in silico* experiments, as is usual in physics and related sciences. For this purpose, one basically has three main approaches: (i) the formulation of continuous models by using differential equations that are solved by means of mathematical and/or numerical techniques [2,3,4], (ii) a discrete approach where the dynamics is implemented at the level of individual cells [5,6,7,30], (iii) multiscale models simultaneously involving continuous and discrete approaches have also been formulated [9,10,11], e.g. nutrients and oxygen diffusion can be treated by means of mean-field differential equations, while the dynamics of cell proliferation and diffusion are undertaken in a discrete fashion. The theoretical approach allows the use of well-established techniques and tools developed in the fields of statistical physics and computer simulations of complex systems, as the dynamic scaling theory cited above.

Within this broad context, the aim of this paper is investigate the growth of linear cell colony fronts from both experimental and theoretical points of view by means of dynamic scaling techniques. We present a review of recent studies on both cultured Vero and HeLa cells for the case of linear and circular interfaces [29,41,47]. Also, we show the main results of a simple discrete model to describe the developing of these interfaces [30]. From the critical exponents associated with the growing interface in both experimental and theoretical models we determine that these cell cultures are in the Kardar-Parisi-Zhang universality class.

The manuscript is organized as follows: in Sections 2 and 3 we present descriptions of the experimental and theoretical models, respectively. Section 4 is devoted to a brief theoretical background for the treatment of interfaces and dynamic scaling theory. Results are presented and discussed in Section 5, and finally our conclusions are stated in Section 6.

2. Experimental Model

2.1. *Cell colony growth procedures*

Vero cells (passages 170-180) and HeLa cells (passages 45-55) were cultured in Petri dishes using Roswell Park Memorial Institute (RPMI 1 640) medium containing 10% fetal bovine serum (FBS) maintained in a 5% carbon dioxide controlled atmosphere at 37^0 C, changing one half of the culture medium every 2 days.

Colonies with linear growth fronts were obtained by first covering the central region of the Petri dish bottom with a 2.2 cm wide and 100 μm thick sterilised Teflon tape. Then, disaggregated cells ($30\,000 - 40\,000$ cell ml^{-1}) were shed and left to grow for about 2 days until confluence in the Teflon-free region was reached. Eventually, the Teflon tape was removed, leaving a cell-free central region with the formation of two facing linearly colony fronts of width L that started to shift in opposite directions perpendicular to L (Fig. 1 **(a)**). From this stage on, the morphology evolution of the colony pattern and the 2D growth front displacement were followed.

Colonies exhibiting radially spreading growth fronts were prepared by shedding disaggregated cells in the Petri dish and were left growing until a

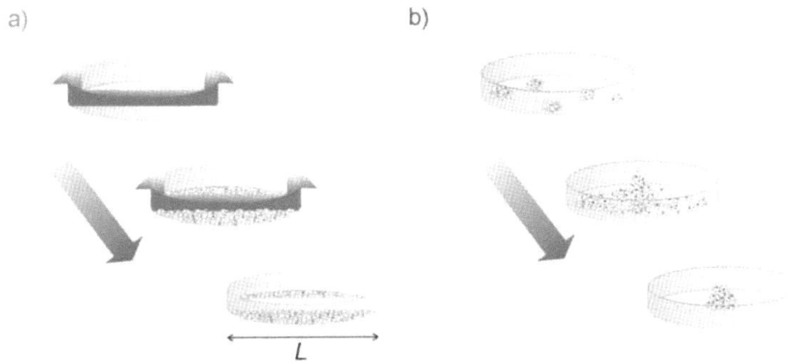

Figure 1. Scheme of the experimental setup for procedures : a) Linear growth fronts: a Petri dish is centrally covered by a Teflon tape. Cells are seeded and left grow until the reach the confluence. After withdrawing the Teflon tape, two linear growth fronts are formed. (b) Very low cell concentration were seeded in a Petri dish. Then 3D clusters were carefully removed and located in a new Petri dish. 2D radially spreading colonies are formed around the cluster.

3D cell cluster with an average radius of about 250–300 μm at the colony center was formed. Subsequently, the 3D cluster was carefully removed with a micropipette and transferred to a second Petri dish containing fresh culture medium. Eventually, the colony continued spreading from the 3D seed rim as a 2D domain.

Each colony growth was followed until it contacted the border of a neighbor colony, a fact that commonly occurred after 5–8 days.

2.2. *Colony imaging*

Colonies were imaged using a Canon digital camera coupled to a Nikon TS100 phase-contrast inverted microscope with a CFI flat field ADL 10× objective at a resolution of 0.88 μm/pixel, where the field covered by the microscope was about 1 000 μm. As the colony grew, an increasing number of partial images to cover the whole colony were taken. Subsequently, these images were stitched to compose the complete image of the colony. Colony fronts were manually traced using a Wacom graphic tablet with a tracing error on the order of the pixel. Typical evolution of colony fronts from 2D radially and linearly spreading colonies are depicted in Fig. 2.

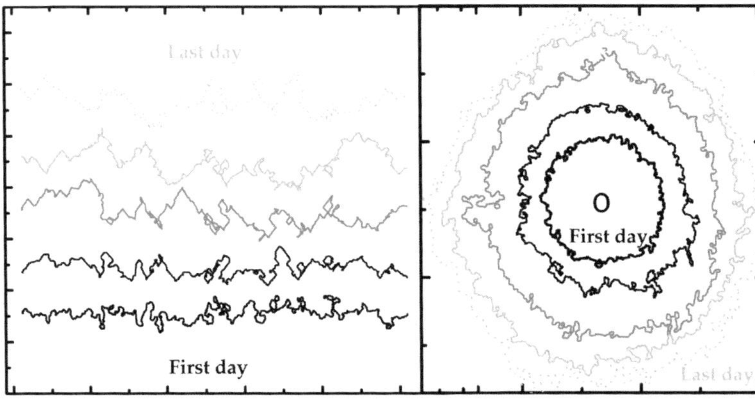

Figure 2. Typical evolution of colony front profiles from cell colonies under linearly (a) and radially (b) spreading. For the sake of clarity, only a section of the linearly fronts is shown.

The analysis of colony fronts was performed employing an in-lab developed program for obtaining the front roughness and other parameters related to the front displacement, with each front consisting of N points (pixels). For the sake of simplicity, both the radial and the linear front distances are denoted as h. For radially spreading colonies, the program allowed us to determine the center of mass (c.m.), the instantaneous distance $h_k(t)$ from c.m. for the kth point at the front ($k = 1, 2, \ldots N$) and the corresponding arc (s_k), whereas for linearly spreading colonies, one obtained the instantaneous colony height $h_k(t)$ from the starting linearly front of width L for each kth front site.

3. Computational Model

3.1. *Description*

In order to simulate the growth of tumor cells *in vitro*, we assume a discrete two-dimensional sample given by a square lattice of sides $L \times H$, which represents a sector of the glass substrate of Petri dishes in actual growth experiments. Along the L-direction (say the vertical direction) we define

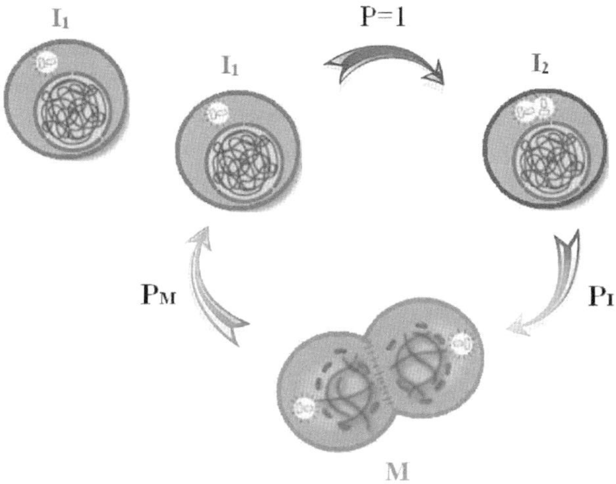

Figure 3. Schematic representation of the proliferation cycle: I_1-state corresponds to the first part of interphase (G_1 phase); I_2-state stands by the second part of interphase (synthesis and the G_2 phase); and $M-$state represents the mitotic phase. A cell of type I_1 spontaneously becomes an I_2-cell ($P = 1$). A cell in the state I_2 can become a cell in state M with probability P_I. A cell of type M splits out into two I_1-cells with probability P_M. More details in the text.

the lines of our system and identify the spatial location of a cell in a given line by the index i ($0 \leq i \leq L-1$). Similarly, the H-direction defines the columns and the positions along them are identified by the index j ($0 \leq j \leq H-1$). We take periodic boundary conditions along the H-direction, while for the L-direction, which corresponds to the cell growth direction, the boundaries are left open. The above choice will become more clear upon the definition of the cell growth rules (see below).

In the proposed model, each site of the lattice with coordinates (i, j) can be occupied by a cell or be empty. Furthermore, the proliferation cycle will be considered in three steps: the first part of interphase (I_1-cells) that represents the G_1 phase; the second part of interphase (I_2-cells) that represents the synthesis (S) and the G_2 phase; and the mitotic phase ($M-$cells) [31]. So, each cell in the lattice can be in three different states, as follows: I_1, I_2, and M, as defined above. Note that the I_1 state represents a cell that has been just divided, so it is ready to evolve into state I_2.

We further assume a homogeneous distribution of nutrients irrespective of the state of the $(i, j)th$ site. The rules for the dynamic evolution of the system, which depend on the state of each particular cell, are the basis of the simulation algorithm and are defined as follows: (i) A cell of type I_1 spontaneously becomes an I_2-cell remaining at the same site. (ii) A cell in the state I_2 can either grow becoming a cell in state M with probability P_I, or diffuse into a nearest-neighbor site with probability $P_I^D = 1 - P_I$. Since an $M-$cell occupies two neighboring sites, the growth process can only proceed if a randomly selected neighboring site of the original I_2-cell is actually empty. In the same fashion, diffusion of an I_2-cell is only allowed to proceed to a neighboring empty site. (iii) Finally, a cell of type M splits out into two I_1-cells, occupying the same sites as the original ones, with probability P_M. Also, this type of cell can diffuse and rotate (in steps of $\pi/2$ degrees in the square lattice) with probabilities P_M^D and P_M^R, with $P_M + P_M^D + P_M^R = 1$, respectively. Of course, these two latter processes involve randomly selected neighboring sites and they are constrained by excluded volume. Transitions between states I_1, I_2, and M are shown in Fig. 3.

In order to implement the simulations a cell is selected at random and, depending on its state, we proceed according to the above-defined dynamic rules. After each random selection the simulation time t_s is increased according to $t_s^{new} = t_s^{old} + 1/N_{tot}$, where $N_{tot} = N_{I_1} + N_{I_2} + N_M$ is the total number of cells in the system. The initial condition is selected by locating I_1-cells filling the first column ($j = 0$) of the substrate. In this way the culture can grow along the H-direction and a growth interface that essentially runs parallel to the L-direction is naturally established.

3.2. *Interfaces definition*

Along this paper we will focus our attention on the properties of the growth interface of the cell culture. For this purpose, it is useful to recall that there are many possible definitions of such an interface. So, we will define two interfaces that are relevant for the present study, namely the so-called single-valuated interface (SVI) and multi-valuated interface (MVI), respectively. In order to locate these interfaces one first determines the biggest cluster of cells (connected by means of nearest neighbors) in contact with the initially filled first column of the culture, i.e. the so-called percolation cluster [32,33,34,35], which is shown by means of black-filled dots in Fig. 4(b).

This cluster is denoted as *land*. On the other hand, empty sites in contact with the last column of the sample, which is never reached by the culture, are linked through both nearest and next-nearest neighbor sites and form a large cluster that is termed the *sea* (notice that sites belonging to this cluster are left in white in Fig. 4(b)). The sites that are not connected with either the largest cluster of *land* or *sea* are identified as *islands* and *lakes* (see also Fig. 4(b)), respectively, but they are irrelevant. In fact, the interface is given by the seashore where *land* and *sea* are in contact [32,33,34,35], which is shown in Fig. 4(c) after removing lakes and islands. We call the set of these sites $h_k, k = 1, ..., N_{int}$, the MVI, where h_k is the horizontal coordinate (H-direction, see Fig. 4) of the *kth* cell belonging to the MVI, and N_{int} is the number of cells of the MVI. The MVI of the snapshot depicted in Fig. 4(a) is shown in Fig. 4(c). On the other hand, for the case of the SVI, the interface has a very simple definition: after disregarding the *islands*, one has to follow a single line of the sample and by starting from the *sea*, the first site of *land* found is just a site of the SVI (for a similar definition see [35]), as shown in Fig. 4(d). Therefore, the SVI is given by the set of sites $(h_k, k = 1, ..., N_{int} = L)$. Note that for the MVI one may have $N_{int} \geq L$. For further technical details on the method used for the location of the interfaces, see e.g. [32,33,34,35].

4. Brief Theoretical Background

A useful tool for the characterization of interfaces is the well established dynamic scaling theory developed by Family and Vicsek [21,36,37]. The methodology relies on the evaluation of the interface width (W) and its rationalization in terms of exponents, which in turn may define a universality class. So, as in the study of second-order phase transitions [38], a set of different models sharing the same critical exponents are said to belong to the same universality class.

In order to define the interface width, first one needs to identify the cells located at such interface. Let h_k ($0 \leq h_k \leq H-1$ and $1 \leq k \leq N_{int}$) be the horizontal coordinate (H-direction, see Figs. 4) of the *kth* cell belonging to the interface, where N_{int} is the number of cells belonging to the interface. Then we define the average interface position $\langle h \rangle$ as the mean value of h_k, namely

$$\langle h \rangle = \frac{1}{N_{int}} \sum_{k=1}^{N_{int}} h_k, \tag{1}$$

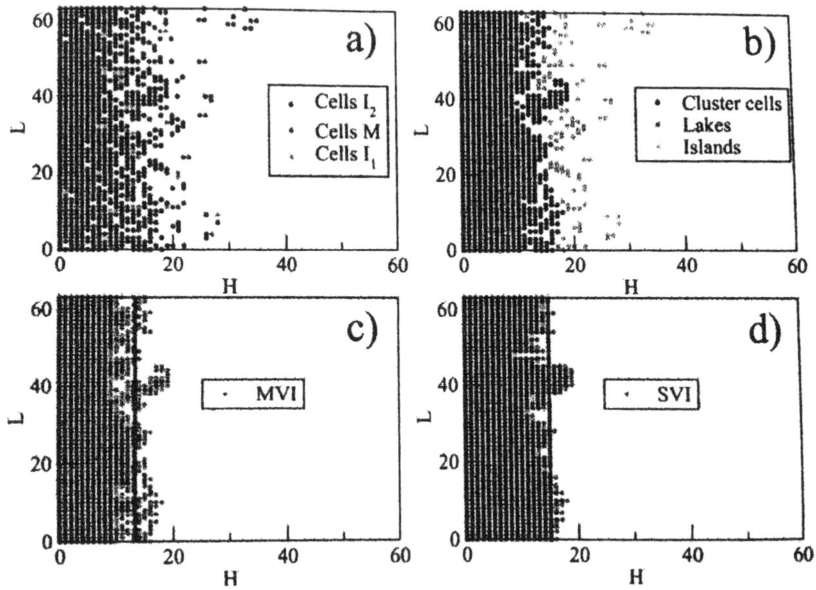

Figure 4. (a) Typical snapshot configuration obtained for the simulation of a system of vertical side $L = 64$. Cells are shown by means of filled dots of different color depending on their type (as indicated), while empty sites are left in white. The subsequent snapshots illustrate the procedure used to locate the interfaces of the aggregate, as follows: (b) Here, the percolating cluster linked to the initially filled column of I_1-cells, on the left-hand side of the sample, is shown by means of black dots. Also, lakes and islands are shown by means of fuchsia and turquoise dots, respectively. (c) For the sake of clearness we remove islands and lakes and show the MVI (green dots), whose average position is shown by the vertical red line. (d) As in (c) but for the SVI. For all Figs. $P_I = P_M = 0.10$, $P_I^D = 0.90$, $P_M^D = 0.89$ and $P_M^R = 0.01$ More details in the text.

where for the SVI one has $N_{int} = L$ while for the MVI one may have $N_{int} \geq L$. Now, the interface width can be evaluated as the rms deviation of $\langle h \rangle$, namely

$$W = \sqrt{\frac{1}{N_{int}} \sum_{k=1}^{N_{int}} (h_k - \langle h \rangle)^2}. \tag{2}$$

According to the experience gained in the study of growth interfaces [21,36], by starting the measurements from a flat interface, as in our case that assumes a line of cells located in the first column of the sample as the initial condition, one has that at early times the width increases according to

$$W \propto t^{\beta}, \tag{3}$$

where β is the growth exponent. During the growth process it is expected that correlations between cells will start to develop. In particular, one has that correlations along the interface can be described by a typical correlation length ($\xi_{||}(t)$), which in turn grows

$$\xi_{||}(t) \propto t^{\frac{1}{z}}, \tag{4}$$

where z is the dynamic exponent. The above paragraphs essentially outline the expected dynamic behavior of a growth interface in an infinite sample. However, both in the experiments and simulations, one has to deal with finite samples of a given width (say L, as shown in Fig. 4). Therefore, $\xi_{||}$ cannot grow indefinitely, as stated by Eq. (4), but now it grows only up to a given crossover time (t_c) according to

$$\xi_{||}(t_c) \propto L \propto t_c^{\frac{1}{z}}. \tag{5}$$

Furthermore, as $t \to t_c$, one also observes deviations from the prediction of Eq. (3) since now the interface growth becomes slower, and eventually the interface width fluctuates around some saturation (or stationary) value (W_s, for $t > t_c$).

Now, in order to give a unified description of the above-outlined arguments as in the case of second-order phase transitions [39], the dynamic scaling theory states that the interface width is a homogeneous function of two characteristic lengths ($W(L, \xi_{||})$), namely $\xi_{||}$ that reflects the properties of the system under study and L that is set by the experimental/simulation conditions. Then, the following scaling Ansatz can be formulated:

$$W(L, \xi_{||}) = W(L, t) = t^{\beta} f(\frac{L}{\xi_{||}}) = t^{\beta} f^{*}(\frac{L}{t^{\frac{1}{z}}}), \tag{6}$$

where $f^{*}(u)$ is a suitable scaling function. In particular, one has that

$$f^{*}(u) = \begin{cases} constant, & u \to \infty \\ u^{\alpha} & u \to 0 \end{cases}, \tag{7}$$

where α is the roughness exponent. In this way in the early time limit $t \to 0$ (i.e. $u \to \infty$), one recovers Eq. (3), while in the long-time regime $t \to \infty$ (i.e. $u \to 0$), one has that the interface width becomes saturated at L-dependent values according to

$$W_s(L) \propto L^{\alpha}, \tag{8}$$

where the above relationship is accomplished when $z = \frac{\alpha}{\beta}$, so that only two of the exponents are actually independent.

In a more general fashion, the scaling behavior of the interface can also be investigated by evaluating the structure factor $S(k,t)$ or power spectrum [21,29,41,40,42]

$$S(k,t) = \langle | \frac{1}{\sqrt{N_{int}}} \sum_{n=1}^{N_{int}} h_n(t) e^{-ikn} |^2 \rangle, \tag{9}$$

where $k = 2\pi n / N_{int}$, with $n = 1, ..., N_{int}/2$. Now, a quite general dynamic scaling hypothesis can also be formulated for the structure factor, yielding [40]

$$S(k,t) = k^{-(2\alpha+1)} s(kt^{1/z}), \tag{10}$$

where the scaling function has the general form

$$s(u) \sim \begin{cases} u^{2(\alpha-\alpha_s)} & \text{if } u \gg 1 \\ u^{2\alpha+1} & \text{if } u \ll 1 \end{cases}, \tag{11}$$

where α_s is the spectral roughness exponent, which can only be evaluated by measuring the structure factor [40]. For the sake of completeness, one can make use of

$$W^2(L,t) = \frac{1}{L} \sum_k S(k,t), \tag{12}$$

in order to obtain a relationship between the structure factor and the width of the interface (cf. Eqs. (2) and (6)).

As already mentioned, the set of exponents that characterize a given interface determines its universality class.

5. Results and Interpretation

5.1. *Velocity of colony front*

Colony front average height, $\langle h \rangle$, versus t, fit reasonable straight lines with constant average front irrespective of the colony geometry. In the case of Vero cells velocity $\langle v \rangle = 0.22 \pm 0.02$ μm min^{-1}, while for the case of HeLa cells the average colony front displacement velocity derived from these plots results in 0.09 ± 0.02 μm min^{-1}, as can be seen in Fig. 5. Thus, for both, different geometries and cell type, one can envision that cell colonies grow linearly with time, even thou, the velocity for HeLa cells is around half the value obtained for Vero cell colonies.

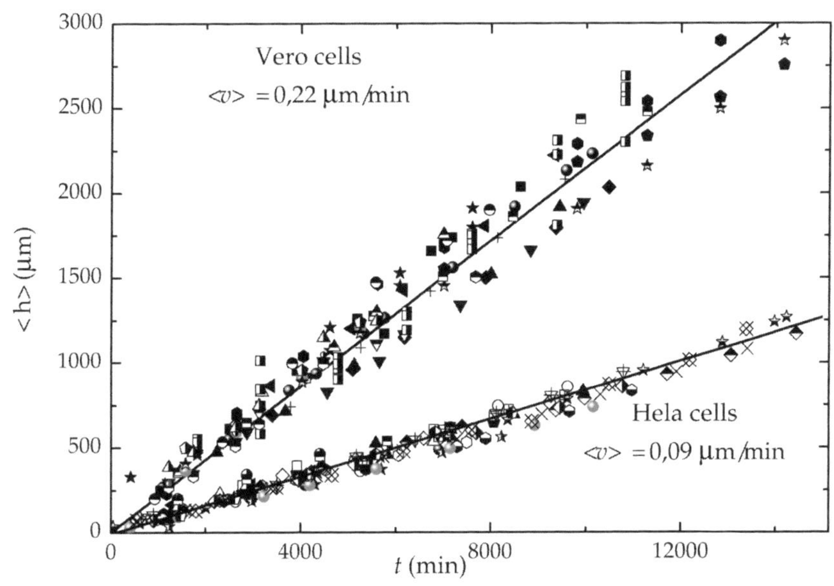

Figure 5. Plots of $\langle h \rangle$ versus t of the colony front profiles of Vero and HeLa cells. Symbols correspond to different radial and linear colonies.

In silico results show that the interfaces in our discrete model also grow at constant speeds. Figure 6 shows plots of the average position of the growth interface $\langle h \rangle$ versus time as obtained for different values of the growth probability P_I.

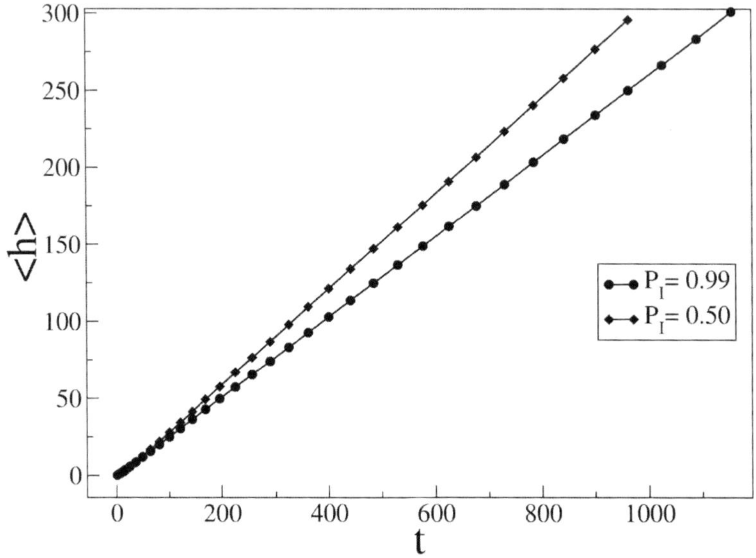

Figure 6. *In silico* results: plots of the average position of the growth interface $\langle h \rangle$ versus time as obtained for different values of the growth probability ($P_I = 0.99$ and $P_I = 0.50$), as indicated. The straight lines obtained indicate that our interfaces grow at constant speeds (arbitrary units).

5.2. *Front roughness*

The instantaneous interface width or colony front roughness was determined from Eq. (2), where L is the width of the growing front, which is constant for linearly fronts and increases with time as $L = 2\pi\langle h(t)\rangle$ for radially expanding fronts. The dependence of the local roughness $w(l, t)$ on the front width l ($l \leq L$) was evaluated from the standard deviation of h_i for different front lengths l at time t. While this procedure is straightforwardly applied for linearly growth fronts, radially spreading colonies require the evaluation of the roughness at different arc lengths s ($s = l < L$)[24,15,41]. For this purpose, the coordinate system was transformed from angle radius into arc radius. Then, the location of each point at the front was determined by coordinates h_i, s_i, with the arc s_i being measured along a circle of radius $\langle h(t)\rangle$.

The experimental roughness was evaluated for both directed and overhang-corrected data. Overhang-corrected colony profiles were obtained by taking the maximum value of $h_i(t)$ at each ith position at the front. The influence of overhang correction, required for a proper scaling of the front roughness, is practically null when the length of the colony front exceeds about 2-3 average cell diameters. Interface width data of both directed and overhang-corrected colony fronts, evaluated from Eq. (2), fulfills linearly log-log plots with time with the slope $\beta = 0.33 \pm 0.04$, irrespective of the colony geometry, for both HeLa and Vero cells (Fig. 7).

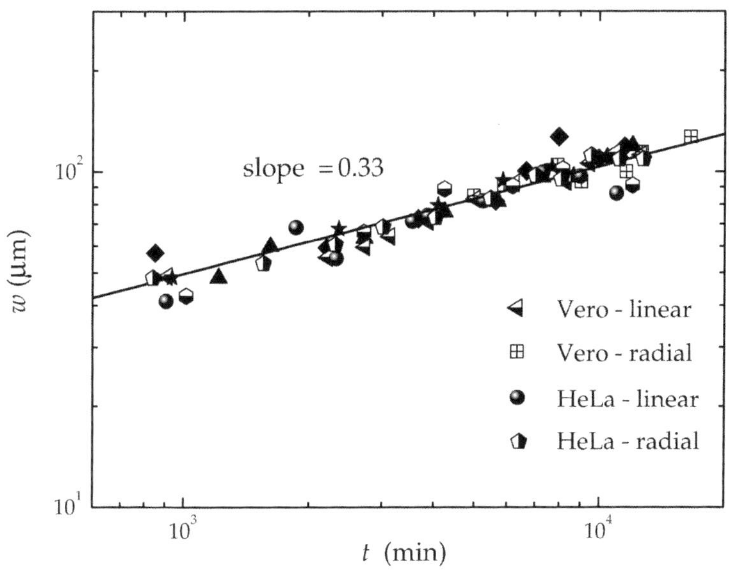

Figure 7. Front roughness or interface width for HeLa and Vero cells versus growth time log-log plots. Linear and radial results are shown together. For the sake of clarity, data have been binned. The straight line corresponds to the slope 0.33 to guide the eye.

Figure 8 shows log-log plots of the interface width (W) versus time as obtained for our *in silico* experiments. We show samples of different sizes (as specified in the plots) and growth probabilities given by $P_I = 0.99$ and 0.50 (inset). It is worth mentioning that Fig. 8 combines results obtained during the early time regime ($t < 2 \times 10^3$) and the stationary regime ($t > 2 \times 10^3$), which were obtained by means of different simulation runs in order to save CPU time and achieve better statistics. We show results obtained by measuring the SVI, which coincide with those corresponding to the MVI, since the formation of overhangs is negligible.

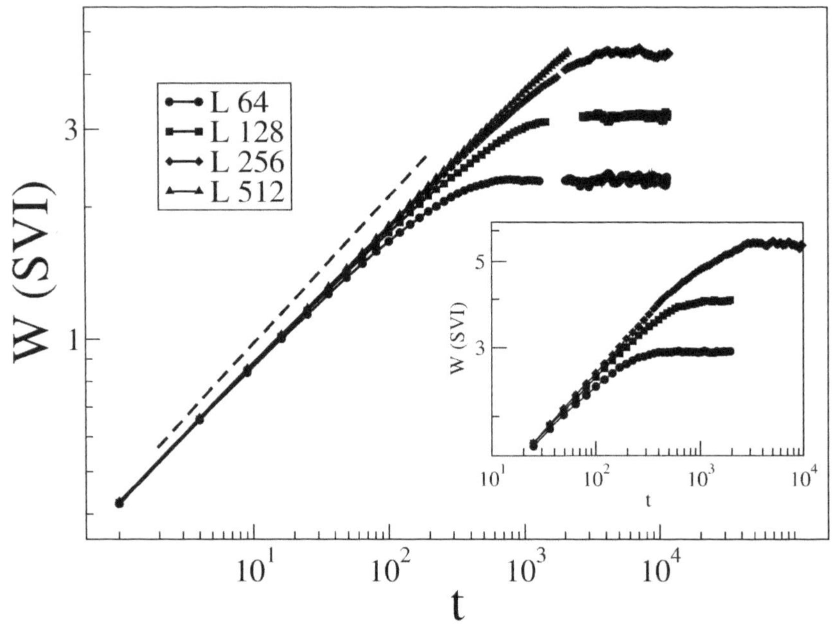

Figure 8. *In silico* results: log-log plots of the interface width (W) versus time as obtained for different sample sizes, as indicated, for $P_I = 0.99$ and $P_I = 0.50$ (inset). The dashed lines have slopes $\beta = 1/3$ and have been drawn for the sake of comparison. More details in the text.

From the cases shown in Fig. 8 we observed an initial increase of W according to Eq. (3), which is followed by a crossover behavior to a saturation value, as already expected according to the discussion of Eq. (8). Then by fitting the early time behavior of the interface width, we obtained $\beta = 0.32(2)$ for the studied values of P_I. Note the agreement between experimental and theoretical results for β, both according to the expected value for the KPZ universality class [43].

5.3. *Structure factor*

For the experimental case, structure factor data obtained from both linearly and radially spreading colonies approach a straight line with a slope close to -2.0 ± 0.1, for both Hela and Vero cells (Fig. 9). Then, in the range $0.02 \leq k \leq 1$, it results in $\alpha_s = 0.50 \pm 0.05$, irrespective of the colony spreading geometry.

From the value of α_s as derived from Fig. 9 and z calculated as $\frac{\alpha}{\beta}$ one can then plot Eq. 10 for both linearly and radially spreading colony data, as can be seen in Fig. 10. For both, different cell lines and geometries, log-log plots exhibit a reasonable collapses into a single curves. Considering the proper noise of biological systems, the above results from the dynamic scaling analysis are fairly good.

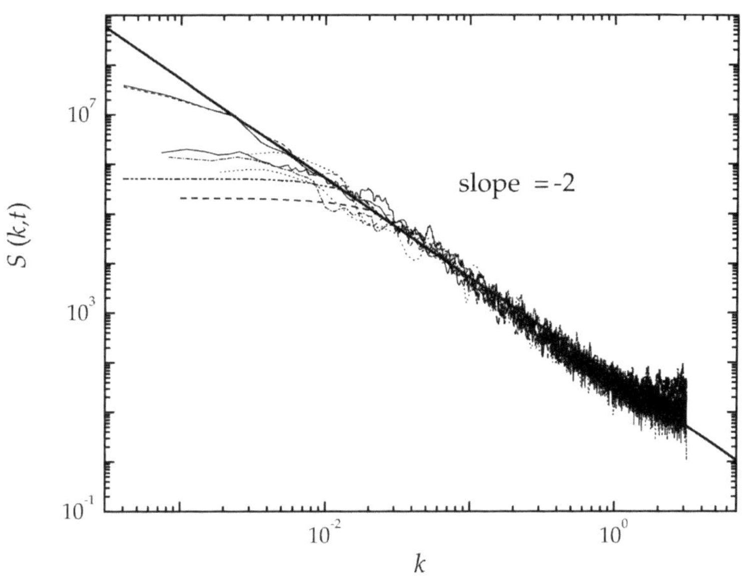

Figure 9. Log-log plots of the structure factor for linearly and radial colony growth fronts of HeLa and Vero cells at different times. Straight lines with slope $-(2\alpha_s+1) = -2$ are drawn to guide the eye.

From the computational point of view, since in order to obtain a reliable value of the roughness exponent by using Eq. (8) one needs additional simulations performed with even larger lattices than those already shown in Fig. 8, we evaluated the structure factor $S(k,t)$ (Eq. (9)) for the stationary regime of the interface, as shown in Fig. 11. In the stationary regime, of

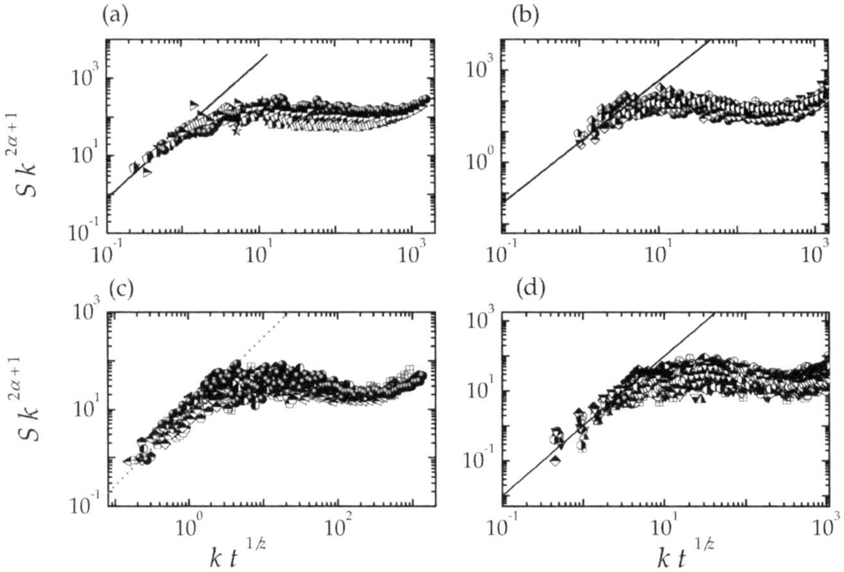

Figure 10. Log-log plots of Family-Vicsek scaling relation (10) for: (a)–(b) linearly and radial growth fronts of Vero cells, and (c)–(d)linearly and radial growth fronts of HeLa cells, considering $\alpha = 0.5$ and $z = 1.5$. Data have been binned for the sake of clarity.

course, the structure factor is independent of time [40], and it scales as

$$S(k, L) \propto k^{-(2\alpha_s+1)} L^{2(\alpha-\alpha_s)}. \tag{13}$$

In Fig. 11 we observe that $S(k)$ is independent of L, so $\alpha = \alpha_s$. Also, from the best fit of the data we obtained $-(2\alpha_s + 1) = -1.96(5)$, i.e. $\alpha_s = \alpha = 0.48(2)$. As discussed in [40], if one has that $\alpha = \alpha_s$, then the structure factor exhibits nonanomalous behavior (cf. Eqs. (10) and (11)), i.e. a characteristic feature of the KPZ universality class. Also, the scaling behavior of $S(k)$, as expected according to Eq. 13, is nicely verified by our data [cf. inset of Fig. 11].

5.4. *KPZ collapses*

After determining both β and α, we are ready to test the dynamic scaling Ansatz given by Eq. (6), since one should obtain data collapse of curves corresponding to different sample sizes just by plotting W/L^α versus t/L^z at log-log scales. These data collapses are shown in Fig. 12 for experimental

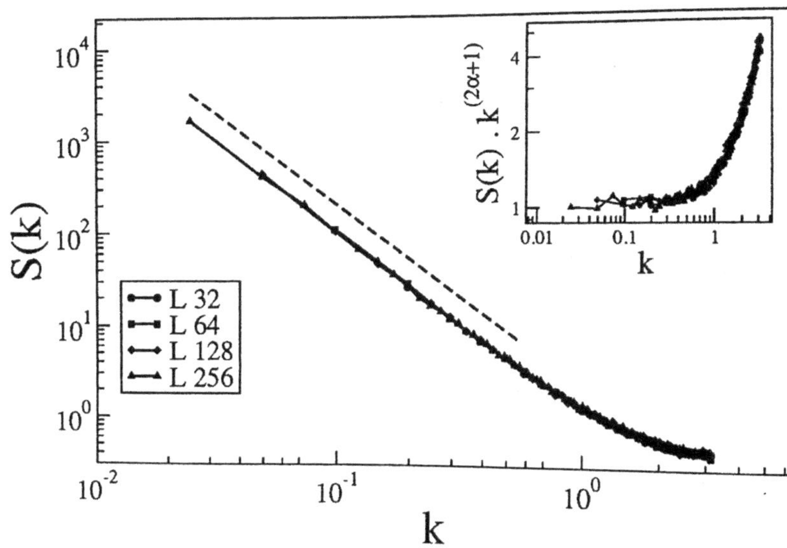

Figure 11. *In silico* results: log-log plots of the structure factor $S(k)$ versus k as obtained during the saturation regime of the interface. Data obtained for different lattice sizes as indicated and taking $P_I = 0.99$. Results averaged over 1000 different samples.The dashed line is the best fit of the data and it has slope $-(2\alpha + 1) = -1.96$. Inset: scaling plot of the data already shown in (a) according to Eqs. (10) and (11) and obtained by taking $\alpha = 1/2$. More details in the text.

results and in Fig. 13, for computational simulations, as obtained for $P_I = 0.99$ and 0.50 (inset). It is worth mentioning that the best collapses are obtained by assuming the exponents corresponding to the KPZ universality class, namely $\alpha = 1/2$, $\beta = 1/3$, $z = 3/2$, as suggested by our previous evaluation of the exponents from Figs. 8 and 11.

6. Discussion

In order to investigate the growth of linear cell colony fronts by means of dynamic scaling techniques, we present a review of recent studies on both cultured Vero and HeLa cells for the case of linear and circular interfaces [29,41,47]. Also, we compare the experimental results with a simple discrete model to describe the developing of these interfaces, that incorporates the effects of excluded volume, cell proliferation, diffusion and rotation [30].

We show that cultures performed *in vitro* and *in silico* exhibit a constant growth velocity. For the experimental case, growing cultures of radial and

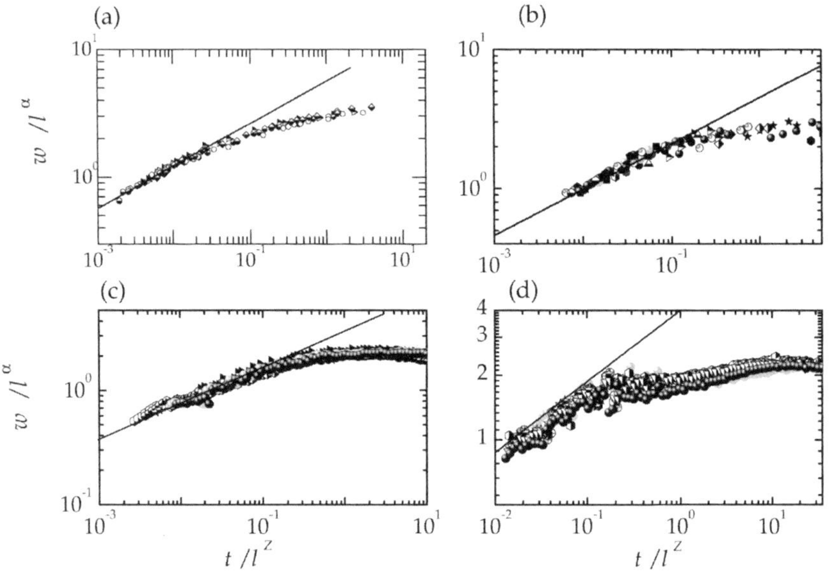

Figure 12. *In vitro* results: log-log plots of the scaled width, $\frac{W}{L^{\alpha}}$ versus $\frac{t}{L^{z}}$, as it follows from Eq. (6), and applied to the data already shown in Fig. 7. The data collapse is achieved, in all cases, by using the exponents of the KPZ universality class, namely $\alpha = 1/2$, and $z = 3/2$.

linear systems present the same velocity, within error bars, for each cell type, Vero and HeLa.

At early times the interface roughness of both, radial and linear growing systems, follows a power law in time given by $w \sim t^{\beta}$, while upon saturation the interface width scales as $w \sim L^{\alpha}$. In that way, conventional dynamic scaling analysis of both direct and overhang-corrected *in vitro* data yields the scaling exponents $\alpha = 0.50 \pm 0.05$, $\beta = 0.33 \pm 0.04$ and $z = 1.5 \pm 0.2$. Also, the comparison between HeLa and Vero cell colony dynamic data, under similar conditions, shows no clear cut dependence of the scaling exponents on either the tumorigenic nature or the genetic background of the cells. For the case of our computational model we have $\alpha = 0.48 \pm 0.02$, $\beta = 0.32 \pm 0.02$ and $z \simeq 1.50 \pm 0.02$.

Therefore, from the critical exponents associated with the growing interface in both *in vitro* and *in silico* models we determine that these cell cultures are in the Kardar-Parisi-Zhang universality class. Finally, while the processes involved in cell culture growth are quite complex, we show

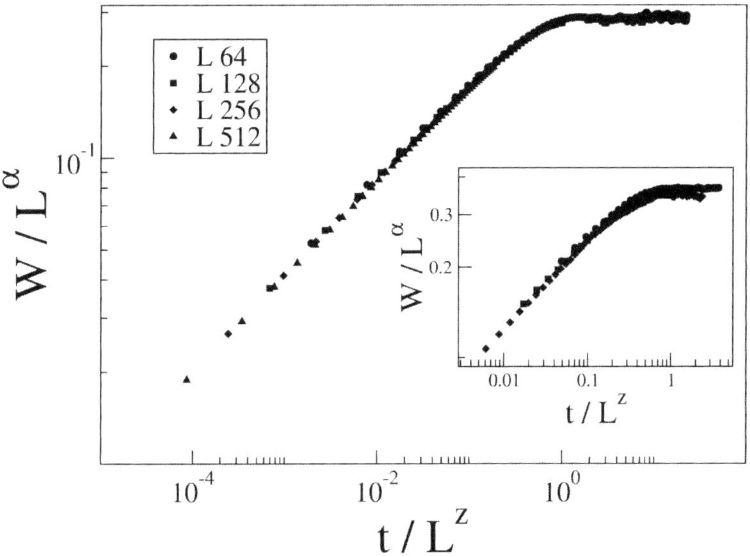

Figure 13. *In silico* results: log-log plots of the scaled width, $\frac{W}{L^{\alpha}}$ versus $\frac{t}{L^{z}}$, as it follows from Eq. (6), and applied to the data already shown in Fig. 8. The data collapse is achieved, in all cases, by using the exponents of the KPZ universality class, namely $\alpha = 1/2$, and $z = 3/2$. More details in the text.

that the complementary analysis of the results obtained by using the dynamic scaling theory is capable of accounting for the universal features of the growth interface by capturing only few but the most relevant underlying ingredients.

References

1. P. Tracqui, *Rep. Prog. Phys.* **72**, 056701 (2009).
2. L.M. Sander and T.S. Deisboeck, *Phys. Rev. E* **66**, 051901 (2002).
3. E. Khain and L.M. Sander, *Phys. Rev. Lett.* **96**, 188103 (2006).
4. S.A. Menchón and C.A. Condat, *Phys. Rev. E* **78**, 022901 (2008).
5. M. Block, E. Schöll and D. Drasdo, *Phys. Rev. Lett.* **99**, 248101 (2007).
6. D. Drasdo and S. Höhme, *Phys. Biol.* **2**, 133 (2005).
7. S. Torquato, *Phys. Biol.* **8**, 015017 (2011).
8. C. Escudero,*Cond. Mat. Stat. Mech.* **arXiv:1001.3323v2**, (2010).

9. T.S. Deisboeck, Z. Wang, P. Macklin, and V. Cristini,*Annu. Rev. Biomed. Eng.* **13**, 127 (2011).
10. P. Ciarletta, L. Foret, and M. Ben Amar,*J R Soc Interface* **56**, 345 (2011).
11. Y. Jiang, J. Pjesivac-Grbovic, C. Cantrell, and J.P. Freyer, *Biophys. J.* **89**, 3884 (2005).
12. P. Meakin,*Cambridge University Press*, (1998).
13. H. E. Stanley, *Introduction to Phase Transitions and Critical Phenomena* (Oxford University Press, Oxford, 1971).
14. A. Brú, J. M. Pastor, I. Fernaud, I. Brú, S. Melle, and C. Berenguer, *Phys. Rev. Lett.* **81**, 4008 (1998).
15. A. Brú, S. Albertos, J. L Subiza, J. L. García-Asenjo, and I. Brú, *Biophys J.* **85**, 2948 (2003).
16. P. Meakin, *J. Theor. Biol.* **118**, 101 (1986).
17. F. Fujikawa and M. Matsushita, *J. Phys. Soc. Japan* **58**, 3875 (1989).
18. M. Matsushita and H. Fujikawa, *Physica A* **168**, 498 (1990).
19. M. Matsushita, J. Wakita, H. Itoh, I. Ráfols, T. Matsuyama, H. Sakaguchi, and M. Mimura, *Physica A: Statistical and Theoretical Physics* **249**, 517 (1998).
20. T. Vicsek, M. Cserzo, and V. K. Horvath, *Physica A* **167**, 315 (1990).
21. A. L. Barabasi and H. E. Stanley, *Fractal Concepts in Surface Growth* (Cambridge University Press, 1995).
22. M. Mandelbrot, *Fractal Geometry of Nature*, (Freeman, 1982).
23. M. T. Santini and G.Rainaldi, *Pathobiology* **67**, 148 (1999).
24. J. Galeano, J. Buceta, K. Juárez, B. Pumariño, J. de la Torre, and J. M. Iriondo, *Europhys. Lett.* **63**, 83 (2003).
25. H. M. Byrne, T. Alarcon, M. R. Owen, and P. K. Maini, *Phil. Trans.* **364**, 1563 (2006).
26. J. M. Pastor and J. Galeano, *Central European J. of Physics* **5**, 539 (2007).
27. H. P. Greenspan, *Studies in Applied Mathematics* **51**, 317 (1972).
28. A. R. A. Anderson, M. A. J. Chaplain, and K. A. Rejniak, *Single-cell-based models in biology and medicine*, (Birkhäuser Verlag AG, Basel,2007).
29. M. A. C. Huergo, M. A. Pasquale, A. E. Bolzán, A. J. Arvia, and P. H. González, *Phys. Rev. E* **82**, 031903 (2010).
30. B. Moglia, N. Guisoni and E. V. Albano, *Phys. Rev. E* **87**, 032713 (2013).
31. D. O. Morgan, *The Cell Cycle, Principles of Control* (New Science Press Ltd, London, 2007).
32. B. Sapoval, M. Rosso, and J.F. Gouyet, *J. Phys. (Paris) Lett.* **46**, L149 (1985).
33. V.C. Chappa and E.V. Albano, *J. Chem. Phys.* **121** , 328 (2004).
34. M.T. Gastner, B. Oborny, D.K. Zimmermann, and G. Pruessner, *Am. Nat.* **174** , E23 (2009).
35. E.S. Loscar, N. Guisoni, and E.V. Albano, *Eur. Phys. J. B.* **85**, 60 (2012).
36. F. Family and T. Vicsek, *J. Phys. A* **18**, L75 (1985).
37. A. Bunde and S. Havlin, *Fractals and Disordered Systems*, (Springer-Verlag, New York, Inc. New York, NY, USA, 1991).

38. M. Henkel and M. Pleimling, *Theoretical and Mathematical Physics, Non-Equilibrium Phase Transitions*, (Volume 2, Ageing and Dynamical Scaling Far from Equilibrium, Springer Dordrecht Heidelberg London New York, 2010). P. Papon, J. Leblond, and P.H.E. Meijer, *The Physics of Phase Transitions, Concepts and Applications*, (2nd Edition Springer Berlin Heidelberg New York, 2006).
39. E.V. Albano, M.A. Bab, G. Baglietto, R.A. Borzi, T.S. Grigera, E.S. Loscar, D.E. Rodriguez, M.L. Rubio Puzzo, and G.P. Saracco, *Rep. Prog. Phys.* **74**, 026501 (2011).
40. J. J. Ramasco, J. M. López, and M. A. Rodríguez, *Phys. Rev. Lett.* **84**, 2199 (2000).
41. M. A. C. Huergo, M. A. Pasquale, P. H. González, A. E. Bolzán, and A. J. Arvia, *Phys. Rev. E* **84**, 021917 (2011).
42. A.B. Kolton, A. Rosso, E.V. Albano, and T. Giamarchi, *Phys. Rev. B.* **74**, 140201R (2006).
43. M. Kardar, G. Parisi, and Y. C. Zhang,*Phys. Rev. Lett.* **56**, 889 (1986).
44. Song-Li, J. L. Guan, and S. Chien, *Ann. Rev. Biomed. Eng.* **7**, 105 (2005).
45. J. Buceta and J. Galeano, *Biophys J.* **88**, 3734 (2005).
46. A. Brú, *Biophys. J.* **88**, 3737 (2005).
47. M. A. C. Huergo, M. A. Pasquale, P. H. González, A. E. Bolzán, and A. J. Arvia, *Phys. Rev. E* **85**, 011918 (2012).
48. S. C.Ferreira and S. G. Alves, *J. Stat. Mech.*, P11007 (2006).

INDEX